HIGHWAY ENGINEERING

Dr. Ing. Alamirew M. Tola

TAP

Toronto Academic Press

HIGHWAY ENGINEERING

Dr. Ing. Alamirew M. Tola

Toronto Academic Press
4164 Lakeshore Road
Burlington ON L7L 1A4
Canada
www.tap-books.com

Email: orders@arclereducation.com

ISBN: 978-1-77956-338-5

ABOUT THE AUTHOR

Dr.-Ing. Alamirew M. Tola is a distinguished civil engineer with a Ph.D. in Highway and Transportation Engineering from the prestigious University of Rostock, Germany. He is an assistant professor at the Kombolcha Institute of Technology, Wollo University, Ethiopia, where he brings a wealth of knowledge and experience to his students and colleagues. Dr.-Ing. Tola's expertise spans a wide array of areas within civil engineering, including Building Information Modeling (BIM), traffic safety, advanced highway design, smart transportation, traffic management and optimization, urban transportation, and machine learning in civil engineering. His dedication to the field is evident through his innovative teaching methods and his commitment to groundbreaking research. Passionate about pushing the boundaries of his field, Dr. Tola continuously seeks to integrate new technologies and methodologies into his work, aiming to improve the sustainability and functionality of civil engineering projects.

Table of Contents

List of Figures

List of Tables

PREFACE

According to the World Health Organization (WHO), road traffic crashes kill at least 1.19 million people and injure 50 million each year. About half of the world's road traffic deaths occur among motorcyclists (21%), pedestrians (23%), and cyclists (6%), collectively known as "vulnerable road users" (WHO 2023). The definition of 'vulnerable road users' includes pedestrians, cyclists, and motorcyclists of all age groups. 'Vulnerable road users' are categorized by the amount of protection they have from other motorized traffic. Pedestrians and cyclists are referred to as 'vulnerable' because they are unprotected. Also, riders of two-wheelers — motorcycles, mopeds, and light mopeds — are largely unprotected; hence, they are referred to as 'vulnerable.' As a result, there can be no thriving local or national economy without a reliable and well-maintained land transportation network. To facilitate economic expansion and social development, society relies on a reliable and convenient land transportation network, and roads have always been. They will always be an integral part of this system. Road's relevance and utility have grown with the development of faster and more efficient forms of transportation and the rapid acceleration of economic activity in modern human civilization. However, when careful consideration is not given to road development at the stages of planning, design, building, and management, the potential for negative consequences has increased in proportion. The discipline of highway engineering must go beyond just satisfying the fundamental necessities of delivering safe and rapid access from one location to another, to an area of study that not only includes "the structural and functional requirements of highways" and city streets but also handles the socio-economic and environmental implications of road network growth, allowing us to maximize the advantages and limit the negative effects of road construction. The traditional engineering curriculum does not fully cover these "softer" elements of "highway engineering" and the social duties of highway engineers.

Contents/Structure of the Book

This book contains ten chapters that comprehensively cover various aspects of highway engineering, providing a thorough understanding of the field.

Chapter 1 lays the foundation by discussing the history, importance, and scope of highway engineering. It introduces key concepts, including the role of highways in economic development and the principles of sustainable and efficient road design. The chapter also touches upon the evolution of highway systems and the current trends and challenges faced in the field.

Chapter 2 covers the principles of geometric design, focusing on the alignment, profile, and cross-section of highways. It explains the standards and guidelines for designing highways that ensure safety, comfort, and efficiency for all road users. Topics such as sight distance, horizontal and vertical alignment, and the design of curves are thoroughly explored.

Intersections are critical points in highway design that require careful planning to minimize accidents and delays. Chapter 3 covers the types of intersections, their design principles, and the application of traffic control devices. It also discusses modern intersection designs such as roundabouts and signalized intersections, along with methods to improve their efficiency and safety.

Focusing on the practical aspects of highway construction, Chapter 4 covers the materials used in road pavements, including their properties and suitability for different conditions. It explains the construction techniques for various types of pavements and the importance of proper drainage systems to ensure longevity and performance. The chapter also discusses the latest advancements in construction materials and methods.

Traffic engineering is crucial for the effective management of highway systems. Chapter 5 introduces the principles of traffic flow, control, and management. It covers topics such as traffic surveys, capacity analysis, traffic signals, and the design of traffic control devices. The chapter also addresses issues related to traffic congestion and the strategies to alleviate them.

Designing roads in hilly terrains presents unique challenges. Chapter 6 focuses on the specific considerations for hill road design, including alignment, stability, drainage, and safety measures. It discusses the techniques for constructing and maintaining roads in mountainous regions, considering the environmental impact and the need for sustainable development.

Chapter 7 contrasts the design principles for urban and rural roadways, highlighting the differences in requirements, challenges, and solutions. It covers the planning and design of road networks in different settings, addressing issues such as land use, traffic density, and the integration of multimodal transportation systems.

The longevity and performance of roads depend on effective maintenance and timely repairs. Chapter 8 outlines the methods for routine and periodic maintenance, identifying common defects and their causes. It also discusses modern techniques and materials used in road repair and rehabilitation.

Chapter 9 emphasizes the role of highway engineers in the overall development process. It covers the planning, design, implementation, and management phases of highway projects. The chapter also explores the regulatory and environmental considerations that engineers must navigate.

Chapter 10 provides an overview of the transportation planning process, highlighting the importance of integrating highway planning with broader transportation and land-use planning. It discusses the steps involved in developing transportation plans, including data collection, modeling, and evaluation of alternatives. The chapter also addresses contemporary issues such as sustainability, smart transportation systems, and the impact of technology on transportation planning.

Audience

This book is designed to serve a diverse audience that includes students, educators, researchers, and government and regulatory bodies.

For students, it provides a comprehensive curriculum in highway engineering, covering fundamental concepts and advanced topics. It is structured to support academic courses in civil engineering, transportation engineering, and related disciplines, serving as a textbook for classes on transportation systems, roadway design, and traffic engineering.

Educators will find the book to be an invaluable resource for teaching highway engineering. It offers detailed explanations, practical examples, and extensive references, making it an excellent tool for developing lecture materials, assignments, and projects.

Researchers involved in research on highway engineering and transportation planning will benefit from the book's coverage of current trends, challenges, and technological advancements in the field. It provides a solid foundation for academic and applied research in this area.

For government and regulatory bodies, the book offers insights into the technical aspects of highway engineering. It can aid in the formulation of policies, standards, and regulations that ensure the safety, efficiency, and sustainability of highway systems.

Learning Features

- Start-of-chapter learning objectives
- Key terms in each chapter
- Summary Points at the end of each chapter
- Review Questions at the end of each chapter
- Case Studies and Activities in each chapter for solving real problems
- References at the end of each chapter

CHAPTER 1

Introduction to Highway Engineering

LEARNING OBJECTIVES

After studying this chapter, you will be able to:

- Discuss the basics of highway engineering
- Examine the road development plan
- Access the road alignment

KEY TERMS FROM THIS CHAPTER

Biodiversity	Civil engineering
Communication networks	Concrete pavements
Economic development	Economic expansion
Highway engineers	Highway systems
Hot-mix asphalt (HMA)	Level of service (LOS)
Pavement quality concrete (PQC)	Portland cement concrete (PCC)
Transport Research Laboratory (TRL)	Socioeconomic

INTRODUCTION

The planning, designing, building, running, and maintaining of roads and highways is the specialty of highway engineering, a subfield of civil engineering. This area is essential to building up the transportation network, enabling the flow of people and products, and boosting economic expansion. Building a roadway network that is sustainable, safe, and able to accommodate growing traffic volumes is the main goal of highway engineering. Comprehensive surveys and studies are conducted during the planning phase to ascertain whether additional roads or the extension of current ones are necessary. This includes analyzing traffic volume, evaluating the effects of the environment, and taking socioeconomic factors into account. The demands of rural and urban areas must be balanced by engineers, who must ensure connectivity while reducing environmental disturbance and land acquisition expenses. Highway engineering's design phase, which includes pavement design, geometric design, and the incorporation of safety features, is crucial. The alignment, profile, cross-section, and intersections of a road are all included in the concept of geometric design. The goal is to create a road that is both comfortable and safe for users, with different vehicle types and speeds being accommodated. The materials and structural layers that comprise the road surface are the main focus of pavement design, as this ensures longevity and durability under a range of loads and environmental conditions.

Designing and building physical structures is the process of construction. To guarantee that the work complies with specifications, budgets, and timelines, this phase necessitates careful project management. Highway engineers have to work in tandem with a range of stakeholders, such as contractors, governmental bodies, and members of the public, to tackle obstacles like site conditions, inclement weather, and unanticipated roadblocks.

Once a highway is built, maintaining and operating it becomes essential to maintaining both its functionality and safety. Road conditions are monitored, control measures are put into place, and traffic flow is managed. In addition to routine tasks like patching potholes, resurfacing, and maintaining drainage systems, maintenance also includes reacting quickly to unplanned events like accidents or natural disasters. Modern innovations and technology are also incorporated into highway engineering to increase productivity and safety.

With the use of sensors, cameras, and communication networks, intelligent transportation systems (ITS) monitor and control traffic in real-time, supplying information that improves incident response, traffic signal optimization, and congestion management. Highway engineers are becoming more concerned with sustainable practices to address resource depletion and climate change. Examples of these practices include using recycled materials and designing for reduced environmental impact.

1.1. BASIC OF HIGHWAY ENGINEERING

Highway engineering, sometimes referred to as roadway engineering or street engineering, is a professional engineering discipline that comes from the transportation engineering subdiscipline of civil engineering. Its tasks include designing, building, operating, and maintaining roads, highways, streets, bridges, and tunnels to guarantee the efficient and safe movement of people and products. In the second half of the 20th century, following World War II, highway engineering gained popularity. Highway engineering standards are always being raised. The design of highway junctions and interchanges, geometric alignment and design, pavement materials and design, pavement thickness structural design, and pavement maintenance are all factors that highway engineers must consider.

1.1.1. History

One could trace the origins of road building back to the Roman era. Road development had to keep up with the technological advancements from two-horse-pulled carriages to vehicles with 100 horses' worth of power. Modern highway building did not start until the late 19th and early 20th centuries. In the UK, the Transport Research Laboratory (TRL) was established in 1930, marking the beginning of highway engineering research. The Federal-Aid Highway Act of 1944, which attempted to connect 90% of cities with a population of 50,000 or more, elevated highway engineering to a prominent field in the US. Pavements need to be improved since they were constantly under stress from cars that got bigger over time. The Preston Bypass, which was built in 1958, was a major factor in the development of new pavement technology, as the previous technology was outdated.

1.1.2. Planning and Development

Estimating present and future traffic volumes on a road network is a part of highway planning. Another essential requirement for highway development is highway planning. The goal of highway engineers is to anticipate and evaluate every potential civil impact of their systems. The negative consequences on the environment, such as air, water, noise, and other ecological effects, are some factors to consider.

1.1.2.1. Financing

Rich nations are continuously confronted with the heavy maintenance costs associated with aging transportation infrastructure. There is a growing need for safer, more efficient, and less congested highways because of the expansion of the motor vehicle industry and the corresponding economic growth. The expansion of business, academia, housing, and defense has historically placed a heavy burden on government coffers, making public transportation financing difficult. The economy, the multipurpose nature of highways, and the developments in highway pricing technology are all ever-evolving. As a result, methods for managing, financing, and maintaining highways are also continuously evolving.

1.1.2.2. Environmental Impact Assessment

The construction of new highways to improve mobility is essential to a community's economic development. But badly thought out, designed, built, and maintained highways can upend a community of any size's social and economic structure. Highway development frequently has negative effects on the environment, including harm to habitat and biodiversity, pollution of the air and water, noise and vibration, destruction of the natural landscape, and disintegration of the social and cultural fabric of a community. Highway infrastructure must be built and maintained in accordance with stringent specifications. Planning, scheduling, building, and maintaining highways can all be done with environmental considerations in mind in three easy steps.

This process is known as an Environmental Impact Assessment, or EIA, as it systematically deals with the following elements:

- Identification of the full range of possible impacts on the natural and socio-economic environment
- Evaluation and quantification of these impacts
- Formulation of measures to avoid, mitigate, and compensate for the anticipated impacts.

1.1.2.3. Highway Safety

Highway systems bear the brunt of human injury and mortality; not counting the 1.19 million fatalities, they inflict injuries on close to 50 million people annually. The primary cause of unintentional death in the first fifty years of life is injuries sustained in traffic accidents. The methodical process of safety management aims to lessen the frequency and seriousness of traffic accidents. Highway safety management is challenged by the unstable human-machine interaction with road traffic systems.

Designing, constructing, and maintaining highway systems to be far more tolerant of the typical range of this human/machine interaction with highways is essential to improving highway safety. Over time, improvements in technology have been made to highway engineering, leading to better design, construction, and maintenance practices. Newer innovations in highway safety have been made possible by these advancements. Every stage of highway planning, design, construction, maintenance, and operation can be evaluated to improve the safety of our highway systems by making sure that all circumstances and opportunities are recognized, taken into account, and implemented as necessary (Mamlouk and Michael, 2005).

1.1.3. Design

During the design phase, a highway's ideal location, alignment, and shape are chosen. In order to provide a safe highway, highway design takes into account the interaction of three main factors: human, vehicular, and roadway. Human factors include following cars, reaction times for braking and steering, and visual acuity for traffic signs and signals.

When choosing design vehicles and determining lane widths and maximum slopes, vehicle dynamics and size are crucial factors to take into account. When designing the geometry of two-lane, two-way roads, highway engineers make sure that cars are stable when navigating curves and grades and that there are enough sight distances to allow for passing maneuvers along curves.

1.1.3.1. Geometric Design

When designing highways for specific site topography, highway and transportation engineers have several safety, service, and performance standards to meet. The term "highway geometric design" mostly describes the components that are visible from the road. When designing a highway's geometry, highway engineers also have to take the surrounding infrastructure's social and environmental impacts into account. (O'Flaherty & Coleman, 2002) (see Figure 1.1).

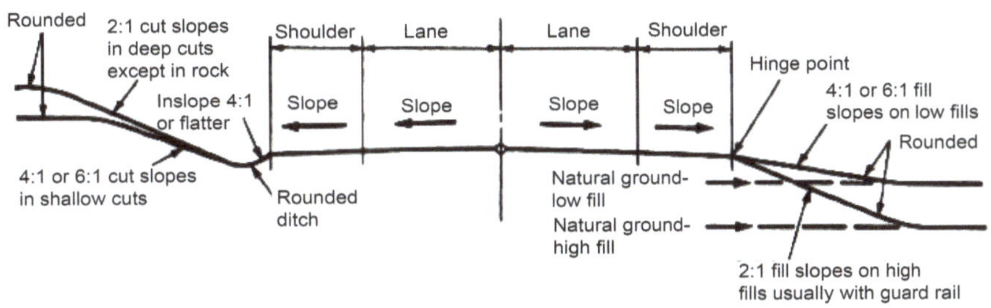

Figure 1.1. *A typical cross-section drawing of a roadway.*

Source: https://upload.wikimedia.org/wikipedia/commons/thumb/a/a1/Typical_roadway_cross-section_sheet_in_transportation_engineering.jpg/1024px-Typical_roadway_cross-section_sheet_in_transportation_engineering.jpg.

There are certain considerations that must be properly addressed in the design process to successfully fit a highway to a site's topography and maintain its safety. Some of these design considerations are:

- Design speed
- Design traffic volume
- Number of lanes
- Level of service (LOS)
- Sight distance
- Alignment, super-elevation, and grades
- Cross section
- Lane width
- Structure gauge, Horizontal and vertical clearance

The operational performance of a highway can be seen through drivers' reactions to the design considerations and their interaction.

1.1.3.2. Materials

The materials used to build roads have advanced over time, going all the way back to the Roman Empire's early years. This advancement in materials has been accompanied by improvements in the techniques used to characterize and apply these materials to pavement structural design. Pavement quality concrete (PQC), Portland cement concrete (PCC), and hot-mix asphalt (HMA) are the three main categories of pavement surfaces. The material layers that provide the pavement system with structural support are located beneath this wearing course. The aggregate base and subbase layers, the treated base and subbase layers, and the underlying natural or treated subgrade are examples of these underlying surfaces. For added support, these treated layers can be treated with lime, asphalt, or cement.

1.1.4. Flexible pavement design

Typically, a flexible asphalt pavement, also known as Tarmac, is composed of three or four layers. A compacted, natural soil subgrade is covered by a surface course, base course, and subbase course in a four-layer flexible pavement. The base course is laid directly on top of the natural subgrade when constructing a three-layer flexible pavement; the subbase layer is not utilized.

The surface layer of a flexible pavement is made of hot-mix asphalt (HMA). The base course is usually composed of unstabilized aggregates, but it can also be stabilized with Portland cement from <Roadstone Recycling>, asphalt, foamed bitumen, or any other stabilizing agent. Local aggregate is typically used to build the subbase, and cement or lime is frequently used to stabilize the top of the subgrade.

The highest stress in a flexible pavement is at the surface, and as the pavement gets deeper, the stress gets less. As a result, the pavement's surface must be made of the highest quality material, but as it gets deeper, lower quality materials may be utilized. The reason asphalt is called flexible is because it can slightly bend and distort before returning to its original shape in response to the application and removal of each traffic load. These minor distortions may eventually become permanent, which could cause rutting in the wheel path over time. A flexible pavement's service life is usually designed to last between 20 and 30 years. The amount of material used, the size, frequency of traffic load repetitions, the environment, and the pavement's intended service life all affect how thick each layer of flexible pavement must be. These kinds of factors are taken into account in the design process to ensure that the pavement will not experience undue distress over its intended life.

1.1.4.1. Rigid Pavement Design

Major highways like the interstate highway system and airports are typically built with rigid pavements. They also frequently function as heavy-duty industrial floor slabs, pavements in port and harbor yards, and pavements in terminals or parks for large vehicles.

Similar to flexible pavements, rigid highway pavements are made to be long-lasting, all-weather structures to handle the high-speed traffic of today. They serve as structural layers to distribute vehicle wheel loads in a way that ensures the induced stresses transmitted to the subgrade soil are of acceptable magnitudes while simultaneously providing excellent riding surfaces for safe vehicular travel. When building stiff pavement slabs, Portland cement concrete (PCC) is most frequently utilized as the building material. Because of its accessibility and the state of the economy, it is very popular. It is necessary to build rigid pavements to withstand repeated, heavy traffic loads. Rigid pavements typically have designed service lives of 30 to 40 years, which is roughly twice as long as flexible pavements.

Reducing fatigue failure brought on by repetitive traffic stresses is a key design factor for rigid pavements. Due to the fact that a typical highway will see millions of wheel passes during its service life, fatigue failure is common among major roads. It is necessary to consider tensile stresses resulting from thermal energy in addition to design criteria like traffic loadings. Many highway engineers have observed as pavement design has advanced that thermally induced stresses in rigid pavements can be as high as those imposed by wheel loadings. Thermal stresses play a major role in the design of rigid pavements because of the relatively low tensile strength of concrete.

A prepared subgrade, a base or subbase, and a concrete slab are the three layers that typically make up a rigid pavement. The concrete slab's construction follows a planned selection of plan dimensions for the slab panels, which has a direct impact on the level of thermal stresses that arise in the pavement. Temperature reinforcements must be planned in addition to the slab panels to regulate the slab's tendency toward cracking. Slab panel dimensions determine joint spacing. In practical application, jointed plain concrete pavement (JPCP), jointed reinforced concrete pavement (JRCP), and continuously reinforced concrete pavements (CRCP) are the three primary varieties of concrete pavements. Contraction joints are used in the construction of JPCPs to guide the pavement's natural cracking process. The steel reinforcing is not used in these pavements.

To prevent pavement cracking, reinforcing steel and contraction joints are used in the construction of JRCPs. The reinforcing steel firmly holds the pavement together when it cracks due to high temperatures and moisture stresses. Dowel bars are commonly positioned at transverse joints to aid in the transfer of the vehicle's load across the fracture. The natural transverse cracks in the pavement are held together by continuous reinforcing steel in CRCPs alone. Although they are less common than the other three, prestressed concrete pavements have also been utilized in the construction of highways. By partially or completely neutralizing thermally induced stresses or loadings, prestretched pavements enable a thinner slab thickness.

1.1.4.2. Flexible pavement overlay design

Accumulated traffic loads over a flexible pavement's service life may result in excessive rutting or cracking, poor ride quality, or insufficient skid resistance. If the pavement is properly maintained, these issues can be avoided; however, maintaining the pavement is usually expensive, and the pavement might not be structurally strong enough to support the anticipated traffic loads. The degree of serviceability of a highway is constantly checked and maintained. Applying an overlay to the pavement's surface is a typical technique for preserving a highway's degree of serviceability. On flexible pavements, overlays made of asphalt and concrete, Portland cement concrete overlay, and ultra-thin Portland cement concrete overlay are the three common types. On top of the flexible surface, the concrete layer of a traditional PCC overlay is applied unbonded. An ultra-thin PCC overlay usually measures no more than 4 inches (10 cm) in thickness. There are two main categories of flexible pavement overlay design procedures:

- Component analysis design
- Deflection-based design

1.1.4.3. Rigid pavement overlay design

When a rigid pavement reaches the end of its useful life, a choice must be made between rebuilding the entire worn pavement and building an overlay layer. Applying overlay layers more frequently is often more financially advantageous because it can be built on a rigid pavement that has not reached the end of its service life.

A rigid pavement that is structurally sound requires a much thinner overlay than one that is nearing the end of its useful life. Rigid pavements like JPCP, JRCP, and CRCP can be rehabilitated using rigid or flexible overlays. Rigid pavement overlays are divided into three subcategories based on the bonding state at the pavement overlay and existing slab interface.

- Bonded overlays
- Unbonded overlays
- Partially bonded overlays

1.1.4.4. Drainage System Design

Proper drainage is essential for the success of highway systems. It is important for highways to be designed and constructed to ensure they remain dry. Without adequate drainage, a highway may experience premature failure due to excess water within the structure. Each highway drainage system is unique and complex, with the methods used depending on the geography of the region.

Highway engineers must determine which design processes are suitable for each situation, often using a combination of methods and materials to divert water away from the structure. Pavement subsurface drainage and underdrains are crucial for extending the life of the pavement and ensuring reliable performance. Excessive moisture beneath a concrete pavement can lead to issues such as pumping, cracking, and joint failure.

Erosion control is also vital when designing highway drainage systems. Proper surface drainage allows precipitation to flow away from the structure. Highways should be designed with a slope or crown to direct runoff water towards the shoulder of the road and into a ditch, away from the site.

Designing a drainage system involves predicting runoff and infiltration, conducting open channel analysis, and designing culverts to direct surface water to a suitable location.

1.1.5. Construction, Maintenance, and Management

1.1.5.1. Highway Construction

Subgrade preparation and thorough surveys typically come before highway construction. The techniques and equipment used to build highways have advanced and become more complex over time. With the development of technology, the level of skill sets needed to oversee highway construction projects has increased. Project complexity and nature, the distinctions between new construction and reconstruction, and the distinctions between projects in urban and rural areas are some of the variables that affect this skill from project to project. The technical and commercial components of the system can be used to categorize various aspects of highway construction.

Some examples of each are listed below:

- Technical elements
  - Materials
  - Material quality
  - Installation techniques
  - Traffic
- Commercial elements
  - Contract understanding
  - Environmental aspects
  - Political aspects
  - Legal aspects
  - Public concerns

Typically, construction begins at the lowest elevation of the site, regardless of the project type, and moves upward. By reviewing the geotechnical specifications of the project, information is given about:

- Existing ground conditions
- Required equipment for excavation, grading, and material transportation to and from the site
- Properties of materials to be excavated
- Dewatering requirements necessary for below-grade work
- Shoring requirements for excavation protection
- Water quantities for compaction and dust control

1.1.5.2. Subbase Course Construction

A subbase course is a layer between the pavement's base course and subgrade that is made of carefully chosen materials. The subbase is intended to support the necessary structural capacity of the pavement section and is typically 4 to 16 inches thick. For a highway subbase, materials like gravel, crushed stone, or subgrade soil stabilized with lime, fly ash, or cement are frequently used. The ability of permeable subbase courses to remove water that has seeped in from the surface is leading to an increase in their use. They also keep subterranean water from rising to the surface of the pavement. Highway engineers can increase the bearing capacity of the underlying soil by adding Portland cement, foamed asphalt, or polymer soil stabilization, such as cross-linking styrene acrylic polymer, which increases the California Bearing Ratio of in-situ materials by a factor of 4–6. This is useful when local material costs are prohibitively high or the material requirements to increase the structural bearing of the sub-base are not easily available.

1.1.5.3. Base Course Construction

The area of the pavement section directly beneath the surface course is known as the base course. If a subbase course exists, the base course is built directly over this stratum. If not, the subgrade is directly covered by the construction. Base course thickness typically varies between 4 and 6 inches, depending on the characteristics of the underlying layer. Pavement surfaces are constantly subjected to heavy loads, most of which are absorbed by the base layer. Crushed stone, slag, or gravel are examples of untreated crushed aggregate that is typically used to build base courses. The base course material will be stable under the weight of construction equipment and have good drainage qualities. The base course materials are frequently treated with fly ash, lime, calcium chloride, bitumen, or cement. These treatments act as a moisture barrier between the base and surface layers, improve support for large loads, and increase frost susceptibility.

1.1.5.4. Surface Course Construction

Portland cement concrete and hot-mix asphalt are the two types of pavement surfaces that are most frequently utilized in the construction of highways. While transferring the heavy traffic loads through the various base courses and into the underlying subgrade soils, these pavement surface courses also offer a safe and smooth riding surface.

1.1.5.5. Hot-mix Asphalt Layers

Flexible pavements are surface courses made of hot-mix asphalt. Since its development in the late 1980s, the Superpave System has allowed for modifications to specifications, mix design, design methodology, and material quality testing. An experienced construction team dedicated to maintaining equipment control and work quality is necessary for the creation of an efficient, long-lasting asphalt pavement.

Construction issues:

- Asphalt mix segregation
- Laydown
- Compaction
- Joints

Before laying the HMA surface course, a low-viscosity asphalt called a prime coat is applied to the base course. Between the base course and the asphalt surface, this coat creates a cohesive layer by binding loose material. A low-viscosity asphalt emulsion called a tack coat is applied to an existing pavement surface to establish a bond with a fresh asphalt overlay. Usually, tack coats are put on nearby curbs to help the HMA and concrete adhere to one another.

1.1.5.6. Portland Cement Concrete (PCC)

Concrete pavements, also known as rigid pavements, consist of three main types: jointed plain, jointed reinforced, and continuously reinforced. Loadings from traffic are transferred between sections either through the interlocking of larger aggregates in the mix or with the use of load transfer devices in transverse joints. Dowel bars are employed as load-transferring devices to effectively transfer loads across transverse joints while ensuring the joints remain aligned horizontally and vertically. Additionally, tie bars, which are deformed steel bars, are placed along longitudinal joints to secure adjacent pavement sections in position.

1.1.5.7. Highway Maintenance

Repairing flaws and maintaining the structural integrity and usability of the pavement are the main goals of highway maintenance. Identifying, comprehending, and documenting defects is necessary to develop a suitable maintenance schedule. Predictive maintenance planning involves addressing optimization problems. When it comes to predictive maintenance planning, data-driven, empirical approaches yield more precise outcomes than mechanical models. Rigid and flexible pavements have different types of defects (Ferragut et al., 2002) (see Figure 1.2).

Figure 1.2. *A U.S. Federal Highway Administration (FHWA) worker inspects a washout on the shoulder of West Virginia Route 150.*

Source: https://upload.wikimedia.org/wikipedia/commons/0/0c/West_Virginia_State_Route_150_shoulder_washout.jpg.

There are four main objectives of highway maintenance:

- repair of functional pavement defects
- extend the functional and structural service life of the pavement
- maintain road safety and signage
- keep road reserve in acceptable condition

Through routine maintenance practices, highway systems and all their components can be maintained to their original, as-built condition.

1.1.5.8. Project Management

From project inception to project completion, project management entails organizing and planning project activities. Building highways and bridges, for example, or performing major and minor maintenance tasks associated with building infrastructure are examples of activities. There must be professionalism throughout the project, and all associated activities must be finished on schedule and within budget. A successful project management strategy also involves minimizing social and environmental impacts.

1.2. ROADS DEVELOPMENT PLAN

A thorough plan that outlines the approach for creating, growing, and maintaining road infrastructure in a particular area or nation. Ensuring the smooth flow of people and goods, fostering economic expansion, and improving connectivity all depend on this plan. It usually includes an evaluation of needs, strategic goals, project prioritization, finance sources, environmental factors, and implementation tactics, among other important elements. A comprehensive needs assessment is a necessary first step in the creation of a road development plan. In this step, traffic patterns are analyzed, bottlenecks are located, road conditions are evaluated, and future demand is predicted using data on population growth, urbanization trends, and economic activity. Road usage statistics, demographic projections, and traffic study data are vital for comprehending the current and future needs of the road network. The needs assessment informs the formulation of strategic objectives. These goals could be boosting sustainable transportation options, lowering traffic, boosting connectivity between important economic zones, boosting road safety, and assisting with regional development. Goals that are specific and quantifiable give subsequent planning and decision-making procedures direction.

Road projects are ranked according to factors like urgency, economic impact, cost-benefit analysis, and alignment with strategic objectives as part of the crucial process of project prioritization. This guarantees that funds are distributed effectively and that the most promising projects are given top priority. This prioritization is frequently determined by variables like accident rates, traffic volume, and accessibility to important hubs. Funding must be secured in order for the Roads Development Plan to be implemented successfully. It is possible to investigate several funding sources, such as public-private partnerships (PPPs), government budgets, foreign loans and grants, and toll collections. Financial sustainability is improved and dependence on a single source is decreased with a diversified funding plan. Cost estimates and financial planning are also essential for controlling spending and guaranteeing cash availability for the duration of the project. Plans for the construction of new roads often take the environment into account. The purpose of environmental impact assessments, or EIAs, is to assess how possible road projects may affect the environment, both natural and human. To reduce adverse effects, mitigation techniques are used, such as creating wildlife corridors, putting erosion control measures in place, and employing environmentally friendly building materials. Utilizing recycled materials and energy-efficient building methods are two examples of sustainable practices that are becoming increasingly important.

The Roads Development Plan's implementation strategies lay out the procedures for carrying them out. This covers the procurement procedures, construction schedules, quality control procedures, and comprehensive project design. To guarantee that projects are finished on schedule, within budget, and in compliance with specifications, effective project management is essential. To solve issues and guarantee smooth progress, coordination with a variety of stakeholders—including governmental organizations, contractors, and local communities—is crucial. Additionally, mechanisms for monitoring and evaluating road projects are included to keep tabs on their advancement and evaluate their results in relation to the strategic goals. Frequent reporting, performance audits, and site inspections aid in the early detection of problems and enable prompt remedial action.

1.2.1. Importance of Roads and Its Characteristics

A road is a crucial piece of infrastructure that facilitates all forms of mobility and helps people get from one place to another. A road facilitates task completion whether it is used for the movement of people or goods. In order to contribute to economic development, growth, and social benefits, roads are essential. Roads are essential for a nation's growth and development because they provide access to social, health, and educational services as well as employment opportunities. This makes them essential for battling poverty and adversity. The market for goods can be expanded and the transportation of fuels, other materials, and raw materials to the sites of production can be greatly facilitated by an efficient transport system. It also makes more isolated areas and production resources available. Additionally, as the need for transportation facilities grows, so does the demand for automobiles, ships, and other means of transportation. A road is a route used to move traffic that is primarily designed with a better surface for use by pedestrians and motorized and non-motorized vehicles. Roads serve as means of transportation, as opposed to streets, which are primarily used as public areas. Parkways, avenues, controlled-access highways (freeways, motorways, and expressways), tollways, interstates, highways, thoroughfares, and local roads are just a few of the various kinds of roads. Lanes, sidewalks (pavement), roads (carriageways), medians, shoulders, verges, bike paths (cycle paths), and shared-use paths are some of the main characteristics of a road.

1.2.2. History

Many times, animals do not follow consistent paths, so it is not widely accepted that the first pathways were the trails left by them. Some people think that some roads were created by people who followed animal trails. The Icknield Way could serve as an example of this kind of road origination, in which both humans and animals chose the same natural line. Around 10,000 BC, human travelers likely followed natural trails and pathways.

- The world's oldest known paved road was constructed in Egypt sometime between 2600 and 2200 BC.
- Stone-paved streets appear in the city of Ur in the Middle East dating back to 4000 BC.
- Corduroy roads (log roads) are found dating to 4000 BC in Glastonbury, England.
- Brick-paved streets appeared in India as early as 3000 BC.

- 1995 BC: an early subdividing of roadways evidenced with sidewalks built in Anatolia.
- The Achaemenid Empire (Persia) had a vast road network when Darius I the Great began building it in 500 BC. Among the best roads of the era was the Royal Road, which connected Sardis, the empire's westernmost major city, to Susa. The road was still in use after the Roman era. As far east as Bactria and India, these road networks were located.
- Considering the cost of building new roads and the fact that river barges have a higher carrying capacity than carts, transportation by river was easier and faster in the past than transportation by land. The horse-drawn boat, which dates back to around 1740, is a combination of ship and road transportation, with the horse following a path that has been cleared along the riverbank.
- From about 312 BC, the Roman Empire built straight strong stone Roman roads throughout Europe and North Africa, in support of its military campaigns. At its peak, the Roman Empire was connected by 29 major roads moving out from Rome and covering 78,000 kilometers or 52,964 Roman miles of paved roads.
- In the 8th century AD, many roads were built throughout the Arab Empire. The most sophisticated roads were those in Baghdad, which were paved with tar. Tar was derived from petroleum, accessed from oil fields in the region, through the chemical process of destructive distillation.
- In Britain, the Highways Act of 1555 gave local parishes control over road maintenance, taking over from the government. Road conditions became poor and inconsistent as a result. In order to address this, the first turnpike trusts were founded circa 1706 to construct decent roads and obtain tolls from passing cars. In the end, there were about 1,100 trusts and 36,800 km (22,900 mi) of engineered roads in Britain. The 1839–1844 Rebecca Riots in Carmarthenshire and Rhayader helped spark a Royal Commission that ultimately resulted in the system's collapse in 1844, coinciding with the growth of the railway network in the United Kingdom.
- In the late-19th century roading engineers began to cater for cyclists by building separate lanes alongside roadways.
- Since the turn of the 20th century, more and more roads have been constructed for both employment and tourism. The Great Dolomite Road is a prime example of how tourism can be stimulated, but the construction of the stunning Strada Costiera, which runs between Duino and Barcola, Italy, in 1928, was primarily done with employment in mind.
- The *Autostrada dei Laghi* ("Lakes Motorway") in Italy, the first controlled-access highway built in the world, connecting Milan to Lake Como and Lake Maggiore, and now parts of the A8 and A9 motorways (Ron Bridle & John Porter (2002)), was devised by Piero Puricelli and was inaugurated in 1924. This motorway, called *autostrada*, contained only one lane in each direction and no interchanges.

1.2.3. Construction and Development

In the field of transportation engineering, the subgrade refers to the original material that lies beneath a road once it is constructed. Building a road involves creating a designated path or foundation that can accommodate vehicles or pedestrians, while also meeting legal requirements. This process typically includes excavating or blasting away earth and rock, building up embankments, bridges, and tunnels, and clearing vegetation. Once this groundwork is complete, pavement material is laid down using various road construction machinery.

Before construction begins, a thorough assessment of the design, approvals, planning, legal considerations, and environmental impact must be conducted. A surveyor then outlines the path of the road, carefully considering the natural terrain to minimize any necessary earthwork. Special attention is paid to maintaining reference benchmarks throughout the construction process (Figure 1.3).

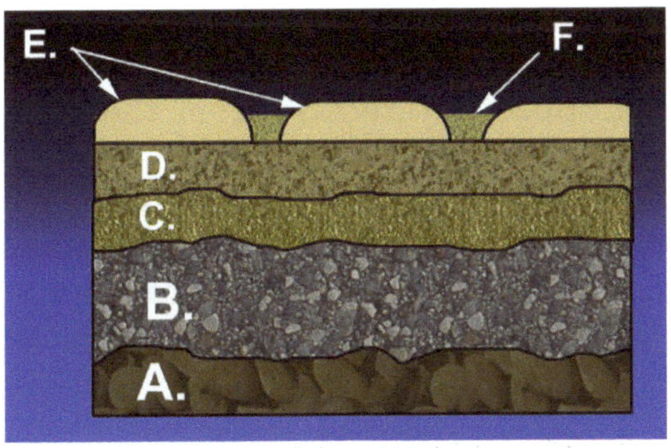

Figure 1.3. *Layers in the construction of a mortarless pavement: A) Subgrade B) Subbase C) Base course D) Paver base E) Pavers F) Fine-grained sand.*

Source: https://upload.wikimedia.org/wikipedia/commons/thumb/d/d6/MortarlessPavement.jpg/1024px-MortarlessPavement.jpg

Roads are constructed with the primary purpose of accommodating both vehicular and pedestrian traffic. It is essential to integrate storm drainage and environmental factors into the design process. Erosion and sediment controls are put in place to prevent any harmful effects. Drainage lines within the road easement are fitted with sealed joints, with runoff coefficients tailored to the land zoning and stormwater system requirements. Approval is needed from the appropriate authority for the drainage discharge into a watercourse, creek, river, or sea.

Road construction sites need to have access to a borrow pit for materials like fill, gravel, and rock, as well as a nearby water source. Local authorities may require permits for water usage or material processing. The topsoil and vegetation from the borrow pit are preserved for future reclamation of the area. Safety guidelines dictate that excavation side slopes should not exceed one vertical to two horizontal.

Before construction commences, existing structures such as old road surfaces, fences, and buildings may need to be dismantled. Trees marked for preservation must be safeguarded during construction, with their root zones protected from disturbance. Compensation or replacement may be necessary if a protected tree is damaged. Vegetation is mulched for reinstatement purposes, while topsoil is stockpiled for embankment rehabilitation. Stump removal and filling of holes are done before earthwork commencement.

Upon completion of road construction, rehabilitation efforts include seeding, planting, watering, and other activities to restore the area to its natural state. During earthwork, procedures include excavation, spoilage material removal, filling, compacting, building, and trimming. If it is found that there is rock or other inappropriate material, it is removed, the moisture content is controlled, and standard fill that has been compacted to meet design requirements is substituted (usually between 90 and 95% relative compaction). Since the intact rock structure makes an ideal road base, blasting is not often used to excavate the roadbed. After the topsoil has been removed, the native bed is compacted when a depression needs to be filled to reach the road grade. The fill is created using the compacted layer method, which involves spreading a layer of fill and then compacting it under saturated conditions to the desired thickness. Until the required grade is obtained, the procedure is repeated. General fill material needs to have a low plasticity index, be free of organic materials, and meet minimum California bearing ratio (CBR) requirements.

Sand or a sand-rich mixture with fine gravel typically makes up the lower fill; this mixture inhibits the growth of plants or other vegetable matter. Additionally, the compacted fill provides lower-stratum drainage. The second fill that has been sieved should contain no big lumps of clay and should primarily consist of gravel, broken rock, or decomposed rock that is smaller than a given particle size. Fill with sand clay is another option. Following the compacting of each layer of fill, the roadbed needs to be proof-rolled. A section is considered to comply if a roller passes over it without causing any noticeable deformation or springing. To enhance the quality of roads, geosynthetic materials like geotextiles, geogrids, and geocells are commonly incorporated into the different layers of pavement. Both public and private roads with little traffic, as well as highways, use these materials and techniques. The four primary roles that geosynthetics play in roads are separation, drainage, filtration, and reinforcement. These roles improve pavement performance, lower construction costs, and require less maintenance.

The finished road is either left with a natural surface, such as gravel, or is finished with paving. The expected usage and economic factors determine the type of road surface. Installations of raised pavement markers, crash barriers, traffic signs, and other types of road surface markings are examples of safety enhancements. A study published in May 2009 by the national transportation research organization TRIP and the American Association of State Highway and Transportation Officials (AASHTO) estimated that driving on rough roads adds about $400 to the annual operating costs of a vehicle for the typical American driver. Due to increased tire wear from bad road conditions, increased maintenance costs, increased vehicle deterioration, and increased fuel consumption, drivers in urban areas with populations over 250,000 are paying upwards of $750 more annually. Duplication, twinning, or doubling is the term used when a single-carriageway road is transformed into a dual carriageway by constructing a second, independent carriageway beside the first. The new carriageway is one-way in the opposite direction from the original two-way carriageway, which was converted to one-way. The new carriageway is not always built directly alongside the existing carriageway, much like when railway lines are converted from single to double track.

1.2.4. Maintenance

Roads deteriorate with time, just like any other structure. Although environmental factors like frost heaves, thermal cracking, and oxidation frequently contribute to degradation, accumulated vehicle damage also plays a role. The effective damage to the road is approximately proportional to the fourth power of axle weight, according to a set of experiments known as the AASHO Road Test, which were conducted in the late 1950s. Compared to a passenger car with 2,000 pounds (0.907 t) on each axle, a typical tractor-trailer weighing 80,000 pounds (36.287 t) with 8,000 pounds (3.629 t) on the steer axle and 36,000 pounds (16.329 t) on both tandem axle groups is expected to cause 7,800 times more damage. Road construction, vehicle braking, and rain damage are the main causes of potholes.

1.2.4.1. Line Marking in a Rural Area

Pavements are made with a planned service life or design life in mind. The standard design life for new bitumen and concrete pavement is forty years in certain regions of the United Kingdom. Upkeep is taken into account in the total cost of ownership for roads with servicing at 10, 20, and 30-year intervals. Roads can be built to last for a number of lifespans, including eight, fifteen, thirty, and sixty years. Pavement that outlasts its planned lifespan could have been overbuilt, or its initial cost could have been excessive.

The owner might incur exorbitant repair and rehabilitation expenses if a pavement fails before its planned design life. Certain asphalt pavements are intended to last longer than 50 years in terms of structural life; these are known as perpetual pavements. Even though they weren't intended to last forever, many asphalt pavements (S.Kameyama et al., 2002) that were constructed more than 35 years ago have held up well after their design life. Many concrete pavements constructed after the 1950s have far exceeded their original lifespans. Certain roads, such as Wacker Drive in Chicago, a significant two-level (and briefly three-level) thoroughfare in the downtown region, are undergoing reconstruction with a 100-year planned service life.

Before their useful lives are over, almost all roads need some kind of maintenance. In order to extend the lifespan of their roads, proactive agencies employ pavement management strategies to regularly assess road conditions and plan preventive maintenance procedures. Technologically advanced organizations use sophisticated tools like laser/inertial profilometers to monitor the state of the road network's surface. Road curvature, cross slope, asperity, roughness, rutting, and texture are among these measurements. This data is used by software algorithms to suggest repairs or new buildings. Thin asphalt overlays, crack sealing, surface rejuvenation, fog sealing, micro milling or diamond grinding, and surface treatments are typical maintenance procedures for asphalt concrete. With thin surfacing, the road's functional condition is preserved, protected, and enhanced while requiring less routine maintenance, resulting in an extended service life without adding to the structural capacity. Dowel bar retrofitting is a technique for fixing older concrete pavements that get cracks. It involves cutting slots in the pavement at each joint, inserting dowel bars into the slots, and then filling the slots with concrete patching material. The concrete pavement's lifespan may be increased by 15 years as a result. Road maintenance issues can have a big financial impact on society. Approximately 50% of US roads are estimated to be in poor condition, with urban areas having worse conditions, according to a 2009 report published by the American Association of State Highway and Transportation Officials. According to the report, the average US driver spends roughly $335 on car repairs annually, compared to $746 for urban drivers. The average driver, on the other hand, pays roughly $171 in road maintenance taxes annually (based on 600 gallons and $0.285 per gallon tax).

1.2.4.2. Slab Stabilization

Loss of support resulting from voids beneath the concrete pavement slabs can lead to distress and a decline in serviceability on concrete roads. The voids are typically caused by surface water infiltration close to joints or fractures. Pumping, consolidation, subgrade failure, and bridge approach failure are the most frequent causes of voids. This issue can be resolved non-destructively with slab stabilization, which is typically used in conjunction with diamond grinding and patching techniques for concrete pavement restoration. By filling in tiny gaps that form beneath concrete slabs at joints, cracks, or the pavement edge, the method restores support to concrete slabs. A cementitious grout or polyurethane mixture is pumped through holes drilled through the slab to complete the process. Small spaces beneath the slab and/or sub-base may be filled in by the grout. After stabilization is finished, the grout also helps to prevent water from saturating and weakening the support beneath the joints and slab edge. It also helps to displace free water. Following the identification of the voids, this method entails three steps: post-testing the stabilized slabs, grout injection, and hole placement and drilling. Slab stabilization doesn't stop erosion, eliminate faulting, raise the design structural capacity, or fix depressions. Nonetheless, it does so by restoring the slab support, which lessens deflections under load. Only where there is a loss of support, such as at joints and fractures, should stabilization be applied. Vacuum can be found most easily by visual inspection. Transverse joint faulting, corner breaks, shoulder drop-off, lines at or near joints, and cracks are indicators that repair is necessary. Another method often used to find voids is deflection testing. Testing should ideally be done at night when temperatures are lower because this is when joints open, aggregate interlock decreases, and load deflections are greatest.

1.2.4.3. Testing

When electromagnetic waves are pulsed into the pavement by ground-penetrating radar, the reflected signal is measured and graphically displayed. This may make holes and other flaws visible. The epoxy/core test uses both mechanical and visual techniques to find voids. Using a dry-bit rotohammer, a 25 to 50-millimeter hole is drilled through the pavement and into the sub-base. After that, a two-part epoxy that has been colored for visibility is poured into the hole. After the epoxy solidifies, experts drill through the opening. The epoxy will adhere to the core and produce tangible proof if there is a void. Polyurethane and grout made of pozzolan cement are common stabilizing materials. Slab stabilization necessitates both strength and the capacity to expand or flow into tiny spaces. Utilizing the pozzolan-cement grouts requires the use of colloidal mixing equipment. A non-pulsing progressive cavity pump or a positive displacement injection pump must be used by the contractor to apply the grout. Drilling is also required, but it has to leave a clean hole free of breakouts or surface spalling. The grout packer in the injection devices needs to be able to seal the hole. In the unlikely event that employees notice slab movement on the uplift gauge, the injection device must also include a return hose or a fast-control reverse switch. Sensitive dial gauges are required for the uplift beam, which aids in slab deflection monitoring.

1.2.4.4. Joint Sealing

This technique, also known as joint and

crack repair, aims to reduce the amount of surface water and incompressible material that seeps into the joint system. Joint sealants are also applied in concrete pavement restoration procedures to lessen dowel bar corrosion. In order to successfully reseal, the reservoir must be shaped and cleaned, the backer rod must be installed, and the sealant must be applied. The old sealant is removed by sawing, hand removal, plowing, and cutting. The reservoir is shaped with saws. After cleaning the reservoir, there should be no dirt, dust, or remnants of the previous sealant. To get rid of any sand, dirt, or dust, it is advised to water wash, sandblast, and then air blow. To install the backer rod to the required depth, a steel roller with two wheels is needed. The sealant is inserted into the joint following the insertion of the backer rod. For this method, a variety of materials can be used, such as silicone, hot pour bituminous liquid, and preformed compression seals.

1.2.5. Safety Considerations

The harm (deaths, injuries, and property damage) caused by traffic collisions on the highway system can be decreased and road traffic safety can be increased with careful design and construction. Bicycle lanes, pedestrian crossings, safety barriers, and traffic calming measures can all help keep drivers, cyclists, and pedestrians safe on neighborhood roads. Some nations and states use Botts' dots or Cat's eyes as lane markers. Botts' dots can be embedded in short, shallow trenches carved in the roadway, as is done in the mountainous regions of California, but they are not used where it is icy in the winter because frost and snowplows can break the glue that holds them to the road. In order to lower the risk of head-on collisions, major roads can benefit from grade-separated junctions, median dividers between opposing-direction traffic, and restricted access from nearby roads and properties.

Positioning energy-attenuation devices (e.g. sand barrels, broad grassy spaces, and guardrails) are also typical. Certain elements of roads, like fire hydrants and road signs, are made to collapse upon collision. Instead of forcefully stopping a car that strikes a light pole, light poles are made to shatter at the base. Larger trees may also be taken down from the direct vicinity of the road by highway authorities. Flooding may occur during periods of intense rainfall if the road's surface is not higher than the surrounding terrain. Limiting speed can increase road safety and lower the number of people killed in traffic accidents. The World Health Organization (WHO) names speed control as one of several interventions likely to help reduce road casualties in their World Report on Road Traffic Injury Prevention Report.

1.2.6. Road Conditions

Road conditions refer to the factors that determine how easy it is to drive on a specific road or in a certain area, such as pavement quality, potholes, road markings, and weather. Transportation issues and road conditions are the main causes of road accidents, with weather and road conditions being closely connected. Certain road conditions are important for different purposes, such as autonomous vehicles which may be affected by changes in shadowing, lighting, road surface texture, and road markings. Government agencies and private entities, like local news services, monitor and inform the public about road conditions to ensure drivers are aware of any hazards in their area. News outlets rely on tips from residents to provide accurate coverage of road conditions in their communities.

1.2.7. Environmental Performance

Adequate planning and building of a road can mitigate any adverse effects on the environment. Roadside pollution can be lessened by utilizing water management systems. Water pollution is often caused by rainwater and snowmelt that run off of roads and pick up pollutants like motor oil, trash, heavy metals, and gasoline. The primary source of polycyclic aromatic hydrocarbons (PAHs), which are produced as combustion byproducts of gasoline and other fossil fuels, as well as nickel, copper, zinc, cadmium, lead, and other elements, is road runoff. Road salts can be poisonous to delicate plants and animals, and de-icing agents and sand can wash up on roadsides, contaminating surface waters and groundwater. When sand is used on icy roads, it can be ground up by passing vehicles into tiny particles, which worsens air pollution. One of the main causes of noise pollution is roads. It was realized in the early 1970s that road design could be used to both influence and reduce noise generation. Near built-up areas, noise barriers can help minimize noise pollution. The use of engine braking may be restricted by regulations. Emissions from motor vehicles cause air pollution. In comparison to certain distances away from the road, there are higher concentrations of air pollutants and detrimental effects on respiratory health close to the road. Vehicle-generated Road dust can cause allergic reactions. Furthermore, scientists claim that greenhouse gas emissions from on-road transportation account for most of the causes of climate change.

1.2.8. Regulation

Depending on the nation, traffic moves on either the left or the right side of the road. When crossing a two-way road, pedestrians should be aware that traffic is approaching from the left first. In nations where traffic flows on the right, traffic signs are typically located on the right side of the road. Roundabouts and traffic circles also rotate counter-clockwise or anti-clockwise.

The opposite is true in nations where traffic moves left. By population, roughly 33% of people drive on the left and 67% drive on the right. Despite the fact that, historically, most traffic traveled on the left, approximately 28% of drivers travel on the left and 72% on the right across road distances. The standard practice of driving on either the right or left side of the road is referred to as road regulation, specifically about right- and left-hand traffic. This practice is essential for upholding safety and order on the roads. This law affects different facets of traffic management, vehicle manufacturing, and road design and varies by nation.

- **Right-Hand Traffic (RHT):** Cars are driven on the right side of the road in nations that allow it. The driver can sit closer to the center of the road because the steering wheel is usually located on the left side of the car. This configuration is meant to make it easier to pass and improve visibility of approaching vehicles. About 65% of the world's countries, including the US, Canada, and the majority of Europe, have right-hand traffic as standard.
- **Left-Hand Traffic (LHT):** On the other hand, countries with LHT laws mandate that cars be driven on the left side of the road. The steering wheel is typically located on the right side of the car in these nations, putting the driver closer to the center of the road.

In a similar vein, this arrangement facilitates improved visibility and safe overtaking. About 35% of people on Earth reside in nations where left-hand traffic is the norm, such as the UK, Japan, Australia, and India.

The decision to drive in left- or right-hand traffic frequently has historical precedent. For instance, left-hand traffic in the UK has its origins in the Middle Ages, when knights favored staying to the left in order to place their sword-wielding right hand closer to a possible adversary. In contrast, the United States and France, who both opted for this system in the 18th and 19th centuries, are credited with influencing the adoption of right-hand traffic in many other nations.

1.2.8.1. Implications of Traffic Regulations

- **Road Design:** The direction of traffic plays a crucial role in road design, influencing the placement of traffic signs, road markings, and the layout of intersections and roundabouts. In countries with right-hand traffic, road signs are positioned on the right side of the road, while in left-hand traffic countries, they are placed on the left. Intersections and roundabouts are carefully designed to facilitate smooth and safe turning and merging based on the traffic flow.
- **Vehicle Manufacturing:** Vehicles are manufactured with the steering wheel located on the opposite side of the driving lane. This means left-hand drive (LHD) vehicles are made for countries with right-hand traffic, and right-hand drive (RHD) vehicles are produced for left-hand traffic countries. This design feature is crucial for safety, as it enhances the driver's visibility of incoming traffic and provides easier access to road controls.
- **International Driving:** Understanding and adjusting to different countries' traffic regulations can be challenging for travelers and imported vehicles. Many countries with opposite traffic directions offer warnings and reminders at border crossings and major roads. Some nations have specific regulations for foreign vehicles to ensure road safety.
- **Safety and Standardization:** Consistent adherence to either right-hand or left-hand traffic within a country is essential to prevent confusion and accidents. Regulatory bodies enforce strict rules to ensure all road users follow the designated traffic directions. This includes driver education, signage, and road design standards that reinforce traffic regulations.

1.2.9. Economics

In order to maximize the advantages and reduce the expenses related to building and maintaining road infrastructure, economics for road development entails the complex analysis and management of financial resources. Making sure that investments in road projects are financially feasible and favorably impact overall economic growth depends on this field.

Cost-benefit analysis (CBA) is the fundamental tool used in the economic analysis of road development to assess a project's financial viability. CBA measures all expenses, including building, purchasing land, mitigating environmental effects, and upkeep, and compares them to expected benefits like shorter travel times, cheaper car operating expenses, increased accessibility, and safer travel. Making sure the benefits' net present value (NPV) is greater than the expenses is the aim to justify the investment. An essential component of the economics of road development is funding. Many strategies are used, such as public-private partnerships (PPPs), where private companies invest in return for future revenue streams, foreign loans, and grants from institutions such as the World Bank, and direct revenue generation from tolls and user fees. Governments may also issue bonds to raise money, paying investors interest as they go. In order to meet the significant financial requirements of road projects and guarantee their sustainability, these various funding strategies are crucial.

The construction of new roads has significant and varied economic effects. By cutting down on travel times and vehicle operating expenses, improved road infrastructure increases productivity by improving the efficiency of the movement of goods and services. Enhanced efficiency leads to increased competitiveness and economic output. The construction of new roads facilitates access to markets, healthcare, education, and employment opportunities while also enhancing connectivity between rural and urban areas, all of which are important components of regional development. Furthermore, because improved accessibility and shorter travel times result from road improvements, property values are frequently raised. Further promoting economic growth are the employment opportunities created by the construction and maintenance phases, which occur both directly in the construction industry and indirectly in related industries. An important factor in the economics of road development is sustainability. Contemporary methods place a strong emphasis on environmental preservation. Examples include using sustainable materials, building roads with little ecological impact, and integrating green infrastructure. Prioritizing social equity means that road projects enhance access to vital services and employment opportunities for marginalized communities as well as other societal segments. Roads' long-term viability is also considered when designing infrastructure, which makes it resistant to natural disasters and climate change and guarantees that it will continue to operate efficiently for the duration of its intended life.

1.2.9.1. The Need for Roads

The road's construction and the infrastructure it supports improve people's living conditions, boost trade, encourage entrepreneurship, and lower the cost of goods, all of which contribute to the reduction of poverty. The population is served by police, fire, rescue, ambulance, repair, and construction services because of the roads. The distribution of healthcare and educational services to previously unreachable areas is made possible by the developed road network. In situations where there are multiple routes and competing modes for both personal and freight (road, rail, air, or ferry), and where induced demand can result in increased or decreased transport levels when road provision is increased by building new roads or decreased (for example California State Route 480), transport economics is used to understand both the relationship between the transport system and the wider economy and the complex effects of the road network structure. The public sector typically builds and maintains roads through taxes, though it occasionally uses road tolls or hires private contractors to carry out the work. Communities can address the rising costs by bringing private funds into the infrastructure through public-private partnerships.

There are four main ones:

- design/build
- design/build/operate/maintain
- design/build/finance/operate
- build/own/operate

Society depends heavily on efficient roads. In the European Union (EU) 44% of all goods are moved by trucks over roads and 85% of all people are transported by cars, buses, or coaches on roads. The term was also commonly used to refer to roadsteads, and waterways that lent themselves to use by shipping.

1.2.9.2. Construction Costs

According to the New York State Thruway Authority, some sample per-mile costs to construct multi-lane roads in several US northeastern states were:

- Connecticut Turnpike – $3,449,000 per mile
- New Jersey Turnpike – $2,200,000 per mile
- Pennsylvania Turnpike (Delaware Extension) – $1,970,000 per mile
- Northern Indiana Toll Road – $1,790,000 per mile
- Garden State Parkway – $1,720,000 per mile
- Massachusetts Turnpike – $1,600,000 per mile
- Thruway, New York to Pennsylvania Line – $1,547,000 per mile
- Ohio Turnpike – $1,352,000 per mile
- Pennsylvania Turnpike (early construction) – $736,000 per mile

1.3. ROAD ALIGNMENT

Selecting the proper alignment for a road involves more than just math; it also involves aesthetic judgment. Maintaining the procedure's uniformity is essential to the design of horizontal alignment so that it can seamlessly integrate into the surrounding topography. It must therefore offer a gentle alignment that validates natural contours, since inflexible or straight alignments damage the environment, ruin naturally occurring slopes, and interfere with plant development. When choosing alignments, keeping the current features intact should be considered. The alignment of a road or highway is determined by the location of its centerline. Road alignment is another name for highway alignment. The road's alignment allows it to travel through the easiest, least complicated, and most cost-effective route possible. Stated differently, it pertains to the orientation of the road.

- To reduce construction costs, the road alignment should be as short as possible. To provide higher traffic speeds and lower construction costs, the alignment should be as straight as possible.
- The alignment of the road must be convenient for construction, maintenance, and traffic flow. At right angles to railways, other roads, and bridges, the alignment should cross them.
- Rivers, canals, and streams must be crossed at places where their width is the smallest. In order to serve the greatest number of people, the alignment should link intermediate towns and a group of villages.
- Natural beauty and scenic areas should not be crossed by the road alignment. There should be a minimum number of bridges, crossing culverts, and embankment places in the alignment.
- The curve must be smooth and the gradient should be easy to follow. In order to avoid earthworks in embankments or cutting, minimal earthworks should be done.
- A reasonable sight distance should be provided by the alignment. Water bodies, lakes, wells, monumental buildings, historical structures, etc., should be removed.
- Ideally, the alignment should pass through areas with easy access to labor and materials for road construction. For the road to run smoothly, it must run on good soil that can bear the weight of traffic safely without damaging the surface.
- It is important that the alignment does not pass through more expensive and cultivated land, as well as through forests.

Usually, roads are built to facilitate the development of an area. The road alignment should not pass through historical and cultural sites.

Road alignments are defined by mandatory points. Road alignments must pass through important towns, villages, and places of religious, social, political, and commercial significance.

1.3.1. Types of Highway Alignment

The alignment of the road is one of the most crucial considerations when constructing or rebuilding any road. Highway alignment is the process of placing a road to its surroundings. It takes into account a number of factors, including topography, traffic volume, and the infrastructure already in place, the environment, and the economy. Highway alignment plays a crucial role in planning in several ways, including lowering construction costs and guaranteeing safety.

In order to guarantee that a road can be constructed in a manner that is safe, efficient, and economical, the alignment of the road must be carefully planned. In addition to taking the environment into account, highway alignment must also take potential risks into account. A road's alignment is determined by several factors, such as the land's contours, the volume of traffic it is expected to handle, the infrastructure already in place, environmental considerations like wetlands, and financial limitations.

The planning process can be finished when these elements are taken into account. The purpose of the road, as well as the environment and terrain it must pass through, determine the type of highway alignment that is selected. The three primary components of highway alignment are vertical alignment, horizontal alignment, and grade alignment.

- **Grade Alignment:** The road surface's slope is referred to as the grade alignment. A road's grade must be appropriate for the intended speed and user comfort. Highway grade alignments can be vertical, with the road having an incline or decline, or horizontal, with the road surface being flat and the highest and lowest points being the same.
- **Horizontal Alignment:** The most typical kind of highway alignment is horizontal alignment. This kind of alignment relates the road's layout to its spatial location. A highway's horizontal alignment is intended to guarantee that, considering the topography of the area, the road takes the most direct route. The right-of-way must be considered when designing a highway's horizontal alignment.
- **Vertical Alignment:** The portion of highway alignment that takes elevation of the road surface into account is called vertical alignment. By making sure that the road's grade is as flat as possible, vertical alignment helps to minimize the energy needed to travel across it. By making sure that slopes are gradual and the road is kept as low as possible, it also helps to reduce the negative environmental effects of the road.

Choosing the right highway alignment is necessary to use any road safely and effectively. When choosing the best kind of highway alignment, it is critical to take the road's intended use, the topography of the surrounding area, and the road's impact on the environment into account. The amount of energy needed to travel a highway can be decreased by choosing the proper highway alignment, which can also lower construction and maintenance costs.

1.3.2. Factors Controlling Highway Alignment

A highway's alignment can vary, and it is vital to take into account the numerous variables that influence highway alignment. These elements typically fall into two groups: economic and geographic characteristics. A highway's alignment is largely influenced by its geographic features. The topography, or the actual shape of the land, is a crucial consideration in determining the best route for a highway. Highways should, in general, follow the path of least resistance, or as closely as possible, the path of the land. For instance, if the land is level, the highway should be as straight as possible. The highway should be curved to follow the land's contours if the terrain is hilly. The current state of the road network has an impact on highway alignment as well. When choosing the alignment of a highway, the existing roads and highways should be taken into account if the highway needs to connect two points. For instance, the highway should be oriented to consider any existing roads that link the two locations. This will lessen the highway's environmental impact while also lowering the cost of construction. The volume of traffic is another crucial consideration when choosing a highway's alignment. A highway's alignment should consider both the anticipated volume of traffic and the speed at which that traffic will be moving. For instance, a highway should be built to handle high traffic volumes and high speeds if that is its intended use. This will guarantee the highway's effectiveness and safety.

The environment plays a significant role in determining a highway's alignment. The effects a highway will have on the environment should be considered when planning its alignment. For instance, the highway design should minimize its impact on the environment if it passes through a naturally beautiful area. Lastly, choosing a highway's alignment can depend on several economic factors. When choosing a highway's alignment, construction costs play a significant role. The alignment that offers the best value for money should be selected. Any grants or subsidies that are available for the highway's construction may be considered in addition to other economic factors. In general, a highway's alignment should be carefully chosen to guarantee that it satisfies the needs of its users. Highway alignment is determined by several factors, including geographic features, traffic volumes, road infrastructure, environmental considerations, and economic considerations. A thorough consideration of each of these elements is necessary before choosing a highway's alignment.

1.3.3. Impact of Factors on Highway Alignment

Highway alignment is a crucial consideration when designing or building a new road since it has a big impact on the project's expenses, efficiency, and safety. It is crucial to take into account the various aspects influencing highway alignment when designing or building a new road.

- **Geographical features.** Geographical features have a major influence on a highway's alignment because they determine the highway's route, the amount of construction required, and the locations where it can be built. The topography of the area, including its slopes and grades, will dictate the kind of alignment required for the highway. The kind of alignment required will also depend on the land's topography.
- **Topographical features.** The topography of the area plays a significant role in choosing a highway's route. In determining the highway's route, slopes and

grades will be important factors. If the terrain is level, the path can be straight and accommodate a grade alignment. But if there are grades or slopes, the alignment needs to be changed appropriately.

- **Volumes of Traffic.** The alignment will also be impacted by the amount of traffic on the highway. It might be necessary to build a wider, more accommodating horizontal alignment if there is a lot of traffic. This will enhance safety and lessen traffic.
- **Infrastructure for roads.** When deciding on a highway's alignment, the current road network must be considered. The location of any existing roads the highway must intersect with must be considered because it will dictate how the highway should be directed. The kind of alignment required will also depend on the current infrastructure, since straight alignments may require assistance from the roads.
- **Environmental Elements.** The environment plays a major role in determining the alignment of highways. In order to make sure that the construction of the highway does not harm or destroy the environment, consideration should be given to the environment. To guarantee that the highway route does not disturb ecosystems or natural habitats, great care must be taken.
- **Economic Factors.** When choosing a highway's alignment, economic factors are also very important. It is necessary to take certain financial limitations into account when building a highway. It is necessary to consider the cost of labor, the cost of materials, and the cost of construction. The kind of alignment that can be achieved will depend on the available funds since certain alignments might cost more and require more resources than others.

The various elements governing highway alignment collectively have a significant influence on a new road's construction and safety. The geographical features, topography, traffic volumes, road infrastructure, environmental factors, and economic factors that will affect the highway alignment must all be taken into account when planning or building a new road. Taking these things into account will guarantee the highway is built effectively and safely.

1.3.4. Benefits of Proper Highway Alignment

An essential component of road planning is highway alignment. Motorists gain from increased traffic flow, lower construction costs, and safer roads when a highway is properly aligned. Moreover, the best possible highway alignment can have additional benefits that can lower travel times and increase the overall efficiency of the transportation system.

- **Enhanced Road Safety:** The security of drivers depends on the alignment of the highway. Road alignment can increase visibility, lower the chance of collisions, and guarantee that all paths and intersections have signage. Road alignment that is appropriate also lessens the need for overpasses, which can be costly and challenging to build. Highway alignment can significantly lower the chance of accidents and increase user safety by avoiding these.

- **Lower Construction Costs:** Proper road alignment can also lower construction costs. Road construction can be simple and affordable when done correctly. Because fewer resources are required to complete the construction process, ensuring that roads are aligned correctly will significantly lower construction costs. Additionally, correctly aligned roads can lessen the need for costly upkeep down the road.
- **Improved Traffic Flow:** Another benefit of highway alignment is an increase in traffic flow. Roads that are properly aligned can ease traffic and make driving more comfortable. Furthermore, since properly aligned roads allow for improved visibility and more effective routing, they can contribute to a decrease in the number of accidents that happen. Travel times can be shortened and traffic flow can be enhanced as a result.
- **Enhanced Efficiency:** The overall efficiency of the transportation system can also be enhanced by optimal highway alignment. Road alignment can lessen the number of vehicles that need to be routed and the need for detours. As a result, the transportation system may operate more efficiently overall and use less fuel.
- **Decreased Travel Times:** Lastly, correctly oriented roadways can also aid in a reduction in travel times. Proper alignment of roads offers drivers more effective and direct routes, resulting in less time spent on the road and shortened travel times.

As is evident, appropriate highway alignment can benefit road design in several ways. Highway alignment can help to make the roads safer, lower construction costs, and give drivers a smoother driving experience, from increased road safety to shorter travel times.

1.3.5. Challenges of Creating Highways

Highway design is a very complicated process because planners have to take a lot of things into account to make sure everything works out. A lot of these variables can be hard to forecast or manage, which can hinder highway alignment success. Taking into account too few or too many factors could lead to a number of potential risks.

- **Geographic Challenges:** When planning to build a highway, it can be challenging to appropriately account for the terrain, landforms, and other geographic features of the area. These characteristics have the potential to seriously affect a road's alignment and to reduce the success of its design and construction.
- **Construction Costs:** Rebuilding or repairing poorly designed or built highways can be expensive. Alignment problems have the potential to significantly raise construction costs and cause completion delays if they are not considered during the initial planning stages.
- **Traffic Flow:** Outdated or ineffective highway alignment can result from unforeseen changes in traffic flow or from shifting traffic patterns. In order to guarantee that a highway is efficient and effective, proper planning takes into account both the traffic patterns of the present and the anticipated future.

- **Impact on the Environment:** Alignment issues with highways can also result from environmental changes. Changes in topography or altered landforms brought about by natural disasters or climate change may cause a highway's intended alignment to be disrupted.
- **Economic Constraints:** The success of highway alignment may also be hampered by a lack of funding. Alignment problems may occur if planners are unable to secure the funding required to construct a highway to the required standards.

In order to ensure that a highway is properly aligned, planners must take into account a variety of factors, some of which can be unpredictable. Planners can design safe, effective, and economical highways by considering all relevant variables.

CASE STUDY

NATIONWIDE DESERT HIGHWAY ASSESSMENT: A CASE STUDY IN CHINA

There are about 7,900 kilometers of desert highways in China, mainly distributed in the Xinjiang Ugur Autonomous Region, Ningxia Hui Autonomous Region, Neimengu Autonomous Region and Gansu Province, and some other areas. The differences between desert highways and general highways are because of the arid climate, the wind sand, and the temperature of highway construction engineering. Another difference is the shortage of suitable filling materials such as soils and stones for highway construction in desert regions, which makes it more difficult to construct a desert highway. A highway exists in the natural environment as a structure. The elements of the natural environment affect highway construction engineering. On another hand, the construction affects the natural environment and puts the ecological environment at a disadvantage. Especially in desert regions the highways can always be blocked by the encroachment of sand dunes [Figure 1(a)], and the highway traffic can cause severe wind erosion of grassland soils [Figure 1(b)]. To satisfy the variety and hierarchy of highway construction and discover the spatiotemporal distribution of the desert natural environment and its effects on highway construction, a desert highway assessment should be carried out, which will effectively pave the way for route planning, design, construction, maintenance, and subsequent management of desert highways, and help formulate standards.

Figure 1. *Interactive effects between highway engineering and the natural environment.*

Source: https://www.mdpi.com/ijerph/ijerph-08-02734/article_deploy/html/images/ijerph-08-02734f1.png.

Desert highway assessment studies have been going on for many years. The earliest desert assessment can be dated back to Lamprey's research in central western Sudan when he introduced the desert encroachment theory, which laid the groundwork for providing a process for measuring the desertification problem. A provisional methodology for assessment and mapping of desertification was formulated and is now used for local and regional assessment and mapping. Most of the studies assess the desert based on the desert topography, vegetation coverage, and wind erosion. The desert is divided into two kinds (fixed and mobile) or three kinds (fixed, partly fixed, or mobile), according to the extent of the activity of the sand dunes. China began a nationwide desertification assessment in the mid-1980s. Historically, China has focused on sandy desertification, which is caused by wind erosion. This classification method was based on geo-morphological changes caused by blowing sand (such as sand dunes and sand mound formations, sand cover, and surface coarseness), the percentage of change in relation to the total area, and the annual rate of increase of desertified land. These classifications have been broadly used in sandy desertification assessment. Based on hundreds of expert questionnaires, sandy desertification intensities were classified as slight, medium, severe, or extremely severe. The main assessment indicators were bare sand ratio, vegetation coverage ratio, and soil texture, and the importance of each indicator was calculated by the Delphi weighted method.

However, a desert highway assessment that comprehensively describes the natural highway distribution or single local highway distribution has not been reported until now at home or abroad. With the increase in highway construction, especially because of the Western Development Plan of China, highway construction in the desert regions has accelerated considerably and the highway mileage has increased yearly. Li Bin had put forward the idea of desert highway assessment in China. He designed the indicators as the moisture coefficient and aimed to determine the differences in the highway subgrade protective level in desert regions. However, he only put forward the theory, and the natural region division map for desert highway assessment was not drawn out. His studies were rough and could not be applied to actual highway construction.

Index System of Highway Desert Assessment in China

The dominant indicator for assessing the desert comprehensive state must be based on a sufficient investigation and a wide collection of field data. Then the advanced technology of GIS and related databases is used. The function equation is established according to the real experience and data files, such as highway engineering standards and protocols. The equation is also combined with the real situation of the highway construction to select the important environmental factors, which are crucial to the highway construction in the desert regions. After the equation is established, every parameter is defined and calculated.

The scale value and weighting coefficients are made clear based on the commensurability theory. Finally, the comprehensive state index of every division region is obtained with the dominant and auxiliary indicators, which form the natural region division map. Based on the order of importance, the natural region division map of the desert comprehensive state index is drawn. According to one dominant indicator (the desert comprehensive state index) and four auxiliary indicators (the sand dune height, the blown sand strength, the vegetation coverage ratio, and the annual average temperature difference), the highway desert assessments are carried out.

The Definition of Important Indicators

The desert comprehensive state index means the effects of the natural environmental factors on highway construction in the desert region. According to years of field experience, the natural environment factors are crucial to the route design, construction, maintenance, and running management of highways, and mainly include the sand dune height, the blown sand strength, the vegetation coverage ratio, and the annual average temperature difference.

- **The sand dune height:** The sand dune height is one of the most significant indicators for analyzing the physiognomy features of blown sand regions. It can reflect the time of formation of the desert physiognomy and the scale of the sand dune growth. Generally, the longer the time of formation of the sand dune is, the more height the sand dune has, due to the sand movement and sand accumulation under wind for a long time. On the contrary, the newly grown sand dunes are short with less sand accumulation. The sand dune height can also reflect the active strength of sand dune movement. Under the same wind flow and natural situation, the height difference is dominant in the active strength of the sand dune movement.

The sand dune height affects the route design, construction, and the later management of the highway engineering operations. The sand dune height directly affects the route layout of the highway and at the same time causes blown sand damage to the highway. During the later running management and maintenance, it is more difficult to maintain the highway in the desert environment, so the effect of the sand dune height on the highway is crucial to recommend the route layout plan and protect the highway from desert damage.

The height of sand dunes is expressed by the average value of all sand dune heights in a region. According to the 1:25,000,000 digital elevation model in China, the sand dune height distribution is not uniform, however, most sand dune heights are within a fixed range, so the average height of sand dunes in a region can be adopted to represent the wave distribution of sand dunes.

- **The blown sand strength:** The blown sand strength is defined as the cumulative velocity record of the blown sand during a rated time (month, season, or year). It depends on the observation times and the observed value of the wind velocity every time, so its unit is m/(s·time unit). For the desert highway assessment in China, the time unit is 30 years and its unit is m/(s·year). The blown sand strength is one of the most important indicators for blown sand movement and is proportional to the wind velocity. It is crucial to the design of the desert highway and also to the subsequent management and maintenance. It will affect the route layout of the highway and cause desert damage. During the construction of highways in desert regions, more attention should be paid to the blown sand strength. It is defined by the sand diameter, the ground characteristics and the water content of the sand. The blown sand strength can indirectly reflect the wind velocity. This is the first time the blown sand strength has been adopted as one of the indicators for desert highway assessment, which can direct the route layout of the highway and its later running management and maintenance.

- **The vegetation coverage ratio:** Vegetation is a general term for the plant life of a region. The vegetation coverage ratio is defined as the percent of the planimetric area of the vegetation (leaves, branches, and stems) on the statistical region area. In desert regions, vegetation is rare, so the ratio is low and in some desert regions there is even no vegetation. However, the vegetation coverage can reflect the stability of a sand dune. Furthermore, it affects the design, construction, and protection of the highway, so it is crucial to evaluate the effect of the vegetation on the highway construction, recommend the optimal route layout, and build different level protective measures into the highway engineering plans.
- **The annual average temperature difference:** The annual average temperature difference means the difference between the maximum and the minimum of the average annual temperature in a region. It reflects the climate effect on the highway construction. The highway will have a route layout exposed to the natural climate, which includes the temperature and rainfall. In a general region, with periodical changes in the temperature, the moisture in the subgrade will undergo frost and thaw cycles, so there can be frost boiling damage, but in the desert regions the rainfall is rare, so the effect of the rainfall on the subgrade is small and the temperature changes affect the pavement more. The asphalt pavement can suffer fatigue cracking under repeated exposure to high and low-temperature cycles. At the same time, the high temperatures will lead to the asphalt aging. The characteristics of the asphalt and its mixtures are different under high or low temperatures. The requirements for asphalt in desert regions are high because the differences between the maximum and minimum temperatures are large. The annual average temperature difference in the desert region directly affects the structure of the highway and its pavement material selection.

Mathematical Equation

The desert comprehensive state can reflect the effects of natural factors on highway construction. It consists of the four natural factors, such as the sand dune height, the blown sand strength, the vegetation coverage ratio, and the annual average temperature difference. The four natural factors have different characteristics, content, and dimensions, so the first step for establishing the function of desert comprehensive state index should be to normalize all the natural factors according to the comprehensive evaluation theory. Then the weight coefficient of every factor should be defined. The established equation is shown in Equation 1:

$$NEP_i = K_h \cdot M_{ih} + K_p \cdot M_{ip} + K_f \cdot M_{if} + K_t \cdot M_{it} \quad (1)$$

where: NEP_i the state for the desired calculation unit;

K_h the weight coefficient of the sand dune height;

K_p the weight coefficient of the vegetation coverage ratio;

K_f the weight coefficient of the blown sand strength;

K_t the weight coefficient of the annual average temperature difference;

M_{ih} the normal contribution of the sand dune height in the desired calculation unit;

M_{ip} the normal contribution of the vegetation coverage ratio in the desired calculation unit;

M_{if} the normal contribution of the blown sand strength in the desired calculation unit;

M_{it} the normal contribution of the annual average temperature difference in the desired calculation unit.

The first step is to normalize the four natural factors, and then calculate the normal contribution of every factor. Following the 1994 United Nations Convention to Combat Desertification (UNCCD), the Chinese Committee for Implementing the UN Convention to Combat Desertification (CCICCD) organized a group of scientists to assess nationwide desertification. The group studied wind erosion, water erosion, temperature and vegetation in addition to degradation by other driving factors. This study determined four desertification intensities (slight, medium, severe, and extremely severe) and the scientists formulated a detailed assessment criteria for classification. Considering the significance of every parameter, the scientists' assessment criteria were based on past research results and extensive expert discussions (Table 1).

Table 1. Scale Value of Every Indicator

Comprehensive characteristics state	Unit	Desertification intensity classification			
		Slight	**Medium**	**Severe**	**Extremely severe**
		1	**3**	**6**	**10**
Sand dune height(H)	m	<=2	2~20	20~60	>60
Vegetation coverage ratio (P)	%	>=40	40~30	30~15	<15
Blown sand strength (F)	m/(s·year)	<=2,500	2,500~3,500	3,500~4,500	>4,500
Annual average temperature difference (T)	°C	<=30	30~35	35~40	>40

Source: https://www.mdpi.com/1660-4601/8/7/2734#:~:text=The%20desert%20highway%20assessment%20divides,not%20conflict%20with%20each%20other.

Then the weighting coefficient of every factor is defined as shown in Table 2. The values of every coefficient in Table 2 are defined by the scores of ten specialists, who are recognized in the highway design and construction area.

Table 2. *Contribution Coefficients to the Desert Comprehensive State Index*

	Sand dune height	Vegetation coverage ratio	Blown sand strength	Annual average temperature difference	Total
Contribution coefficient	0.3	0.3	0.28	0.12	1

Source: https://www.mdpi.com/1660-4601/8/7/2734#:~:text=The%20desert%20highway%20assessment%20divides,not%20conflict%20with%20each%20other.

Calculation Process of the Four Auxiliary Indicators

- **The sand dune height:** The dedicated analysis tool for the sand dune height is developed based on the ArcInfo9.1 software and GIS module. The unit is selected as 5 kilometers square. The Digital Elevation Model of the 1:25 desert region in China is traversed one unit by one unit. The maximum value and average difference of elevation of every unit are calculated, which forms the distribution data of the sand dune height in China. According to Table 1., the scale value is calculated and the map of the sand dune height is drawn.
- **The blown sand strength:** The weather data from the local weather stations in the desert regions are collected. The cumulative annual average strength of blown sand was calculated for 30 years. With the software ArcInfo9.1 and its "Geostatistical Analysis" module, the spatial distribution of blown sand strength is analyzed by the Cregeen interpolation method, which borrows from the spatial variation theory. Further, the isopleth map of the blown sand strength is drawn. The average strength of blown sand in every region unit can then be calculated. According to the calculated results and the criteria of Table 1., the level value of every region unit is obtained and the map of the blown sand strength is drawn.
- **The vegetation coverage ratio:** According to the 1:1,000,000 map of vegetation coverage in China, the percentage of vegetation coverage in every region unit can be calculated using the ArcInfo9 software. According to the calculated results and the criteria of Table 1., the level value of every region unit is obtained and the map of the vegetation coverage is formed.
- **The annual average temperature difference:** The weather data are collected from the local weather stations in the desert regions. The annual average temperature difference is also calculated for 30 years. The maximum and minimum annual average temperatures in every region unit are obtained. With the software ArcInfo9.1 and its "Geostatistical Analysis" module, the spatial distribution of the annual average temperature in every region unit is analyzed. Further, the isopleth map of the annual average temperature difference is drawn. The temperature difference in every region unit can be calculated. According to the

calculated results and the criteria of Table 1., the level value of every region unit is obtained and the map of the annual average temperature difference is formed.

Natural Region Division Map for Desert Highway Assessment in China

By Equation 1, Tables 1 and 2, and the four auxiliary indicators calculation method, the desert comprehensive state index is calculated for every regional unit. Considering the maximum and minimum value of the index, the index is divided into three levels, as shown in Table 3.

Table 3. *Effect of the Desert Environment on Highway Construction*

Desert Comprehensive state index	**≤3**	**3~6**	**7~10**
Level of effect on highway construction	slight	medium	severe

Source: https://www.mdpi.com/1660-4601/8/7/2734#:~:text=The%20desert%20highway%20assessment%20divides,not%20conflict%20with%20each%20other.

The dominant indicator is defined as the desert comprehensive state index. The four auxiliary indicators are defined as the sand dune height, the blown sand strength, the vegetation coverage ratio, and the temperature annual average difference. The division boundary of desert highway assessment is mainly drawn according to the dominant indicator and revised in fine detail by the four auxiliary indicators. By the three-section name method (geographical position + affection strength + title) the Chinese deserts are classified into three regions, as shown in Table 4. The first region is named as the east medium effect region, the second as the middle medium-severe effect region and the third as the west slight-medium effect region.

Table 4. *Desert Highway Natural Division in China*

Name	**Main Criteria and index value**	**Range**
I the east medium effect region	The comprehensive state index is within the region [0 6] and the region [4 6] is dominant.	Hulun Beir Sandy Land Nenjiang sandy land Horqin Sandy Land Hunshandake Sand Land

II the middle medium-severe effect region	The comprehensive state index is within the region [3 10] and the region [4 6] is dominant.	Mu Us Sandy Land Kubuqi Desert Ulan Bub Sandy Land Tengger Desert Badan Jilin desert Desert of Gansu Corridor Kumtag Desert Desert of Qaidam Basin Eastern of Gurbantunggute Desert
III the west slight-medium effect region	The comprehensive state index is within the region [0 6] and the region [0 3] is dominant.	Middle and Western of Gurbantunggute Desert Taklimakan Desert

Source: https://www.ncbi.nlm.nih.gov/pmc/articles/PMC3155326/.

As shown in Figure 2, in the east medium effect region, there are continuous sand dunes. The height difference is small among all the sand dunes. During the route design, the long straight lines should be dominant, the vertical section should follow the natural geology and the low-filling embankments should be dominant.

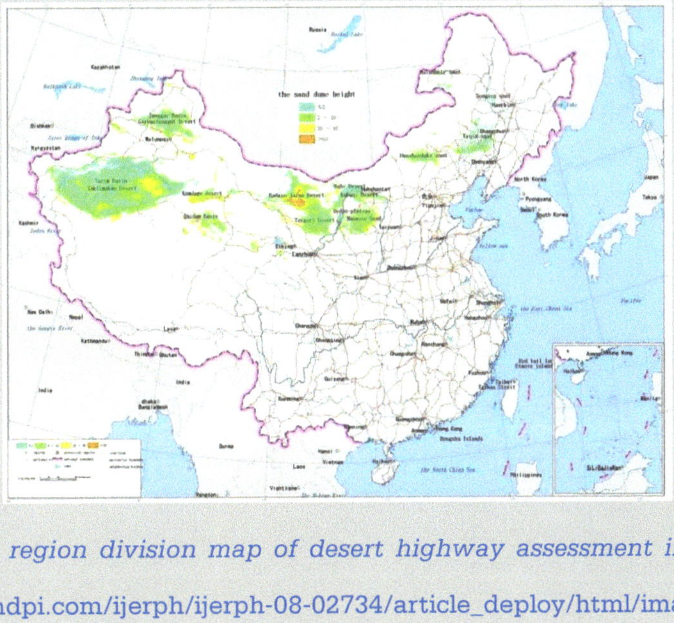

Figure 2. *The natural region division map of desert highway assessment in China.*

Source: https://www.mdpi.com/ijerph/ijerph-08-02734/article_deploy/html/images/ijerph-08-02734f6.png.

In the middle medium-severe effect region, the height differences of the sand dunes are big. The topographic features of the sand dunes are complex. The route design is limited by the longitudinal and transverse distribution plane of sand dunes. In the plane, the route trend should be parallel to the wind direction. In the longitudinal section, the route should follow the natural geology. If the route reaches the high and big sand dunes, the route should be arranged in front of the windward slope, not on the back of the lee slope.

The route must not be arranged along the top of the sand-sliding slopes. It is best for the route not to cross the lee slope of sand dunes. If this cannot be avoided, the route length should be as short as possible. In the west slight-medium effect region, there are low sand dunes and rough sand land. For the route design, the vegetation should be protected as much as possible. The low-filling embankments should be dominant for the subgrade. It is better not to excavate the land and protect the vegetation.

Project Application

The Yuling-Jingbian Expressway is the first highway in the desert regions of China. The length of the route is 115.98 km. The design standards are a totally closed, double-direction and four-lane expressway. The designed maximum vehicle speed is 100 km/h and the width of the expressway is 26 m. The route crosses the Yulin wind-sand grass shoal area, which is defined as part of the west slight-medium effect region according to the desert assessment. The terrain is flat, and sparsely populated, with good vegetation. The route design should avoid vegetation and protect the natural environment, so the design route length is lengthened to 127 km from 115.98 km. The whole investment of the Yuling-Jingbian expressway is 2.8 billion U.S. dollars. The investment for the afforestation project and sediment control structure is nearly 0.15 billion dollars. The subgrade is mainly a low-filled embankment, with a maximum excavation of 6.26 m and a maximum filling of 2.93 m. If the embankment slope height is less than 3 meters, the slope ratio is selected as 1:3 to 1:8. If the slope height is more than 3 meters and less than 8 m, the slope ratio is selected as 1:2. The shelter forest is constructed along both sides of the embankment, which serves as a “green corridor” for the desert highway.

CLASS ACTIVITY

ROAD DESIGN:

Divide students into small groups and assign each group the task of designing a segment of a highway or road. They should consider factors such as horizontal and vertical alignment, sight distance, superelevation, and lane width.

Objective: To understand the principles of road design and the factors that influence it.

Materials: Graph paper, rulers, compasses, calculators, and design guidelines.

ROLE MODEL

DUFF A. ABRAMS (1880–1965)

Duff A. Abrams (1880–1965) was an American researcher in the field of composition and properties of concrete. He developed the basic methods for testing concrete characteristics that remain in use. A professor with the Lewis Institute, he studied the component materials of concrete in the early 20th century.

Abrams was researcher, professor, and director of the research laboratory of the Portland Cement Association in Chicago. He was elected in 1915 a fellow of the American Association for the Advancement of Science.[1] He was also president of the American Concrete Association (ACI) from 1930 to 1931. He was awarded the Frank P. Brown Medal in 1942.

Abrams investigated the influence of the composition of concrete mixes on the strength of the end product. Some of the results of his research were: the definition of the concept of fineness modulus; the definition of the water–cement ratio; a test method for the workability of a concrete mix by using what the Abrams cone. In a comprehensive research program, Abrams established the relationship between the water–cement ratio and the compressive strength of concrete. The results were first published in 1918 in D. A. Abrams, Design of Concrete Mixtures, Bulletin 1, Structural Materials Research Laboratory, Lewis Institute, Chicago, 1918.

SUMMARY

- Highway engineering is a specialized branch of civil engineering that focuses on the planning, design, construction, operation, and maintenance of roads and highways.
- Design is a critical aspect of highway engineering, involving geometric design, pavement design, and the integration of safety features. Geometric design pertains to the layout of the road, including alignment, profile, cross-section, and intersections.
- Highway engineering (also known as roadway engineering and street engineering) is a professional engineering discipline branching from the civil engineering subdiscipline of transportation engineering that involves the planning, design, construction, operation, and maintenance of roads, highways, streets, bridges, and tunnels to ensure safe and effective transportation of people and goods.
- The beginning of road construction could be dated to the time of the Romans. With the advancement of technology from carriages pulled by two horses to vehicles with power equivalent to 100 horses, road development had to follow suit.
- Highway planning involves the estimation of current and future traffic volumes on a road network. Highway planning is also a basic need for Highway development.
- The economic growth of a community is dependent upon highway development to enhance mobility. However, improperly planned, designed, constructed, and maintained highways can disrupt the social and economic characteristics of any size community.
- Highway and transportation engineers must meet many safety, service, and performance standards when designing highways for certain site topography. Highway geometric design primarily refers to the visible elements of the highways.
- The materials used for roadway construction have progressed with time, dating back to the early days of the Roman Empire. Advancements in methods with which these materials are characterized and applied to pavement structural design has accompanied this advancement in materials.
- Designing for proper drainage of highway systems is crucial to their success. A highway should be graded and built to remain "high and dry." Regardless of how well other aspects of a road are designed and constructed, adequate drainage is mandatory for a road to survive its entire service life. Excess water in the highway structure can inevitably lead to premature failure, even if the failure is not catastrophic.
- Highway construction is generally preceded by detailed surveys and subgrade preparation. The methods and technology for constructing highways have evolved over time and become increasingly sophisticated.

REVIEW QUESTIONS

1. What are the key components involved in the planning and development of road infrastructure?
2. How is flexible pavement designed, and what are its main considerations?
3. Why are roads important, and what are their essential characteristics?
4. What safety considerations must be considered in highway engineering?
5. What are the different types of highway alignment, and how are they determined?

REFERENCES

1. Bridle, R., & Porter, J. (2002). *The Motorway Achievement: Frontiers of Knowledge and Practice*. Thomas Telford.
2. Estache, A., Romero, M., & Strong, J. (2000). The Long and Winding Path to Private Financing and Regulation of Toll Roads. *The World Bank*, Washington, DC.
3. Ferragut, T. R., Harrington, D., & Brink, M. (2002). Road Map to the Future. *United States Department of Transportation–Federal Highway Administration*.
4. Kameyama, S., Kato, M., Kawamura, A., Himeno, K., & Kasahara, A. (2002). Effects of Frost Heave on the Longitudinal Profile of Asphalt Pavements in Cold Regions. *ISAP 9th Conference Titles & Abstracts*, 09044.
5. Newcomb, D. E., Willis, R., & Timm, D. H. (2010). *Perpetual Asphalt Pavements: A Synthesis*. Lanham, MD: Asphalt Pavement Alliance.
6. O'Flaherty, C. A. (2002). *Highways: The Location, Design, Construction & Maintenance of Road Pavements* (4th ed.). Butterworth-Heinemann.
7. Rogers, M. (2002). *Highway Engineering*. Blackwell Science.
8. World Health Organization. (2004). *World Report on Road Traffic Injury Prevention*. Geneva: World Health Organization.

CHAPTER 2

Geometric Design of Highway

LEARNING OBJECTIVES

After studying this chapter, you will be able to:

- Describe various terms used in highway and its standards
- Focus on demonstrate the basic requirements of curves
- Explain the sight distance – definition, types IRC – recommendations
- How to Calculate minimum and maximum superelevation

KEY TERMS FROM THIS CHAPTER

Vertical alignment
Sight distance
Vertical curves
Stopping sight distance
Superelevation
Geometric roadway design
road surface
Intersection sight distance

INTRODUCTION

The geometric design of highways is a critical aspect of civil engineering that focuses on optimizing the physical layout and alignment of roadways to ensure safety, efficiency, and comfort for road users. This multifaceted process involves the careful planning and implementation of several key elements, including horizontal alignment, vertical alignment, cross-sectional elements, sight distance, and intersections.

Horizontal alignment refers to the path of the road in the horizontal plane, including straight sections, curves, and transitions. Properly designed curves are essential to accommodate the centrifugal forces experienced by vehicles, reducing the risk of skidding and ensuring a smooth driving experience. Factors such as curve radius, superelevation (banking of the roadway), and transition curves are meticulously calculated to provide balance between comfort and safety.

Vertical alignment, on the other hand, deals with the road's profile in the vertical plane, which includes gradients (slopes) and vertical curves. The design of vertical curves ensures that changes in gradient are gradual, providing adequate sight distances for drivers and facilitating water drainage. The coordination between horizontal and vertical alignments is crucial to prevent sudden changes that could pose hazards to drivers.

The cross-sectional elements of a highway encompass the width and layout of lanes, shoulders, medians, and roadside areas. Lane width is typically standardized to accommodate vehicles safely, while shoulders provide space for emergency stops and recovery of errant vehicles. Medians serve as a physical barrier between opposing traffic flows, enhancing safety. Roadside design also includes clear zones free from obstacles, ensuring that any vehicle leaving the roadway has a chance to recover without hitting a hazardous object.

Sight distance is a vital safety consideration, ensuring that drivers have adequate visibility to react to obstacles or changes in the roadway. This involves ensuring unobstructed views along straight sections, around curves, and over hills. Intersection design integrates sight distance requirements with efficient traffic control to manage vehicle movements safely and effectively. The design of intersections must consider traffic volume, turning movements, and potential conflict points to minimize accidents and improve flow.

Modern highway geometric design also integrates considerations for non-motorized users, such as pedestrians and cyclists, ensuring their safety and accessibility. With advancements in technology, intelligent transportation systems (ITS) are increasingly incorporated into highway design to enhance traffic management and provide real-time information to drivers. The geometric design of highways is a comprehensive and dynamic field that balances safety, efficiency, and comfort. By addressing the intricate details of road alignment, cross-sectional elements, sight distance, and intersections, civil engineers create roadways that meet the diverse needs of all users while promoting a smooth and secure driving environment.

2.1. DESCRIBE VARIOUS TERMS USED IN HIGHWAY AND ITS STANDARDS

The geometric design of roads is the branch of highway engineering concerned with the positioning of the physical elements of the roadway according to standards and constraints. The basic objectives in geometric design are to optimize efficiency and safety while minimizing cost and environmental damage. The geometric design also affects an emerging fifth objective called "livability," which is defined as designing roads to foster broader community goals, including providing access to employment, schools, businesses, and residences, accommodating a range of travel modes such as walking, bicycling, transit, and automobiles, and minimizing fuel use, emissions, and environmental damage.

Geometric roadway design can be broken into three main parts: alignment, profile, and cross-section. Combined, they provide a three-dimensional layout for a roadway.

- The alignment is the route of the road, defined as a series of horizontal tangents and curves.
- The profile is the vertical aspect of the road, including crest and sag curves, and the straight grade lines connecting them.
- The cross-section shows the position and number of vehicle and bicycle lanes and sidewalks, along with their cross slope or banking. Cross sections also show drainage features, pavement structure, and other items outside the category of geometric design.

2.1.1. Design Standards

Roads are designed in conjunction with design guidelines and standards. These are adopted by national and sub-national authorities (e.g., states, provinces, territories, and municipalities). Design guidelines take into account speed, vehicle type, road grade (slope), view obstructions, and stopping distance. With proper application of guidelines, along with good engineering judgment, an engineer can design a roadway that is comfortable, safe, and appealing to the eye.

2.1.1.1. Profile

The profile of a road consists of road slopes, called grades, connected by parabolic vertical curves. Vertical curves are used to provide a gradual change from one road slope to another, so that vehicles may smoothly navigate grade changes as they travel. Sag vertical curves are those that have a tangent slope at the end of the curve that is higher than that of the beginning of the curve. When driving on a road, a sag curve would appear as a valley, with the vehicle first going downhill before reaching the bottom of the curve and continuing uphill or level.

Crest vertical curves are those that have a tangent slope at the end of the curve that is lower than that of the beginning of the curve. When driving on a crest curve, the road appears as a hill, with the vehicle first going uphill before reaching the top of the curve and continuing downhill.

The profile also affects road drainage.

2.1.1.2. Symbols Used

- BVC = beginning of vertical curve
- EVC = end of vertical curve
- g_1 = initial roadway grade, expressed in percent
- g_2 = final roadway grade, expressed in percent
- A = absolute value of the difference in grades (initial minus final), expressed in percent
- h_1 = height of eye above the roadway, measured in meters or feet
- h_2 = height of an object above the roadway, measured in meters or feet
- L = curve length (along the x-axis)
- PVI = point of vertical interception (intersection of initial and final grades)
- S = headlight sight distance
- Tangent elevation = elevation of a point along the initial tangent
- x = horizontal distance from BVC
- Y (offset) = vertical distance from the initial tangent to a point on the curve
- Y' = curve elevation = tangent elevation—offset

2.1.1.3. Sag Curves

Sag vertical curves are curves which, when viewed from the side, are concave upwards. This includes vertical curves at valley bottoms, but it also includes locations where an uphill grade becomes steeper, or a downhill grade becomes less steep. The most important design criterion for these curves is headlight sight distance. When a driver is driving on a sag curve at night, the sight distance is limited by the higher grade in front of the vehicle.

This distance must be long enough that the driver can see any obstruction on the road and stop the vehicle within the headlight sight distance. The headlight sight distance (S) is determined by the angle of the headlight and angle of the tangent slope at the end of the curve. By first finding the headlight sight distance (S) and then solving for the curve length (L) in each of the equations below, the correct curve length can be determined. If the $S < L$ curve length is greater than the headlight sight distance, then this number can be used. If it is smaller, this value cannot be used. Similarly, if the $S > L$ curve length is smaller than the headlight sight distance, then this number can be used. If it is larger, this value cannot be used.

Units	**Sight distance < curve length (S < L)**	**Sight distance > curve length (S > L)**
Metric (S in meters)	$L = \frac{AS^2}{120 + 3.5S}$	$L = 2S - \frac{120 + 3.5S}{A}$
US customary (S in feet)	$L = \frac{AS^2}{400 + 3.5S}$	$L = 2S - \frac{400 + 3.5S}{A}$

These equations assume that the headlights are 600 millimeters (2.0 ft) above the ground, and the headlight beam diverges 1 degree above the longitudinal axis of the vehicle.

2.1.1.4. Crest Curves

Crest vertical curves are curves which, when viewed from the side, are convex upwards. This includes vertical curves at hill crests, but it also includes locations where an uphill grade becomes less steep, or a downhill grade becomes steeper. The most important design criterion for these curves is stopping sight distance. This is the distance a driver can see over the crest of the curve. If the driver cannot see an obstruction in the roadway, such as a stalled vehicle or an animal, the driver may not be able to stop the vehicle in time to avoid a crash. The desired stopping sight distance (S) is determined by the speed of traffic on a road. By first finding the stopping sight distance (S) and then solving for the curve length (L) in each of the equations below, the correct curve length can be determined. The proper equation depends on whether the vertical curve is shorter or longer than the available sight distance. Normally, both equations are solved, then the results are compared to the curve length.

sight distance > curve length ($S > L$)

$$L = 2S - \frac{200(\sqrt{h_1} + \sqrt{h_2})^2}{A}$$

sight distance < curve length ($S < L$)

$$L = \frac{AS^2}{200(\sqrt{h_1} + \sqrt{h_2})^2}$$

US standards specify the height of the driver's eye is defined as 1080 mm (3.5 ft) above the pavement, and the height of the object the driver needs to see as 600 mm (2.0 ft), which is equivalent to the taillight height of most passenger cars. For bicycle facilities, the cyclist's eye height is assumed to be 1.4 m (4.5 ft), and the object height is 0 inches since a pavement defect can cause a cyclist to fall or lose control.

2.1.2. Cross Section

The cross-section of a roadway can be considered a representation of what one would see if an excavator dug a trench across a roadway, showing the number of lanes, their widths and cross slopes, as well as the presence or absence of shoulders, curbs, sidewalks, drains, ditches, and other roadway features.

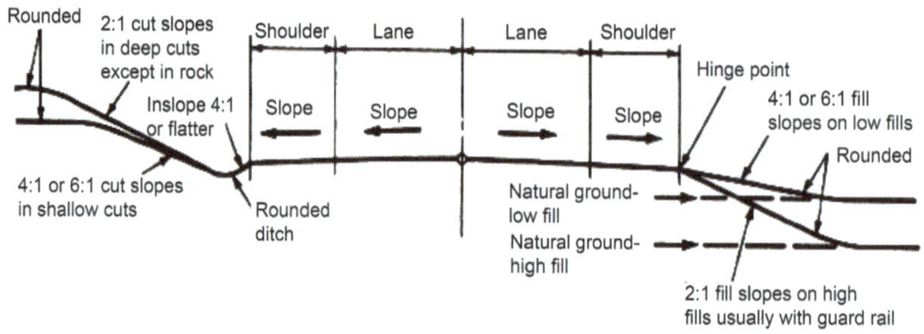

Figure 2.1. *Typical cross-section for two-lane highway.*

Source: https://image.slidesharecdn.com/geometricdesign3-230508125144-f1c78207/75/Geometric-Design-3-ppt-14-2048.jpg

The cross-sectional shape of a road surface, in connection to its role in managing runoff, is called a "crown".

2.1.2.1. Lane Width

The selection of lane width affects the safety, maximum capacity, and cost, of a highway. Safety is best at a width of 3.0 to 3.1 meters (9.8 to 10.2 ft) in urban settings, where both narrow (less than 2.8 meters (9 ft 2 in)) and wide (over 3.1 meters (10 ft)) lanes have higher crash risks. Lanes wider than 3.3~3.4m (10.8–11.2 ft)) are also associated with 33% higher impact speeds and more severe collisions, as well as higher crash rates.

Carrying capacity is also optimal at a width of 3.0 to 3.1 meters (9.8 to 10.2 ft), both for motor traffic and for bicycles. Throughput is maximal at 18 miles per hour (29 km/h); as lane width decreases to 3.0 to 3.1 meters (9.8 to 10.2 ft), traffic speed diminishes, and so does the interval between vehicles.

The pedestrian volume also increases as lanes narrow, and intersections with narrower lanes provide higher capacity for bicycles. Narrow lanes typically cost less to build and maintain. They lessen the time needed to walk across and reduce stormwater runoff.

Lane widths in Northern America are typically greater than optimum, ranging from 3 meters (9.8 ft) to 3.6 meters (12 ft). Wider lanes and shoulders are usually used on roads with higher speeds and higher volume traffic and with significant numbers of trucks and other large vehicles.

2.1.2.2. Cross Slope

Cross slope describes the slope of a roadway perpendicular to the centerline. If a road were completely level, water would drain off it very slowly. This would create problems with hydroplaning, and ice accumulation in cold weather. In tangent (straight) sections, the road surface cross slope is commonly 1–2% to enable water to drain from the roadway. Cross slopes of this size, especially when applied in both directions of travel with a crown point along the centerline of a roadway are commonly referred to as "normal crown" and are generally unnoticeable to traveling motorists.

In curved sections, the outside edge of the road is *superelevated* above the centerline. Since the road is sloped down toward the inside of the curve, gravity draws the vehicle toward the inside of the curve. This causes a greater proportion of centripetal force to supplant the tyre friction that would otherwise be needed to negotiate the curve.

Superelevation slopes of 4 to 10% are applied in order to aid motorists in safely traversing these sections while maintaining vehicle speed throughout the length of the curve. An upper bound of 12% was chosen to meet the demands of construction and maintenance practices, as well as to limit the difficulty of driving a steeply cross-sloped curve at low speeds. In areas that receive significant snow and ice, most agencies use a maximum cross slope of 6 to 8%. While a steeper cross slope makes it difficult to traverse the slope at low speed when the surface is icy, and when accelerating from zero with warm tyres on the ice, a lower cross slope increases the risk of loss of control at high speeds, especially when the surface is icy. Since the consequence of high-speed skidding is much worse than that of sliding inward at a low speed, sharp curves have the benefit of greater net safety when designers select up to 8% superelevation, instead of 4%. A lower slope of 4% is commonly used on urban roadways where speeds are lower, and where a steeper slope would raise the outside road edge above adjacent terrain.

The equation for the desired radius of a curve, shown below, takes into account the factors of speed and superelevation (e). This equation can be algebraically rearranged to obtain the desired rate of superelevation, using the input of the roadway's designated speed and curve radius.

$$R = \frac{u^2}{15(e + f_s)}$$

The American Association of State Highway and Transportation Officials (AASHTO) provides a table from which desired superelevation rates can be interpolated, based on the designated speed and radius of a curved section of roadway. This table can also be found in many state roadway design guides and manuals in the U.S.

2.1.3. Safety Effects of Road Geometry

The geometry of a road influences its safety performance. While studies of contributing factors to road accidents show that human factors predominate, roadway factors are the second most common category, with vehicle factors last.

2.1.3.1. Design Consistency

Collisions tend to be more frequent in locations where a sudden change in road character violates the driver's expectations. A common example is a sharp curve at the end of a long tangent section of road. The concept of *design consistency* addresses this by comparing adjacent road segments and identifying sites with changes the driver might find sudden or unexpected.

Locations with large changes in the predicted operating speed are likely to benefit from additional design effort. A horizontal curve with a significantly smaller radius than those before it may need enhanced curve signs. This is an improvement on the concept of design speed, which only sets a lower limit for geometric design. In the example given above, a long tangent followed by a sharp curve would be acceptable if a 30-mph design speed was chosen. Design consistency analysis would flag the decrease in operating speed at the curve.

2.1.3.2. Safety Effects of Alignment

The safety of a horizontal curve is affected by the length of the curve radius, whether spiral transition curves are used, and the superelevation of the roadway. For a given curve deflection, crashes are more likely on curves with a smaller radius. Spiral transitions decrease crashes, and insufficient superelevation increases crashes.

A safety performance function to model curve performance on two-lane roads is:

$$AMF = \frac{1.55L_c + \frac{80.2}{R} - 0.012S}{1.55L_c}$$

where,

AMF = Accident modification factor, a multiplier that describes how many more crashes are likely to occur on the curve compared to a straight road.

L_c = Length of the horizontal curve in miles.

R = Radius of the curve in feet.

S = 1 if spiral transition curves are present.

= 0 if spiral transition curves are absent.

2.1.3.3. Safety Effects of Cross-Section

Cross slope and lane width affect the safety performance of a road. Certain types of crashes, termed "lane departure crashes," are more likely on roads with narrow lanes. These include run-off-road collisions, sideswipes, and head-on collisions. For two-lane rural roads carrying over 2000 vehicles per day, the expected increase in crashes is:

Lane width	Expected increase in crashes
12 feet (3.7 m)	0%
11 feet (3.4 m)	5%
10 feet (3.0 m)	30%
9 feet (2.7 m)	50%

The effect of lane width is reduced on urban and suburban roads and low-volume roads. Insufficient superelevation will also increase the crash rate.

The expected increase is shown below:

Superelevation deficiency	Expected increase in crashes	
	Cars	Heavy trucks
<0.01	0%	< 5%
0.02	6%	10%
0.03	9%	15%
0.04	12%	20%
0.05	15%	25%

2.1.4. Sight Distance

Road geometry affects the sight distance available to the driver. Sight distance, in the context of road design, is defined as "the length of roadway ahead visible to the driver." Sight distance is how far a road user (usually a vehicle driver) can see before the line of sight is blocked by a hill crest or an obstacle on the inside of a horizontal curve or intersection. Insufficient sight distance can adversely affect the safety or operations of a roadway or intersection. The sight distance needed for a given situation is the distance traveled during the two phases of a driving maneuver: perception-reaction time (PRT), and maneuver time (MT). Perception-reaction time is the time it takes for a road user to realize that a reaction is needed to a road condition, decided what maneuver is appropriate, and start the maneuver. Maneuver time is the time it takes to complete the maneuver. The distance driven during perception-reaction time and maneuver time is the sight distance needed. During highway design and traffic safety investigations, highway engineers compare the available sight distance to how much sight distance is needed for the situation. Depending on the situation, one of following types of sight distances will be used.

2.1.4.1. Stopping Sight Distance

Stopping sight distance is the distance traveled during perception-reaction time (while the vehicle driver perceives a situation requiring a stop, realizes that stopping is necessary, and applies the brake), and maneuver time (while the driver decelerates and comes to a stop). Actual stopping distances are also affected by road conditions, the mass of the car, the incline of the road, and numerous other factors. For design, a conservative distance is needed to allow a vehicle traveling at design speed to stop before reaching a stationary object in its path. Typically the design sight distance allows a below-average driver to stop in time to avoid a collision.

2.1.4.2. Decision Sight Distance

Decision sight distance is used when drivers must make decisions more complex than stop or don't stop. It is longer than stopping sight distance to allow for the distance traveled while making a more complex decision. The decision sight distance is "distance required for a driver to detect an unexpected or otherwise difficult-to-perceive information source or hazard in a roadway environment that may be visually cluttered, recognize the hazard or its threat potential, select an appropriate speed and path, and initiate and complete the required maneuver safely and efficiently." Ideally, roads are designed for the decision sight distance, using 6 to 10 seconds for perception-reaction time and 4 to 5 seconds to perform the right maneuver.

2.1.4.3. Intersection Sight Distance

Intersection sight distance is the sight distance needed to safely proceed through an intersection. The distance needed depends on the type of traffic control at the intersection (uncontrolled, yield sign, stop sign or signal), and the maneuver (left turn, right turn, or proceeding straight). All-way stop intersections need the least, and uncontrolled intersections require the most. Intersection sight distance is a key factor in whether no control or yield control can be safely used, or whether more restrictive control is needed.

2.1.4.4. Corner Sight Distance

Corner sight distance (CSD) is the road alignment specification which provides a substantially clear line of sight so that the driver of a vehicle, bicyclist or pedestrian waiting at the crossroad may safely anticipate the driver of an approaching vehicle. Corner sight provides an adequate time for the waiting user to either cross all lanes of through traffic, cross the near lanes and turn left, or turn right, without requiring through traffic to radically alter their speed.

2.1.4.5. Uncontrolled and Yield-Controlled Intersections

Uncontrolled and yield (give way) controlled intersections require large sight triangles clear of obstructions in order to operate safely. At uncontrolled intersections, the basic right-of-way rules apply (either yield to the vehicle on the right, or the boulevard rule, depending on the location). Vehicle drivers must be able to see traffic approaching on the intersecting road at a point where they can adjust their speed, or stop, if need be, to yield to the other traffic before reaching the intersection. It is not the only criterion for allowing these types of intersection control. Changing an intersection to stop control is a common response to poor safety performance.

2.1.4.6. Two-Way Stop Control

When determining corner sight distance, a setback distance for the vehicle waiting at the crossroad must be assumed.

Setback for the driver of the vehicle on the crossroad has been standardized by some state MUTCDs and design manuals to be up to a minimum of 10 feet plus the shoulder width of the major road but not less than 15 feet.

However, the Federal MUTCD requires that a stop line, if used, shall be at least 4 feet from the nearest travel lane. Line of sight for corner sight distance is to be determined from a 3 and 1/2-foot eye height at the vehicle driver's location on the minor road to a 4 and-1/4-foot object height at the center of the approaching lane of the major road. Corner sight distance, D_{CSD}, is equivalent to a specified time gap, t_g, at the design speed, V_{DS}, required for a stopped vehicle to turn right or left:

$$D_{CSD} = V_{DS} t_g$$

For passenger vehicles at two-lane intersections, this time gap equivalence is commonly a distance of 7.5 seconds away at the design speed. Longer gaps are required for trucks and buses and multilane roads. Generally, the public right-of-way should include and maintain this line of sight.

2.1.4.7. All-Way Stop Control and Signalized Intersections

Drivers at intersections with all-way stop control or traffic signals need the least sight distance. At all-way stops, drivers need to be able to see vehicles stopped at other approaches. At signals, drivers approaching the intersections need to see the signal heads. In jurisdictions that allow right turn on red, drivers in the right lane stop control need the same sight distance as two-way stop control. Although not needed during normal operations, additional sight distance should be provided for signal malfunctions and power outages.

2.1.4.8. Effects of Insufficient Sight Distance

Many roads were created long before the current sight distance standards were adopted, and the financial burden on many jurisdictions would be formidable to: acquire and maintain additional right-of-way; redesign roadbeds on all of them; or implement future projects on rough terrain, or environmentally sensitive areas. In such cases, the bare minimum corner sight distance should be equal to the stopping sight distance. While a corner sight distance that far exceeds the braking distance at the design speed should be afforded to the driver, he or she is still generally required to maintain such control and safe speed as to be able to stop within the assured clear distance ahead (ACDA), and the basic speed rule always applies. Jurisdictions often provide some level of design immunity against government claims actions, in such cases. Warning signs are often used where sight distance is insufficient. The Manual on Uniform Traffic Control Devices requires Stop Ahead, Yield Ahead, or Signal Ahead signs at intersections where the traffic control device is not visible from a distance equal to the stopping sight distance at the speed of approaching traffic. Hill Blocks View signs can be used where crest vertical curves restrict sight distance. However, many jurisdictions still expect drivers to use ordinary care regarding conditions readily apparent to a driver, without the prompting of a sign. The care and focus ordinarily required of a driver against certain types of hazards may be somewhat amplified on roads with lower functional classification. The probability of spontaneous traffic increases proportionally to the density of access points, and this density should be readily apparent to a driver even when a specific access point is not. For this reason, full corner sight distance is almost never required for individual driveways in urban high-density residential areas, and street parking is commonly permitted within the right-of-way.

2.1.5. Camber – Definition, Purpose, Types

Camber, often referred to as cross slope, is a critical geometric design feature of highways and roads. It refers to the transverse inclination provided to the road surface, typically sloping downwards from the centerline towards the edges. The primary purpose of camber is to facilitate effective drainage of surface water, enhancing the safety and longevity of the pavement.

2.1.5.1. Definition and Purpose

Camber is the angle or slope provided across the road width to ensure that water does not accumulate on the road surface. Proper drainage is crucial because standing water can cause hydroplaning, leading to skidding and accidents. Moreover, water seepage can damage the road base, reducing the pavement's lifespan and increasing maintenance costs. By guiding water towards the sides of the road, the camber prevents such issues, thereby maintaining the structural integrity of the roadway and ensuring safer driving conditions.

2.1.5.2. Types of Camber

There are several types of camber used in highway design, each tailored to specific conditions and requirements:

- **Straight Line Camber**: This type of camber involves a uniform slope from the centerline to the edges. It is simple to construct and is often used in regions with moderate rainfall. The straight-line camber is effective for two-lane roads where the traffic flow is relatively straightforward.
- **Parabolic Camber**: Here, the slope decreases gradually from the centerline towards the edges, forming a parabolic curve. This type of camber is particularly effective in heavy rainfall regions as it provides a more gradual and smoother drainage path, reducing the risk of erosion at the edges. Parabolic camber is preferred for highways with high-speed traffic as it offers better vehicle stability.
- **Composite Camber**: A combination of straight and parabolic camber, composite camber is used in areas with varying rainfall intensity. The central part of the road might have a parabolic camber for efficient

drainage, while the outer parts have a straight camber to simplify construction and maintenance.

- **Sloped or Tangent Camber**: In this design, the camber consists of two straight slopes meeting at the centerline. It is often used in low-speed roads and urban areas where aesthetic and functional requirements necessitate a simpler design.

2.1.5.3. Geometric Design Considerations

Incorporating camber into the geometric design of highways involves careful consideration of several factors:

- **Rainfall Intensity**: The amount and frequency of rainfall influence the type and degree of camber. Regions with heavy rainfall require more pronounced camber to ensure rapid drainage.
- **Road Surface Material**: Different materials, such as asphalt or concrete, have varying capacities for water absorption and drainage, affecting camber design.
- **Traffic Speed and Volume**: Higher speeds necessitate smoother transitions and less steep camber to avoid discomfort and ensure vehicle stability. High-volume roads require robust drainage solutions to manage increased runoff.
- **Maintenance and Durability**: The chosen camber type should facilitate easy maintenance and long-term durability of the road surface, considering local environmental conditions and available maintenance resources.

2.1.6. IRC – Recommendations

The Indian Roads Congress (IRC) provides comprehensive guidelines for the geometric design of highways in India. These recommendations aim to ensure safety, efficiency, and comfort for road users while considering the diverse climatic, geographical, and traffic conditions prevalent in the country. Here is an overview of key IRC recommendations for various aspects of highway geometric design.

2.1.6.1. Design Speed

Design speed is a fundamental parameter in highway design, influencing all geometric elements. According to IRC, design speeds should be selected based on the functional classification of the road, terrain, and anticipated traffic volume. For example:

- **Expressways**: 120 km/h
- **National Highways (NH)**: 80-100 km/h in plain areas, 40-60 km/h in hilly areas
- **State Highways (SH)**: 80 km/h in plain areas, 40-60 km/h in hilly areas

2.1.7. Kerbs, Road Margin, Road Formation, and Right of Way

Understanding the key components of road design is essential for civil engineering and urban planning. Here is a detailed explanation of kerbs, road margin, road formation, and right of way:

- Kerbs: **Kerbs** (or curbs in American English) are the boundaries between the roadway and the sidewalk. They serve several purposes:
  - **Structural support**: They provide a structural edge to the pavement, preventing the

lateral spread of the roadway materials.

- **Safety**: They help to delineate the edge of the road, improving safety for vehicles and pedestrians.
- **Drainage**: Kerbs guide water runoff to drainage inlets, helping to manage surface water and reduce flooding on the roadway.
- **Aesthetics and functionality**: They enhance the visual appeal of the road and can accommodate various roadside features such as planting strips and parking spaces.

Types of kerbs include:

- **Barrier kerbs**: High kerbs designed to prevent vehicles from leaving the roadway.
- **Mountable kerbs**: Lower kerbs that can be driven over, often used in residential areas.
- **Semi-barrier kerbs**: Intermediate height, providing a balance between the functions of barrier and mountable kerbs.

- Road Margin: The **road margin** refers to the areas adjacent to the carriageway (the part of the road used by vehicular traffic). This includes:
  - **Shoulder**: The strip of land immediately adjacent to the traveled way, providing space for stopped vehicles, emergency use, and lateral support for the pavement.
  - **Verges**: The green areas or strips of land beyond the shoulder, which can include landscaping, sidewalks, and utilities.
  - **Sidewalks**: Paths for pedestrian use, often found within the road margin.
  - **Bike lanes**: Dedicated lanes for bicycles, which may be within the road margin.
- Road Formation: **Road formation** involves the preparation and construction of the base layers of a road. This includes:
  - **Subgrade**: The natural soil prepared to support the loads of the road and traffic. Proper compaction and drainage are crucial.
  - **Sub-base**: A layer of material laid on the subgrade, providing additional support and aiding drainage. It distributes loads to prevent the subgrade from deforming.
  - **Base course**: The layer between the sub-base and the surface course, providing structural capacity to the pavement.
  - **Surface course**: The top layer of the road, made from materials like asphalt or concrete, provides a smooth and durable driving surface.
- Right of Way (RoW): **Right of Way** refers to the legal right to pass through property owned by another. In road construction and planning, it includes:
  - **Width of the road corridor**: The total width of the land acquired for road construction, including the carriageway, shoulders, footpaths, and verges.

- **Acquisition and easements**: Legal processes to acquire land for road development, including compensating property owners and establishing easements for utilities.
- **Usage rights**: Regulations that dictate who can use the road and under what conditions, such as vehicles, pedestrians, cyclists, and public transport.

These components are crucial in road design and infrastructure planning, ensuring functionality, safety, and efficient use of space and resources. Properly designed roads enhance connectivity, improve traffic flow, and contribute to the overall urban and rural infrastructure.

2.1.8. Design Speed IRC – Recommendations

The Indian Roads Congress (IRC) provides guidelines for the design speed of various categories of roads to ensure safety, efficiency, and comfort. Design speed is the maximum safe speed that can be maintained on a road under favorable conditions. The key recommendations from IRC for different types of roads:

2.1.8.1. Design Speed Recommendations by IRC

- Urban Roads
  - Arterial Roads: 50-80 km/h
  - Sub-Arterial Roads: 50-60 km/h
  - Collector Streets: 40-50 km/h
  - Local Streets: 30 km/h
- Rural Roads
- National Highways and State Highways:
  - Plain and Rolling Terrain: 100 km/h
  - Hilly and Mountainous Terrain (with sparse development): 60-80 km/h
  - Steep Terrain (with restricted development): 40 km/h
- Major District Roads:
  - Plain and Rolling Terrain: 80 km/h
  - Hilly and Mountainous Terrain: 40-50 km/h
- Other District Roads and Village Roads:
  - Plain and Rolling Terrain: 60 km/h
  - Hilly and Mountainous Terrain: 30-40 km/h

2.1.8.2. Factors Affecting Design Speed

- **Terrain**: Design speeds are adjusted for different types of terrain (plain, rolling, hilly, and mountainous) due to the varying difficulties and safety considerations.
- **Road Function**: Roads serving different functions (e.g., highways, local roads) have different design speed requirements.
- **Traffic Volume**: Higher design speeds are recommended for roads with higher traffic volumes to ensure smooth flow and safety.
- **Safety Considerations**: Safety features such as sight distance, curve radius, and road width are designed based on the recommended design speeds.
- **Urban vs. Rural**: Urban roads generally have lower design speeds

due to higher pedestrian activity and intersections, while rural roads have higher design speeds because of fewer interruptions.

2.1.8.3. Implementation and Adjustments

- **Geometric Design Standards**: The geometric design of the road, including alignment, cross-section, and intersections, should be based on the designated design speed.
- **Consistency**: To avoid confusion and potential accidents, the design speed should be consistent along the road, except where natural constraints necessitate a change.
- **Regulatory Speed Limits**: The posted speed limit on the road may be lower than the design speed to account for traffic conditions, road environment, and safety considerations.

The IRC design speed recommendations aim to ensure that roads are designed to accommodate safe and efficient travel for the intended road users. These recommendations consider various factors such as terrain, road function, and traffic volume to provide a framework for consistent and safe road design across India. Properly implementing these guidelines helps in achieving a road network that supports economic growth, enhances connectivity, and improves safety for all road users.

2.2. DEMONSTRATE THE BASIC REQUIREMENTS OF CURVES

Curves are an essential element of road design, ensuring smooth transitions and safe navigation for vehicles. Properly designed curves enhance safety, comfort, and efficiency on roadways. Here are the basic requirements and considerations for designing curves.

2.2.1. Basic Requirements of Curves

Types of Curves

Curves in road design are primarily classified into two types: horizontal and vertical curves.

- **Horizontal Curves**: These curves are used to change the direction of a road in a horizontal plane. Key parameters include the radius, length, and deflection angle. The design of horizontal curves must consider factors such as sight distance, super-elevation, and transition curves to ensure safety and comfort.
- **Vertical Curves**: These curves are used to change the gradient of a road in a vertical plane. They provide a smooth transition between different grades, enhancing visibility and ride quality. Vertical curves are characterized by their length and the difference in grade at the ends of the curve.

Sight Distance

Sight distance is a critical factor in curve design. It ensures that drivers have a clear line of sight to perceive and react to any obstacles or changes in the road ahead.

The two main types of sight distances considered are:

- **Stopping Sight Distance (SSD)**: The minimum distance required for a driver to perceive an obstacle and bring the vehicle to a complete stop.
- **Overtaking Sight Distance (OSD)**: The distance required for a vehicle to safely overtake another vehicle on a two-lane road.

Super-Elevation

Super-elevation is the banking of the roadway at a curve to counteract the lateral acceleration produced by a vehicle negotiating the curve. It helps in maintaining vehicle stability and passenger comfort. The rate of super-elevation is determined based on the design speed and radius of the curve. Proper drainage should be ensured to avoid water accumulation on the curved sections.

Radius and Curvature

The radius of the curve directly impacts the safety and comfort of the road users. A larger radius provides a gentler curve, which is safer and more comfortable, but may require more land. The choice of radius depends on the design speed, terrain, and available right of way. The degree of curvature is the angle subtended by a 100-meter arc of the curve, affecting the design and alignment of the road.

Transition Curves

Transition curves (e.g., spirals) are introduced between straight sections and circular curves to provide a gradual change in curvature and super-elevation. This ensures a smooth and comfortable transition for vehicles, reducing the risk of skidding and improving ride quality.

Gradients and Vertical Alignment

The vertical alignment of a highway is generally defined as the presence of heights and depths in the vertical axis with respect to the horizontal axis of alignment. These heights and depths in roads may be in the form of gradients (straight lines in a vertical plane) or vertical curves (Figure 2.2).

Vertical curves must be designed considering the gradient of the approaching and departing road segments. The length of the vertical curve affects the comfort and safety of the ride.

Figure 2.2. *Vertical alignment of highway.*

Source: https://i0.wp.com/theconstructor.org/wp-content/uploads/2016/04/clip_image002.jpg?resize=600%2C419&ssl=1.

Vertical alignment of highways consists:

- Gradients
- Grade compensation
- Vertical curves (valley curve, summit curve)

2.2.2. Gradients

A gradient in road design refers to the slope of the road surface along its longitudinal direction. It is expressed as a percentage or ratio, representing the vertical rise or fall per unit of horizontal distance. For example, a 5% gradient means a 5-meter rise or fall over a 100-meter horizontal distance. Gradients are critical in road design for ensuring vehicle safety, managing vehicle speed, and providing adequate drainage.

A gradient is defined as the rise or fall corresponding to some horizontal distance. Rise with respect to horizontal distance is called Upward gradient (+n %). Fall with respect to horizontal distance is called Downward Gradient (-n %). The gradient is represented in the figure below:

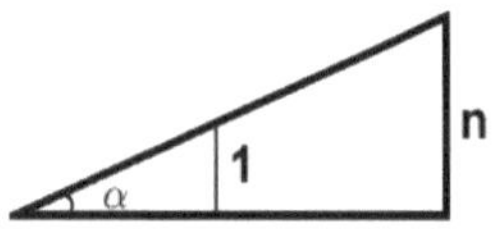

2.2.2.1. Types of Gradients

1. *Ruling Gradient:* This is the maximum gradient which is generally used to design the vertical profile of the highway. So, it is also called a designer gradient. Ruling gradient depends on the terrain, length of the grade, speed, pulling power of the vehicle, and the presence of the horizontal curve. It is adopted by considering a particular speed as the design speed and for a design vehicle with standard dimensions. In flat terrains it is possible to provide flat gradients and in hilly terrains it is very costly and sometimes it is difficult to provide ruling gradient in hilly terrains.
2. *Limiting Gradient:* This gradient is provided as shorter stretches in highways. Whenever ruling gradients costs high for the hilly terrains then limiting gradient is provided which will reduce the cost. This gradient is adopted frequently in terrains with limited stretches.
3. *Exceptional Gradient:* These are very steeper gradients given at unavoidable situations and they are adopted for stretches not exceeding 100m of length.
4. *Minimum Gradient:* To drain rainwater from the road minimum gradient is needed. Generally, for lateral drainage Camber is provided.

   But for the longitudinal drainage along the side drains require some slope for smooth flow of water. For concrete drains minimum gradient of 1 in 500 and open soil drains gradient of 1 in 200 is suitable.

2.2.2.2. Grade Compensation

When a horizontal circular curve lies in a vertical curve there will be an increased resistance offered by the circular curve in the form of curve resistance in addition to the component of gravity. IRC specifications for grade compensations are as follows.

- For grades flatter than 4% grade compensation is not required due to negligible loss of tractive force.
- Grade compensation is $\frac{30+R}{R}\%$
- Maximum value of gradient compensation =75/R%,

  R= radius of horizontal curve.

2.2.2.3. Vertical Curves

Generally, two types of vertical curves are there to adopt for the vertical alignment. They are:

- Summit curve; and
- Valley curve.

The summit curve is a vertical curve adopted mainly when the gradient is upward. In the case of the summit curve simple parabola is considered as best curve shape.

2.2.3. IRC Recommendations

The Indian Roads Congress (IRC) provides specific recommendations for gradients based on the type of road and the terrain it traverses. These recommendations are intended to ensure safety, comfort, and operational efficiency.

1. *National Highways and State Highways*
   - **Plain and Rolling Terrain**:
     - Ruling Gradient: 2.5%
     - Limiting Gradient: 3.3%
     - Exceptional Gradient: 5%
   - **Hilly Terrain (without snow)**:
     - Ruling Gradient: 5%
     - Limiting Gradient: 6%
     - Exceptional Gradient: 7%
   - **Hilly Terrain (with snow)**:
     - Ruling Gradient: 6%
     - Limiting Gradient: 7%
     - Exceptional Gradient: 8%
2. *Other Roads*
   - **Major District Roads**:
     - **Plain and Rolling Terrain**:
       a. Ruling Gradient: 3%
       b. Limiting Gradient: 4%
       c. Exceptional Gradient: 5%
     - **Hilly Terrain**:
       a. Ruling Gradient: 5.5%
       b. Limiting Gradient: 6%
       c. Exceptional Gradient: 7%
   - **Other District Roads and Village Roads**:
     - **Plain and Rolling Terrain**:
       a. Ruling Gradient: 4%
       b. Limiting Gradient: 5%
       c. Exceptional Gradient: 6%
     - **Hilly Terrain**:
       a. Ruling Gradient: 6%
       b. Limiting Gradient: 7%
       c. Exceptional Gradient: 8%

2.2.3.1. Application of Gradients

- **Safety Considerations**: Steep gradients can pose safety risks, especially for heavy vehicles which may have difficulty ascending and descending. Proper gradient design ensures vehicles can maintain control and speed.
- **Comfort**: Gradients influence the comfort of the ride. Excessive gradients can cause discomfort due to the need for frequent gear changes and braking.
- **Drainage**: Adequate minimum gradients are essential to ensure proper drainage, preventing water accumulation that can lead to hydroplaning and road damage.
- **Construction Feasibility**: Practical site constraints often dictate the need for using different gradients. Balancing construction costs and design standards is crucial for effective road planning.

Proper gradient design, adhering to IRC recommendations, is vital for safe and efficient roadways. By understanding the types of gradients and their appropriate applications, road designers can create road networks that balance safety, comfort, and cost-effectiveness, while addressing the challenges posed by various terrains.

2.3. SIGHT DISTANCE – DEFINITION, TYPES IRC – RECOMMENDATIONS

Sight distance refers to the distance along a roadway visible to a driver. It is a critical factor in road design and traffic safety, allowing drivers to perceive and react to potential hazards, obstacles, or changes in road conditions. Adequate sight distance is essential for safe driving, overtaking, merging, and navigating intersections. The visibility of the road ahead of the driver will help in the safe and efficient operation of the vehicles. This will hence demand the geometric design to be highly efficient so that the length of the road is highly visible to the driver even from a distance ahead. This distance is hence termed as the sight distance.

2.3.1. Sight Distance and Their Types

The actual distance that is observed along the road surface which is visible for a driver from a specified height above the carriage way is called as the sight distance at a point. This distance will let the driver see all the stationary and the moving objects in front of the vehicle. Mainly in the geometric design of road construction, mainly three sight distances are taken into consideration. They are:

- **SSD** – Stopping Sight Distance or Absolute Minimum Sight Distance
- **ISD** – Intermediate Sight Distance: This is twice the value of SSD
- **OSD** – Overtaking Sight Distance or Passing Sight Distance (PSD)

Other than these we have:

- Head Light Sight Distance
- Safe Sight Distance

The distance that is available for the driver during the night travel is called as the headlight sight distance. At night the drive is facilitated by the illumination of the headlights. The sight distance available to enter an intersection is called the Safe sight distance. Whatever be the stopping distance taken into consideration, it is necessary for the driver who is traveling at the design speed to possess sufficient carriageway distance. This distance will be within the line of vision to stop the vehicle from colliding with a moving or stationary object in the traffic lane (Figure 2.3).

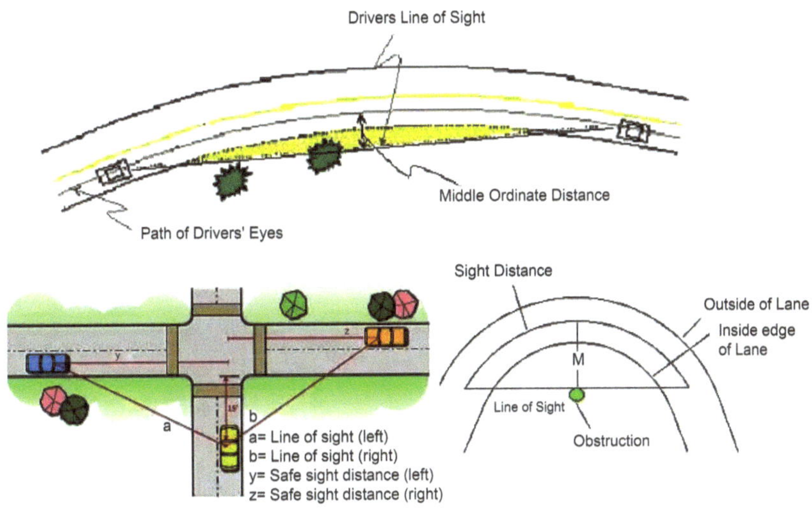

Figure 2.3. *The actual distance that is observed along the road surface.*

Source: https://i0.wp.com/theconstructor.org/wp-content/uploads/2018/02/highway-sight-distance.jpg?resize=675%2C407&ssl=1.

2.3.2. Stopping Sight Distance

This is defined as the sight distance that is available for moving the vehicle on the highway that will enable the driver to stop the vehicle safely without collision with any other obstacle. As mentioned above, the parameter safe stopping distance is the most important feature in traffic engineering. Safe stopping distance is the distance from the point it first perceives to the time the deceleration is complete. Adequate time is necessary for the drivers for them to react to the obstacle spontaneously. This demand the sight distance used in the geometric design to be equal to the safe stopping distance. The Stopping distance can be defined as the sum of the lagging distance to the brake distance. The lagging distance is the distance that is moved by the vehicle in a time 't' at a velocity of 'v' in m/s. Hence lag distance is '**vt**.' The distance that is traveled by the vehicle during the operation of braking is called the braking distance. In the case of a level road, the work that is done in stopping the vehicle is equated to the kinetic energy of the vehicle which will give us the braking distance. Let 'F' be the maximum frictional force that is developed and 'l' be the braking distance. Hence, the work that is done against friction is given by:

Fl = fWl --> (1)

Here W = the weight of the vehicle. The Kinetic energy attained at the design speed of the vehicle

= (1/2g) mv^2 = (1/2g)Wv^2 --> (2)

(1) is equated to (2), we get:

$(1/2g)\ Wv^2 = fWl$

$l = (1/(2fg))v^2$

Hence,

The Stopping Sight Distance (SSD) = Lag Distance + Braking Distance

-> $SSD = vt + (1/(2fg))v^2$

Here, v is the speed in m/s^2, t is the reaction time taken, f is the coefficient of friction, and g is the acceleration due to gravity.

Speed kmph	<0	40	50	60	>80
f	0.4	0.38	0.37	0.36	0.5

If the road possesses an ascending gradient in an amount equal to +n%, to the braking action the component factor of gravity will be added. This will decrease the braking distance. This component of gravity acts along the direction of the braking force which is given by:

$W\sin F \sim W\tan F = Wn/100$

After equating the kinetic energy and the work done we get,

$l = v^2 / (2g\ (f + n/100\ ))$

We can also derive the braking distance for a descending gradient which is performed similarly, and we get:

$SSD = vt + (v^2/\ (2g\ (f \pm 0.01n)))$

2.3.2.1. Overtaking Sight Distance (OSD)

The minimum distance available for the driver to safely overtake the slow vehicle in front of him by considering the traffic in the opposite direction is called the overtaking sight distance. This distance will make us see whether the road is clear enough to undergo an overtaking movement. The overtaking sight distance is also called as the passing sight distance that will be measured along the center line of the road. This is the line level over which the driver keeping an eye level of 1.2 m above the road level can easily see the top of the object 1.2 m above the road surface (Figure 2.4).

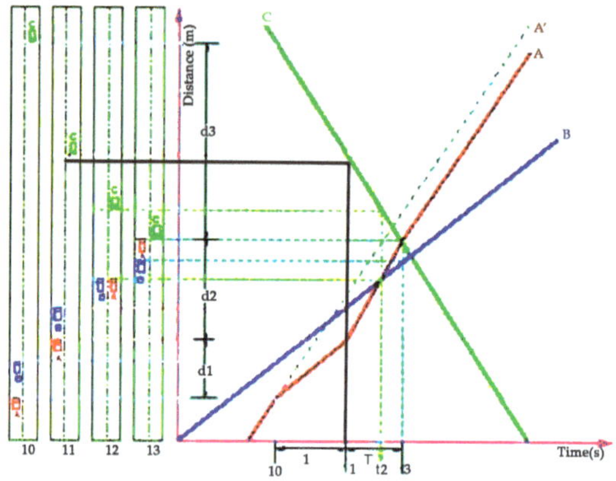

Figure 2.4. *Representing Overtaking Sight Distance through Time Space Diagram*

Source: https://i0.wp.com/theconstructor.org/wp-content/uploads/2018/02/overtaking-sight-distance.jpg?resize=427%2C319&ssl=1.

2.3.3. Factors Affecting Overtaking Sight Distance

The main factors that affect the OSD are:

- Spacing Between the vehicles
- Speed of the vehicles
- The gradient of the road
- The acceleration rate of the overtaking vehicle
- The velocities of the vehicle which is overtaking, overtaken, and coming in the opposite direction
- The driver's skill
- The reaction of the driver

2.3.3.1. Computation of Sight Distance

The computation of the sight distance mainly depends on the:

- Driver's Reaction Time
- Vehicle's Speed
- Efficiency of Brakes
- The frictional Resistance Existing Between the Tire and the Road
- Gradient of Road

2.3.3.2. Brake Efficiency

Many factors like the age of the vehicle, and the characteristics of the vehicle will affect the brake efficiency of the vehicle. An efficiency of 100% implies that the vehicle will stop at the moment the brake is applied. Obtaining 100% is not practicable. This is an ideal condition for the vehicle. This means, for a lower value of brake efficiency, it is necessary to obtain a higher value of sight distance. In the process of determining a safe geometric design, it is required to assume 50% brake efficiency.

2.3.3.3. Friction between the Road and the Tire

The stopping of a vehicle is also dependent on the frictional resistance between the tire and the road. Having a higher value of frictional resistance will result in efficient stopping of the vehicle when applied. Here, the sight distance required will be less. During the computation of sight distance, no special provision or consideration is given for the brake efficiency. This is considered along with the factor of longitudinal friction. In India, the value of longitudinal friction is between 0.35 and 0.4. This is as per the Indian Road Congress.

2.3.3.4. Speed of the Vehicle

The speed of the vehicle clearly affects the sight distance. If the speed employed is high, the time required to stop the vehicle will be high. This means that with the increase in the speed, there is an increase in the sight distance.

2.3.3.5. Driver's Reaction Time

The time from the moment the driver observes the obstacle in front of him to the moment he applies the brake is called the reaction time of the driver. Based on the PIEV theory, the reaction time can be divided into 4 components. All these times will be combined to form a total perception of reaction time while undergoing actual practice and design of the highway. From studies conducted it is reviewed that the drivers will require about 1.5 to 2 seconds for normal conditions. As this value may vary based on the characteristics of the vehicle, a higher value for the reaction time can be employed for the design purpose. In India, as per IRC, the reaction time is set to be 2.5 seconds.

2.3.3.6. Gradient of Road

The sight distance is greatly influenced by the gradient of the road. There are chances for the vehicles to stop suddenly when the vehicle is climbing suddenly. This situation will ask for a small sight distance. When the vehicle is moving down, the movement is supported by the gravity action. This will hence ask for more time to stop the vehicle.

2.3.4. Curves – Necessity, Types- Horizontal, Vertical and Transition Curves, Widening of Roads

Curves play a crucial role in road design, allowing roads to navigate terrain, accommodate changes in direction, and provide smooth transitions for vehicles.

2.3.4.1. Road Curve

Road curves are irregular bends in roads to bring a graduation change of direction. Similar curves are on railways and canals. Curves provided in the horizontal plane are known as horizontal curves and are generally circular or parabolic. Curves provided in the vertical plane are known as vertical curves.

There are five types of horizontal curves on roads and railways and two types of vertical curves on roads.

2.3.4.2. Horizontal Curves

1. *Simple curve:* A simple curve has the same radius throughout and is a single arc of a circle, with two tangents meeting at the intersection (B in this diagram).
2. *Compound curve:* A compound curve has two or more simple curves with different radii that bend the same way and are on the same side of a common tangent. In this diagram, MN is the common tangent.
3. *Reverse curve:* Also called a serpentine curve, it is the reverse of a compound curve, and two simple curves bent in opposite directions are on opposite sides of the common tangent.
4. *Deviation curve:* A deviation curve is simply a combination of two reverse curves. It is used when it is necessary to deviate from a given straight path to avoid intervening obstructions such as a building, a body of water, or another significant site.
5. *Transition curve:* Is a curve with a gradual change in elevation on the outside of the curve to help drivers comfortably take turns at faster speeds

2.3.4.3. Vertical Curves

1. *Valley curve:* Also called a sag curve, this curve dips down and then rises back up. These are placed at the base of hills. The opposite of the summit curve.
2. *Summit curve:* Also called the crest curve, this curve rises and then dips down. At the peak of hills. The opposite of the valley curve.

2.3.4.4. Widening of Roads

Widening of roads is a strategic approach in transportation infrastructure development aimed at improving traffic flow, enhancing safety, and accommodating increasing vehicle volumes. This process involves expanding the width of existing roadways to provide additional lanes, shoulders, or auxiliary facilities. Here is a detailed exploration of the widening of roads.

2.3.4.5. Importance of Widening Roads

- **Traffic Congestion Relief**: As populations grow and economies expand, the demand for transportation infrastructure increases. Widening roads helps alleviate traffic congestion, particularly during peak travel times, by increasing the capacity of existing roadways.
- **Enhanced Safety**: Additional lanes and wider road shoulders provide more space for vehicles to maneuver, reducing conflicts between vehicles and improving safety conditions. Widening also facilitates better visibility and accommodates emergency vehicles, leading to a decrease in accidents and improved response times.
- **Accommodation of Growth**: Widening roads is essential to accommodate the increasing traffic volumes resulting from population growth, urbanization, and economic development. It ensures that road networks can efficiently handle current and future demands, supporting continued socio-economic growth.
- **Improved Accessibility**: Widening roads enhances accessibility for various modes of transportation, including private vehicles, public transit, pedestrians, and cyclists. It promotes multimodal mobility and facilitates smoother interactions between different modes of transportation, leading to more efficient urban mobility.

2.3.4.6. Methods of Road Widening

- **Lane Addition**: Adding lanes to existing roadways is a common method of road widening. This involves expanding the width of the roadway to accommodate additional travel lanes, such as adding a new lane in each direction or creating dedicated turning lanes at intersections.
- **Shoulder Widening**: Widening road shoulders provide space for emergency stops, breakdowns, and pedestrian or cyclist accommodations. It enhances safety by providing a buffer zone between moving traffic and roadside hazards, reducing the risk of collisions, and improving overall comfort for road users.
- **Intersection Improvements**: Widening at intersections involves expanding the width of the roadway to accommodate additional turning lanes, queuing space, and pedestrian facilities. This helps reduce congestion, minimize conflicts between vehicles and pedestrians, and improve the efficiency of traffic movements at busy junctions.
- **Bridge Widening**: Widening bridges is essential for accommodating additional lanes or improving pedestrian and cyclist facilities. It involves expanding the width of existing bridge structures or constructing new bridge decks adjacent to the existing ones to provide additional capacity and accommodate wider roadways.

2.3.4.7. Planning and Implementation Considerations

- **Traffic Analysis**: Thorough traffic analysis is conducted to assess current and projected traffic volumes, identify congestion hotspots, and determine the most effective widening strategies to address capacity constraints.
- **Environmental Impact Assessment**: Environmental impact assessments are conducted to evaluate potential impacts on land use, habitats, air and water quality, and cultural resources. Mitigation measures are developed to minimize adverse effects and ensure compliance with environmental regulations.
- **Community Engagement**: Public input and stakeholder engagement are essential throughout the planning and implementation stages of road widening projects. Community feedback helps identify concerns, preferences, and potential conflicts, leading to more informed decision-making and project outcomes that reflect the needs of local residents and businesses.
- **Construction Phasing and Traffic Management**: Road widening projects often require staged construction to minimize disruption to traffic flow and ensure safety for road users and construction workers. Effective traffic management strategies, such as temporary lane closures, detours, and traffic signal adjustments, are implemented to maintain traffic flow and minimize delays during construction.

Widening of roads is a critical aspect of transportation infrastructure development, aimed at enhancing traffic flow, safety, and accessibility. By expanding roadways, adding lanes, widening shoulders, and improving intersections, transportation agencies can alleviate congestion, enhance mobility, and promote economic growth. Effective planning, stakeholder engagement, and implementation strategies are essential to the success of road widening projects, ensuring that they meet the needs of communities while minimizing environmental impacts and maximizing benefits to society.

2.4. CALCULATE MINIMUM AND MAXIMUM SUPERELEVATION

The minimum and maximum superelevation of a highway are determined based on various factors including design speed, curve radius, and terrain conditions. Superelevation is the banking or tilting of the roadway provided on curves to counteract the centrifugal force experienced by vehicles when traversing the curve.

2.4.1. Super Elevation: What are the Benefits and Uses?

When a driver approaches a curve, the road is frequently slanted or racked up at an angle, allowing him or her to maneuver the curve at a reasonable speed without skidding or tipping. This is a superelevation in action (Figure 2.5).

On curved roads, superelevation assists motorists in maintaining both safety and optimal speed. Many cars will indeed slide or skid through curves without super elevation, particularly in wet or cold weather or at full speed, or even tip and roll over. It also allows traffic to maintain a certain speed on curves, avoiding unnecessary slowdowns when the road bends.

Figure 2.5. *Superelevation is essential for a driver to maneuver curved roads.*

Source: https://housing.com/news/wp-content/uploads/2022/11/super-elevationT-compressed.jpg.

2.4.1.1. Superelevation: What Is It?

Superelevation is a method of constructing infrastructure in roadway curves in which the edges of the road are raised above the inner edge. Since it is a component of the vertical position or "profile" of a road viewed in cross-section, it is an important security element in the design specifications of any road with curves.

A super-elevated road must be meticulously designed. Its construction necessitates the purchase of heavy machinery, extensive materials, and a large labor force. However, if done correctly, most people will not even notice it is there. Superelevation is also known as "canting" or "banking," as in a racing speedway's banked track.

2.4.1.2. Super Elevation: Working of Superelevation

Superelevation works with several interacting physics forces to help drivers maintain speed and stay on the road safely through a curve. The need for super elevation on a road or highway is determined by a complex conjunction of rotational and centripetal force, friction, inertia, weight, and velocity or speed.

2.4.1.3. Forces Acting on Super Elevation

As a vehicle moves through a curve, the centripetal force pulls towards the inside edge of the pavement. It may feel like you are sliding, tipping, or both. It is the inverse of gravitational force that also pulls you away from the center and toward the outside edge of your vehicle as it curves (and tries to push you into your seat). A super-elevated curve in a highway balances these two forces, as well as several others that act in different directions.

2.4.1.4. Superelevation: Advantages and Purpose

- When a vehicle travels on a curved path, it experiences an outward force known as centrifugal force. To mitigate the effects of centrifugal force, super elevation is provided.
- It aids in maintaining a constant speed while remaining comfortable on a curved path.
- It lowers the number of accidents.
- It aids in the drainage of rainwater towards the inner side. So, there is no need for a camber on this stretch of road.
- It also aids in keeping vehicles on the correct side of the road, preventing head collisions between vehicles traveling in opposite directions.

2.4.1.5. Super Elevation: Disadvantages

- High engineering, geological, and concrete work costs result in increased construction costs.
- Complexity may cause instability or a loss of vehicle control.
- Increased tyre maintenance is the cause of excessive maintenance. This raises the price.
- The construction project needs more room.
- Increases the vehicle's weight, which results in a 20% increase in traction force on the base course.

CASE STUDY

IMPROVING SIGHT DISTANCE ON A CURVED SEGMENT OF A HIGHWAY

Background

A curved segment of a highway located in a hilly terrain experienced a high number of accidents. The primary cause identified was inadequate sight distance around the curves, preventing drivers from having sufficient reaction time to obstacles or oncoming vehicles.

Objectives

1. Assess the existing sight distance.
2. Identify critical points with insufficient sight distance.
3. Propose and implement design improvements.
4. Evaluate the effectiveness of the improvements.

Methodology

1. Site Analysis

- **Data Collection:** Gathered accident reports, traffic volume, and speed data.
- **Field Measurements:** Conducted field surveys to measure the current sight distances at various points along the curve.

2. Sight Distance Calculation

- **Stopping Sight Distance (SSD):** Calculated using the formula:

$$SSD = v \bullet t_r + \frac{v^2}{2a}$$

Where:

- v = Speed of the vehicle (m/s)
- tr = Reaction time (seconds)
- a = Deceleration rate (m/s²)
- **Passing Sight Distance (PSD):** Calculated using the formula considering the time for overtaking maneuvers and safe clearance.

3. Identification of Critical Points

- Plotted the measured sight distances against the required SSD and PSD to identify sections where the sight distance was below standard.

4. Design Improvements

- **Geometric Modifications:** Increased the curve radius and adjusted the road alignment to improve sight lines.
- **Clear Zone Enhancements:** Cleared vegetation and other obstructions from the roadside.
- **Signage and Markings:** Installed warning signs and improved road markings to alert drivers.

5. Implementation

- Executed the proposed design modifications during a scheduled maintenance period to minimize traffic disruption.

6. Evaluation

- **Post-Implementation Survey:** Conducted follow-up measurements of sight distance and collected data on traffic speed and volume.
- **Accident Analysis:** Compared accident rates before and after the improvements.

Results

Before Improvements

- **Measured SSD:** 80 meters (average)
- **Required SSD:** 120 meters (based on design speed)
- **Accidents:** 15 per year

After Improvements

- **Measured SSD:** 130 meters (average)
- **Required SSD:** 120 meters
- **Accidents**: 4 per year

CLASS ACTIVITY

For a class activity on the geometric design of highways, students could participate in a virtual simulation or case study where they analyze and propose design solutions for a hypothetical highway segment. They could work in groups to assess factors such as sight distance, horizontal and vertical alignment, curve design, intersection layout, and roadside features. Through this hands-on exercise, students would apply theoretical knowledge to real-world scenarios, enhancing their understanding of the complexities and considerations involved in highway design. Additionally, discussions following the activity could focus on the trade-offs and challenges encountered during the design process, promoting critical thinking and problem-solving skills.

SUMMARY

- The geometric design of highways is a critical aspect of civil engineering that focuses on optimizing the physical layout and alignment of roadways to ensure safety, efficiency, and comfort for road users
- The geometric design of roads is the branch of highway engineering concerned with the positioning of the physical elements of the roadway according to standards and constraints.
- Sag vertical curves are curves which, when viewed from the side, are concave upwards. This includes vertical curves at valley bottoms, but it also includes locations where an uphill grade becomes steeper, or a downhill grade becomes less steep.
- Road geometry affects the sight distance available to the driver. Sight distance, in the context of road design, is defined as "the length of roadway ahead visible to the driver."
- Camber, often referred to as cross slope, is a critical geometric design feature of highways and roads. It refers to the transverse inclination provided to the road surface, typically sloping downwards from the centerline towards the edges.
- Curves are an essential element of road design, ensuring smooth transitions and safe navigation for vehicles. Properly designed curves enhance safety, comfort, and efficiency on roadways.
- Sight distance refers to the distance along a roadway visible to a driver. It is a critical factor in road design and traffic safety, allowing drivers to perceive and react to potential hazards, obstacles, or changes in road conditions.
- This is defined as the sight distance that is available for moving the vehicle on the highway that will enable the driver to stop the vehicle safely without collision with any other obstacle.
- Road curves are irregular bends in roads to bring a graduation change of direction. Similar curves are on railways and canals.
- Widening of roads is a strategic approach in transportation infrastructure development aimed at improving traffic flow, enhancing safety, and accommodating increasing vehicle volumes.

REVIEW QUESTIONS

1. What are the primary factors considered in the geometric design of highways?
2. Explain the concept of sight distance and its importance in highway design.
3. What are the key elements of horizontal alignment design in highways?
4. Describe the types of curves commonly used in highway design.
5. How is superelevation used in highway curves, and what factors influence its determination?
6. Discuss the factors affecting the design of vertical curves in highways.

7. Explain the concept of stopping sight distance and its role in highway design.
8. What are the different types of intersections found in highway design, and how are they designed to ensure safety and efficiency?
9. Describe the principles of interchange design in highways.
10. Discuss the importance of clear zones and roadside design in highway safety.

REFERENCES

1. American Association of State Highway and Transportation Officials. (2004). *A Policy on Geometric Design of Highways and Streets*. Washington, DC: American Association of State Highway and Transportation Officials.
2. Garber, N. J., & Hoel, L. A. (2001). *Traffic and Highway Engineering* (3rd ed.). Brooks/Cole Publishing.
3. Homburger, W. S., Hall, J. W., Reilly, W. R., & Sullivan, E. C. (2001). *Fundamentals of Traffic Engineering* (15th ed.). ITS Course Notes UCB-ITS-CN-01-1.
4. Kobry⊠, A. (2017). Sample applications of transition curves in vertical alignment. In *Transition Curves for Highway Geometric Design* (Springer Tracts on Transportation and Traffic, Vol. 14, pp. 109–131). Springer International Publishing.

CHAPTER 3

Design of Highway Intersections

LEARNING OBJECTIVES

After studying this chapter, you will be able to:

- Define deriving design reference flows from baseline traffic figures
- Explain the roundabout intersections
- Understand the basics of traffic signal control: optimization and delays

KEY TERMS FROM THIS CHAPTER

Signalized intersections
Grade-separated interchanges
Pedestrian crossings
Conflict points
Traffic flow
Annual average hourly traffic (AAHT)
Turbo roundabouts
Roundabouts
Traffic patterns
Traffic calming
Visibility
T-junctions
Mini roundabouts

INTRODUCTION

Designing highway intersections is a critical aspect of civil engineering that significantly impacts traffic flow, safety, and overall transportation efficiency. A well-designed intersection facilitates the smooth and safe movement of vehicles and pedestrians, minimizing delays and the risk of accidents. The design process involves a thorough analysis of traffic patterns, volume, and speed, as well as careful consideration of the surrounding environment and land use. By integrating these factors, engineers can create intersections that accommodate current demands and are adaptable to future changes.

The primary goal in the design of highway intersections is to manage the conflicting movements of vehicles efficiently. This is achieved through various configurations such as signalized intersections, roundabouts, and grade-separated interchanges. Each type has its specific applications depending on traffic volumes and the nature of intersecting roads. For instance, roundabouts are effective in reducing conflict points and improving safety at lower traffic volumes, while signalized intersections can handle higher traffic volumes with controlled signal timings. Grade-separated interchanges, which involve the construction of overpasses or underpasses, are used to eliminate conflict points entirely, ensuring the uninterrupted flow of traffic on major highways.

Safety is paramount in intersection design. Engineers employ various measures to enhance visibility, reduce vehicle speeds, and protect vulnerable road users like pedestrians and cyclists. This includes the strategic placement of signage, implementation of pedestrian crossings, and the use of traffic calming measures. Additionally, advanced technologies such as intelligent transportation systems (ITS) are increasingly being integrated into intersection design. These technologies enable real-time monitoring and management of traffic, improving both safety and efficiency.

Sustainability and environmental considerations also play a crucial role in the modern design of highway intersections. Efforts are made to minimize land use and environmental disruption through compact and efficient designs. Landscaping and green spaces are often incorporated to mitigate the visual and ecological impact of the infrastructure. Moreover, the use of eco-friendly materials and construction methods helps reduce the environmental footprint of intersection projects.

3.1. DERIVING DESIGN REFERENCE FLOWS FROM BASELINE TRAFFIC FIGURES

3.1.1. Existing Junctions

It will be feasible to precisely estimate daily and peak-hour traffic flows at existing intersections, along with all turning movements. Ideally, the measurements should be taken during a regular weekday (Monday to Thursday) in a neutral month (April, May, June, September, or October) to ensure that the results are as representative of overall peak flow levels as possible. The National Road Traffic Forecasts' indices can be used to factor up the observed morning and evening peak hour flows, which can result in the derivation of DRFs for the design year—typically 15 years after opening. It is possible to extrapolate flow patterns and turning proportions from the base year to forecast movement patterns in subsequent years (Burrow, 1987).

3.1.2. New Junctions

A traffic modeling process that produces estimates of 12, 16, or 24-hour link flows for a future design year must be used to determine flows when designing a junction for a new road or when changes to the general network are expected to alter the flow patterns through an existing junction significantly. Then, by factoring the 12, 16, or 24-hour flows, AADT flows are obtained. After calculating the AAHT (AADT ÷ 24), derived factors are used to factor the appropriate highest hourly flow. Next, tidal flow is considered; typically, a 60/40 split is assumed, with the peak hour direction receiving more of the flow. Additionally, turning proportions are estimated in order to design the junction.

3.1.3. Short-Term Variations in Flow

Typically, traffic does not approach a junction steadily or consistently. There may be times when traffic arrives faster than the DRF and times when it arrives slower. Such deviations may be permitted by using a "flow profile" when conducting the junction analysis for a priority junction/roundabout using one of the Transport Research Laboratory's computer programs (PICADY/ARCADY). An example of a typical profile would be to input peak-time flows every 15 minutes. Such short-term variations can be considered when manually calculating a priority junction/roundabout by using an hourly flow equal to 1 point 125 times the DRF.

This modification should be made to the design flows on both the minor and major arms in the event of a priority junction. This factored flow will affect both the circulating flows inside the intersection and the entry flows to the roundabout in the case of a roundabout intersection.

3.1.4. Major/Minor Priority Intersections

There is a priority intersection where two roads meet, one known as the "major" road and the other as the "minor" road. The primary route is always given preference for traffic flow over secondary routes. Traffic from the minor road may only enter the major road during suitable gaps, giving the major road priority over the minor road. This kind of junction's main benefit is that it does not cause delays for traffic on the main route.

Priority intersections are designed with the idea that they should mirror the way that traffic moves through them. The easiest paths ought to be available to the traffic flows with the highest volumes. A priority intersection's layout must prioritize visibility, especially for traffic leaving the minor junction. In addition to decreasing the intersection's inherent capacity, poor visibility can raise the frequency of major collisions.

Priority intersections can take the shape of straightforward T-junctions, staggered junctions, or crossroads; the latter should be avoided whenever feasible since it may lead to misunderstandings about traffic priorities amongst drivers leaving the minor road. Accidents may arise as a result of this. Diagrammatic representations of the three forms are given in Figure 3.1.

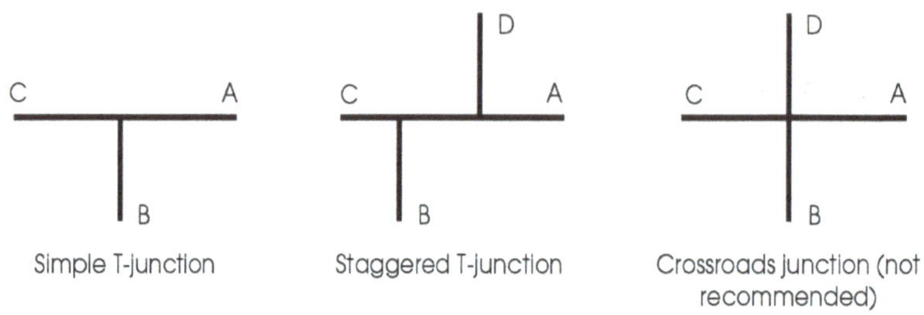

Figure 3.1. *Three forms of priority intersection.*

Source: Martin Rogers, Highway Engineering, ISBN 0-632-05993-1.

Priority intersections can take the shape of straightforward T-junctions, staggered junctions, or crossroads; the latter should be avoided whenever possible since it may lead to misunderstandings about traffic priorities amongst drivers leaving the minor road. Accidents may occur as a result of this.

- **Simple junctions:** A T-junction or staggered junction that does not have any channeling islands in the minor road approach or ghost or physical islands in the major road (see Figure 3.2).

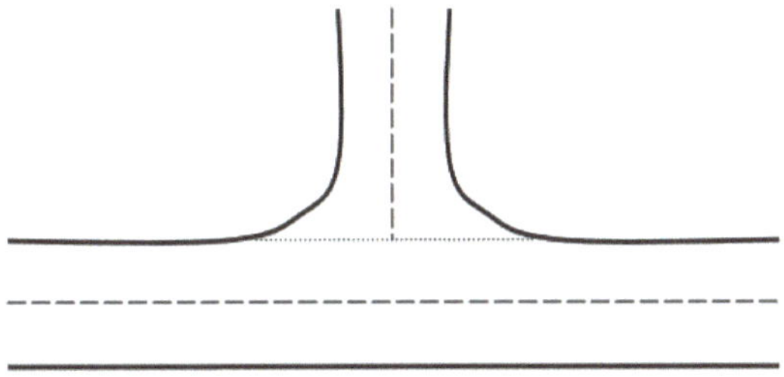

Figure 3.2. *Simple T-junction.*

Source: Martin Rogers, Highway Engineering, ISBN 0-632-05993-1.

- **Ghost island junctions:** These are typically T-junctions or staggered junctions where a designated area is marked on the carriageway, positioned, and shaped to guide traffic flow (refer to Figure 3.3).

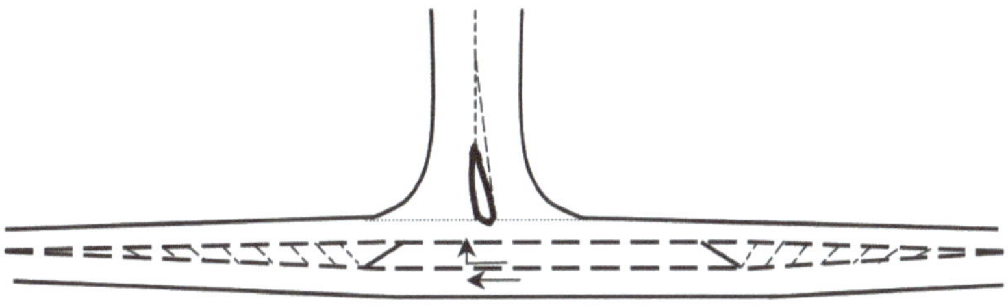

Figure 3.3. *Ghost Island junction.*

Source: Martin Rogers, Highway Engineering, ISBN 0-632-05993-1.

- **Single-lane dualling:** Typically, a T-intersection or staggered intersection with central reservation islands positioned and shaped to guide traffic flow (refer to Figure 3.4).

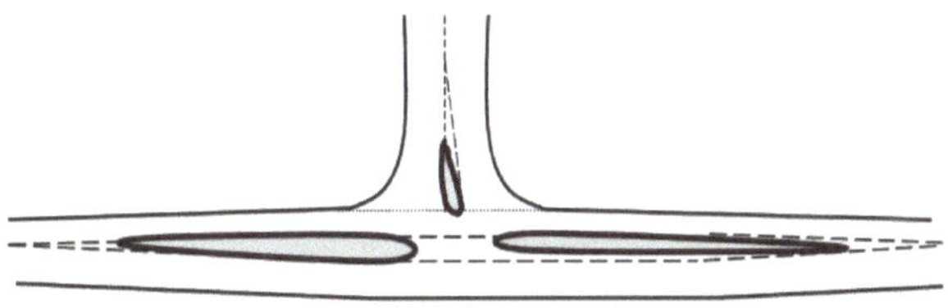

Figure 3.4. *Single lane dualling.*

Source: Martin Rogers, Highway Engineering, ISBN 0-632-05993-1.

The flows from major and minor roads determine the kind or level of junction that is used. The "simple" layout is suitable for new intersections in rural areas where the major road's 2-way AADT is expected to be no more than 13,000 vehicles, and the minor road's 2-way AADT is not expected to exceed 300 vehicles.

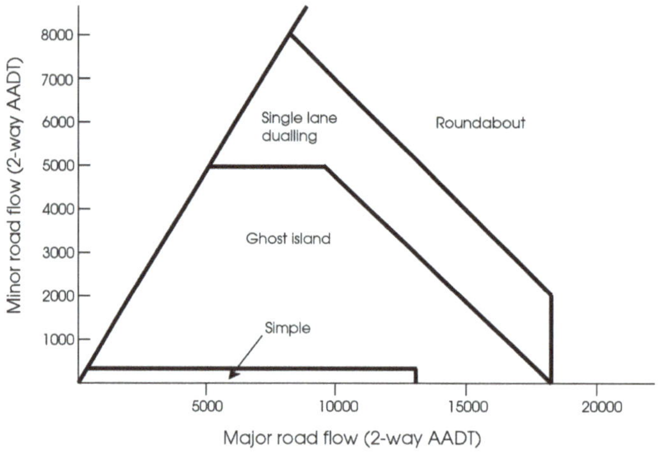

Figure 3.5. *Different levels of priority intersections for various major and minor road flows in the design year.*

Source: https://www.hzu.edu.in/engineering/highway-engineering-blackwell-science.pdf.

Figure 3.5 shows the various T-junction levels appropriate for a variety of flow combinations on single-carriageway roads. The data accounts for entry and turning capacities, accident costs, and geometric and traffic delays. (Kimber, R.M. & Hollis, E.M. (1979))

3.1.4.1. Example 5.1 – Determination of Flows at a Priority Intersection

An existing simple major/minor priority junction at the intersection of two single-carriageway 2-lane roads is to be upgraded. The major route is a major interurban highway. Figure 3.6 shows the annual average daily traffic (AADT) flow ranges predicted for each of the arms to the intersection in the design year, assumed in this case to be 15 years after opening.

Taking the upper AADT flows of 10 250 for the major carriageway and 3000 on the minor carriageway, Figure 3.5 indicates that the junction must be upgraded to a ghost island junction. Using these AADT values, the 2-way annual average hourly traffic (AAHT) on all three arms can be computed by dividing each AADT value by 24, giving the results shown in Figure 3.7.

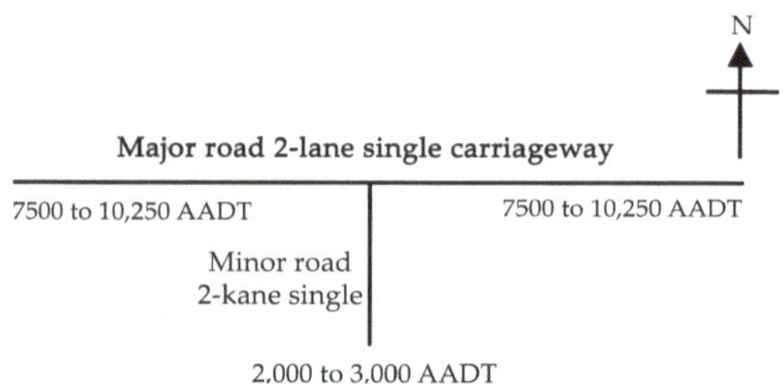

Figure 3.6. *AADT 2-way flows.*

Source: Martin Rogers, Highway Engineering, ISBN 0-632-05993-1.

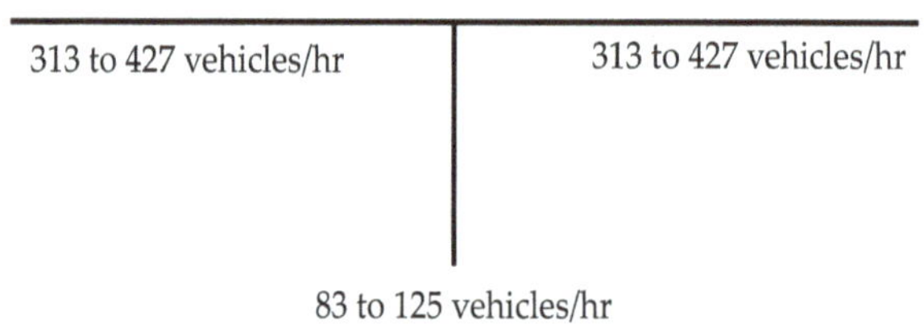

Figure 3.7. *AAHT flows.*

Source: Martin Rogers, Highway Engineering, ISBN 0-632-05993-1.

It is decided on economic grounds to design the junction to cater to the thirtieth highest annual hourly flow. Based on local traffic count data, it is found that the

AAHT flow on each of the approach roads is translated into a design peak hour flow by multiplying by a factor of 3.0, with the results shown in Figure 3.8.

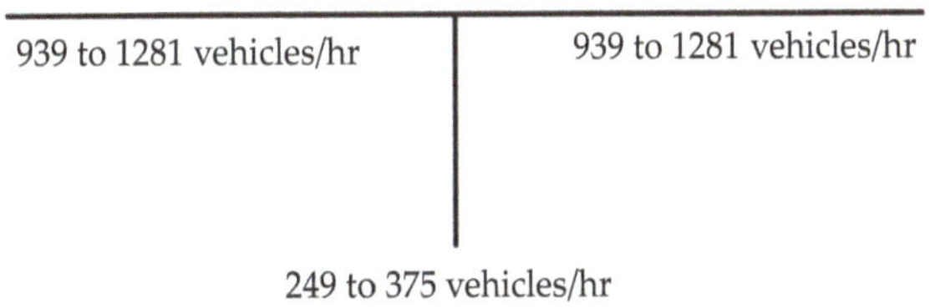

Figure 3.8. *Design peak hour flows (2-way).*

Source: Martin Rogers, Highway Engineering, ISBN 0-632-05993-1.

Directional splits derived from local traffic data are applied to these design flows in order to obtain the final directional flows, in this case, a 70:30 split in favor of flows from the southern and eastern approaches, as shown in Figure 3.9.

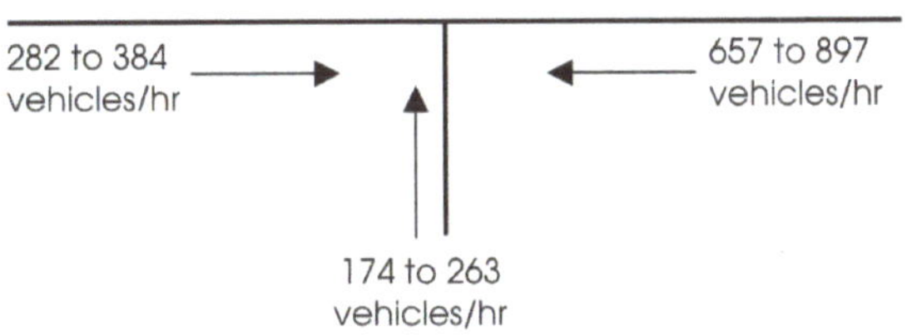

Figure 3.9. *Directional flows from all arms.*

Source: Martin Rogers, Highway Engineering, ISBN 0-632-05993-1.

Turning movements were assessed based on present-day traffic flow data from the junction, which indicated the proportions in Figure 3.10.

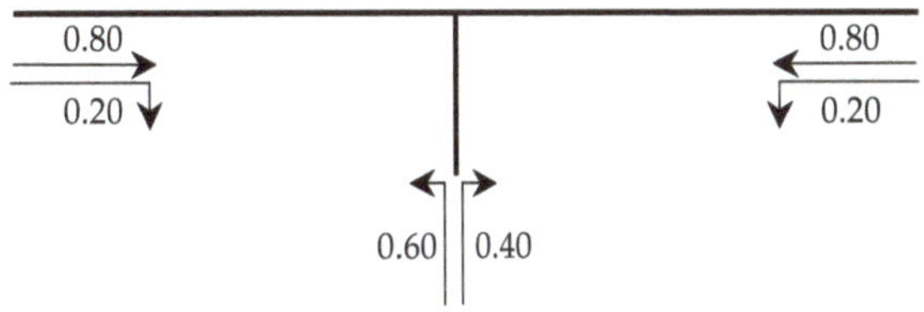

Figure 3.10. *Turning movements.*

Source: Martin Rogers, Highway Engineering, ISBN 0-632-05993-1.

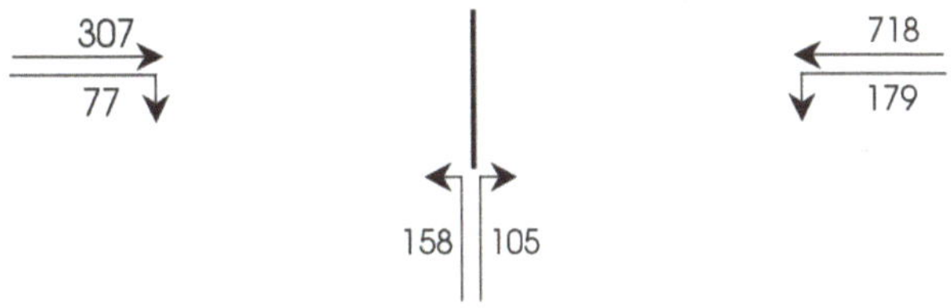

Figure 3.11. *Final upper range flows (vehicles per hour).*

Source: Martin Rogers, Highway Engineering, ISBN 0-632-05993-1.

Taking the upper flow ranges from Figure 3.9, combined with the turning proportions indicated in Figure 3.10, a set of flows is derived which can then be used to analyze the junction, as shown in Figure 3.11.

3.2. ROUNDABOUT INTERSECTIONS

In order to control merging and conflicting traffic flows at an intersection, a roundabout performs the following two major functions:

- It defines the priority between traffic streams entering the junction, usually on the basis that traffic wanting to join the circulatory flow must give way to the traffic to their right already circulating in the roundabout. (In the UK and Ireland traffic circulates in a clockwise direction.)
- It causes the diversion of traffic from its preferred straight-line path, requiring drivers to slow down as they enter the junction.

In order to work efficiently, sufficient gaps must appear in the circulating flows on the roundabout that drivers then accept. Traffic on the entry arms can thus enter, circulate, and then leave at their desired exit arm. Its operation has, therefore, certain similarities to that of a priority intersection and, as we shall see later in the chapter, the design procedure in both has certain similarities. The situation is more complex in the case of a roundabout intersection as there is no clearly identifiable major road traffic flow that can be used as a basis for designing the junction, with the circulating flow depending on the operation of all entry arms to it.

If properly designed, the angles at which traffic merges/diverges will be small. This, combined with the relatively slow traffic speeds on the roundabout, will help reduce accident rates.

As seen from Figure 3.5, when traffic flows at an intersection are relatively low, adequate control can be attained using the priority option. As flow levels increase, however, with this intersection type, delays/queue lengths become excessive and some alternative form is required. While grade-separated junctions may be the preferred option at high flow levels, the expense involved may be prohibitive. For this reason, particularly in an urban setting, at-grade roundabouts or signalized intersections become viable junction options at levels of flow above those suitable for priority control. If the cost of land is an important factor, traffic signals will be preferred as land requirements for a standard 3 or 4-arm conventional roundabout would be greater. However, right-turning vehicles can cause operational difficulties at signal-controlled junctions, particularly where volumes within this phase are large. Roundabouts have difficulty dealing with unbalanced flows, in which case signalization may be preferable. In situations where flows are relatively well balanced, and where three or four entry arms exist, roundabouts cope efficiently with the movement of traffic. Where the number of arms exceeds four, however, efficiency may be affected by the failure of drivers to understand the junction layout. It may also prove difficult to correct this even with comprehensive direction signing. In addition to their ability to resolve conflicts in traffic as efficiently as possible, roundabouts are often used in situations where there are:

- A significant change in road classification/type
- A major alteration in the direction of the road
- A change from an urban to a rural environment.

3.2.1. Types of Roundabouts

3.2.1.1. Mini Roundabouts

Mini roundabouts are small, compact circular intersections with a central island that is typically traversable by larger vehicles. The diameter of the central island is usually less than 4 meters, and the overall diameter of the roundabout is generally between 13 to 28 meters (Figure 3.12).

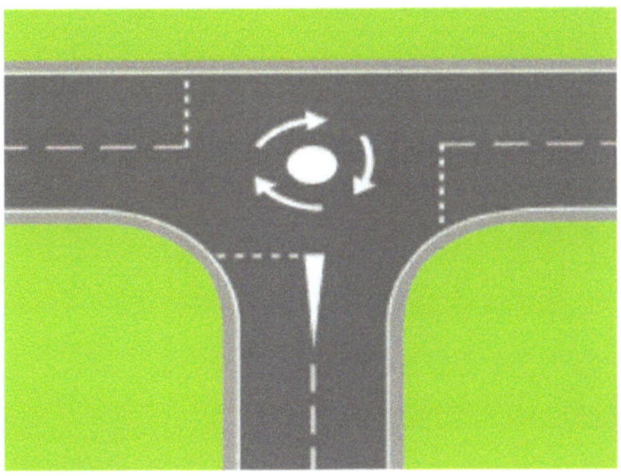

Figure 3.12. *Mini roundabout configuration.*

Source: https://encrypted-tbn0.gstatic.com/images?q=tbn:ANd9GcS2QotQylg4XwpBNA1bMUNsOhca1E4ZzvZLAoYAk3A4oA&s.

3.2.1.1.1. Applications

- They are used in urban and suburban settings where space is limited and traffic volumes are relatively low.
- Ideal for retrofitting existing intersections without requiring extensive land acquisition.

3.2.1.1.2. Benefits

- Enhance safety by reducing speeds and minimizing severe collisions.
- Cost-effective as they require minimal changes to existing infrastructure.

3.2.1.1.3. Considerations

- Not suitable for locations with high volumes of large trucks or heavy traffic.
- The central island must be designed to accommodate occasional large vehicle overruns.

3.2.1.2. Single-Lane Roundabouts

Single-lane roundabouts have one lane entering and exiting each leg of the roundabout. The central island is non-traversable, and often landscaped to improve aesthetics and discourage drivers from cutting across it (Figure 3.13).

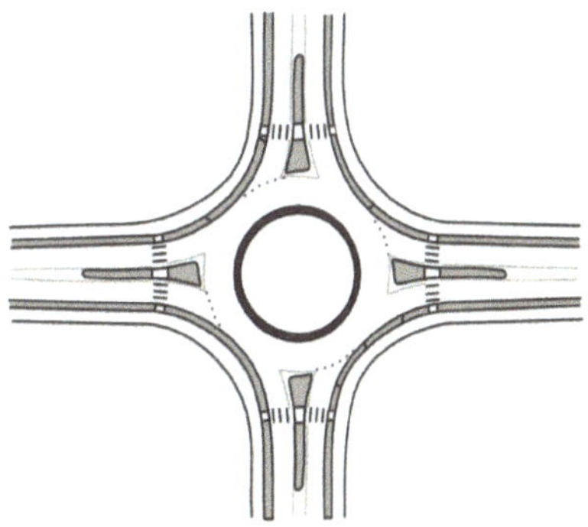

Figure 3.13. *Single-lane roundabouts configuration.*

Source: https://www.researchgate.net/publication/263585407/figure/fig1/AS:1088914408378368@1636628834114/Single-lane-roundabout-A-type.jpg.

3.2.1.2.1. Applications

- Commonly used in suburban and rural settings.
- Suitable for moderate traffic volumes and where space permits a larger footprint compared to mini-roundabouts.

3.2.1.2.2. Benefits

- Improve safety by reducing the likelihood and severity of crashes.
- Efficient handling of traffic with minimal delay.

3.2.1.2.3. Considerations

- Requires sufficient right-of-way to accommodate the larger diameter compared to mini-roundabouts.
- May require more extensive construction and land acquisition.

3.2.1.3. Multi-Lane Roundabouts

Multi-lane roundabouts are larger and more complex than single-lane or mini roundabouts, designed to handle higher traffic volumes and multiple lanes of traffic (Figure 3.14).

Figure 3.14. *Multi-lane roundabouts configuration.*

Source: https://media.licdn.com/dms/image/C4E12AQG6b6Y7pUV7kg/article-cover_image-shrink_600_2000/0/1520043812888?e=2147483647&v=beta&t=kXFlBiACZhMDIfSiL9Q7C34cnrdcKBCgwczdHj3Wv84.

3.2.1.3.1. Design and Structure

- **Central Island:** The central island is raised and landscaped, often with features to improve aesthetics and ensure it is not driven over.
- **Multiple Lanes:** Two or more lanes encircle the central island, allowing for higher traffic capacity and smoother flow.

- **Lane Markings and Signage:** Clear Lane markings, signs, and sometimes road markings (such as arrows) are essential to guide drivers on lane usage and exits.
- **Splitter Islands:** These are raised or painted islands on the approaches to the roundabout, designed to slow down traffic and guide vehicles into the correct lanes.
- **Pedestrian Crossings:** Typically located a short distance away from the roundabout entrances and exits to enhance safety.

3.2.1.3.2. Rules and Usage

- **Yield on Entry:** Vehicles must yield to traffic already circulating within the roundabout.
- **Lane Discipline:** Drivers must choose the correct lane before entering based on their intended exit, following the signage and road markings.
- **Left Lane:** Generally used for left turns or continuing straight.
- **Right Lane:** Typically used for right turns or continuing straight.
- **Indicating:** Drivers should use their indicators when exiting the roundabout to inform other road users of their intentions.

3.2.1.3.3. Benefits

- **Improved Traffic Flow:** Facilitates continuous movement of traffic, reducing stop-and-go situations and improving flow compared to traditional signalized intersections.
- **Safety:** Lower speeds within the roundabout and fewer conflict points can reduce the frequency and severity of accidents.
- **Capacity:** Can handle higher traffic volumes efficiently, making them suitable for busy urban areas or major intersections.

3.2.1.3.4. Considerations

- **Lane Markings and Signage:** Clear and consistent lane markings and signage are crucial to guide drivers correctly and prevent confusion.
- **Driver Education:** Familiarity with how to navigate multi-lane roundabouts is essential for safety and efficiency. Public education campaigns can help in areas where such roundabouts are new.
- **Non-Motorized Users:** Safe and accessible crossings for pedestrians and cyclists should be well-designed and clearly marked.

3.2.1.3.5. Challenges

- **Driver Confusion:** The complexity of multi-lane roundabouts can confuse some drivers, leading to improper lane usage and potential accidents.
- **Large Vehicles:** Accommodating larger vehicles like trucks and buses requires careful design, including wider lanes and more gradual curves.
- **Space Requirements:** Multi-lane roundabouts require more space than single-lane roundabouts or traditional intersections, which can be a constraint in urban areas.

3.2.1.3.6. Examples of Use

- **Urban Intersections:** Common in cities to manage high traffic volumes at major intersections.

- **Highway Interchanges:** Used to connect highways with local roads, facilitating smooth transitions and reducing congestion.
- **Commercial Areas:** Effective in areas with high commercial activity and traffic, helping to manage the flow efficiently.

3.2.1.4. Turbo Roundabouts

Turbo roundabouts, also known as spiral roundabouts or super roundabouts, are a variation of the traditional roundabout design that aims to improve traffic flow and capacity even further (Figure 3.15).

Figure 3.15. *Turbo roundabouts configuration.*

Source: https://media.arcadis.com/-/media/project/arcadiscom/com/blogs/north-america/united-states/brian-moore/2020/bringing-the-turbo/bringing-the-turbo-roundabout-to-the-us--header.jpg?rev=d41d4910e487411b8043a9a46d8d56df.

3.2.1.4.1. Design and Structure

- **Spiral Design:** Instead of a traditional circular layout, turbo roundabouts feature a spiral or helical shape, with gradually decreasing radii as vehicles approach the center.
- **Multiple Entry Lanes:** Turbo roundabouts often have multiple entry lanes, allowing for smoother merging and increased capacity.
- **Guidance Markings:** Clear lane markings, arrows, and signage guide drivers to choose the correct lane and navigate the spiral effectively.
- **Raised Islands**: Raised islands between lanes help to delineate traffic flows and discourage lane-changing within the roundabout.
- **Dedicated Lanes for Exiting:** Each lane typically has a designated exit, reducing weaving and conflicts within the roundabout.

3.2.1.4.2. Rules and Usage

- **Yield on Entry:** Like traditional roundabouts, vehicles entering the turbo roundabout must yield to circulating traffic.
- **Lane Discipline:** Drivers should choose their entry lane based on their desired exit, following guidance from signs and markings.
- **Smooth Merge:** The multiple entry lanes and spiral design facilitate the smooth merging of traffic into the circulating flow.
- **Indicating:** Drivers should use turn signals to indicate their intended exit, enhancing predictability for other road users.

3.2.1.4.3. Benefits

- **Increased Capacity:** The spiral layout and multiple entry lanes allow turbo roundabouts to handle higher traffic volumes more efficiently than traditional roundabouts.

- **Improved Flow:** Vehicles can enter and exit the roundabout with minimal disruption, reducing congestion and delays.
- **Safety:** By reducing conflict points and encouraging smoother merging, turbo roundabouts can enhance safety compared to conventional intersections.

3.2.1.4.4. Considerations

- **Driver Familiarity:** Turbo roundabouts may be unfamiliar to drivers accustomed to traditional roundabouts or standard intersections, requiring education and awareness campaigns.
- **Design Challenges:** Creating a well-functioning turbo roundabout requires careful consideration of lane widths, entry and exit geometries, and traffic signal coordination.
- **Pedestrian and Cyclist Safety:** Designing safe crossings and pathways for pedestrians and cyclists is essential, especially given the higher speeds and complex traffic flows.

3.2.1.4.5. Examples of Use

- **High-Traffic Intersections:** Turbo roundabouts are often deployed at intersections with heavy traffic volumes, where conventional roundabouts may struggle to handle demand.
- **Urban Arterials:** They can be effective at intersections along major urban thoroughfares, providing efficient traffic movement while minimizing disruptions to cross streets and adjacent properties.

3.2.1.5. Double Roundabouts

Double roundabouts, or "twin roundabouts," consist of two closely spaced roundabouts connected by a short roadway segment. They are designed to manage traffic at intersections where a single large roundabout might not be feasible (Figure 3.16).

3.2.1.5.1. Applications

- Suitable for complex junctions with multiple intersecting roads.
- Effective in urban areas where space constraints prevent the construction of a single large roundabout.

3.2.1.5.2. Benefits

- Improve traffic distribution and reduce congestion.
- Enhance safety by breaking up traffic flows into smaller, manageable segments.

Figure 3.16. *Double roundabouts configuration.*

Source: https://nomorecorners.wordpress.com/types-of-roundabouts/.

3.2.1.5.3. Considerations

- Can be confusing for drivers unfamiliar with the design.
- Requires clear signage and road markings to guide drivers through the two roundabouts.

3.2.1.6. Signalized Roundabouts

Signalized roundabouts incorporate traffic signals at one or more entry points to manage traffic flow, particularly during peak periods. The signals can be timed or actuated based on traffic conditions (Figure 3.17).

Figure 3.17. *Signalized roundabouts configuration.*

Source: https://asyatrafik.com/wp-content/uploads/2020/01/Sinyalize-Kavsak-Turleri-Nelerdir-1024x576.jpg.

3.2.1.6.1. Applications

- Suitable for high-traffic intersections where traditional roundabouts may become congested.
- Effective in locations where pedestrian and cyclist movements need to be controlled.

3.2.1.6.2. Benefits

- Combines the benefits of roundabouts and traffic signals.
- Can significantly improve capacity and reduce delays during peak hours.

3.2.1.6.3. Considerations

- More expensive to install and maintain compared to unsignalized roundabouts.
- Requires synchronization of signals to optimize traffic flow.

3.2.1.7. Grade-Separated Roundabouts

Grade-separated roundabouts, also known as "stacked roundabouts," incorporate bridges or tunnels to separate traffic flows vertically. These designs can include single-lane or multi-lane roundabouts on different levels (Figure 3.18).

Figure 3.18. *Grade-separated roundabouts configuration.*

Source: https://bicycledutch.wordpress.com/wp-content/uploads/2011/05/houten-roundabout.jpg.

3.2.1.7.1. Applications

- Used in areas with extremely high traffic volumes and where land space is limited.
- Suitable for complex intersections with significant crossing movements.

3.2.1.7.2. Benefits

- Eliminates conflict points associated with crossing traffic streams.
- Significantly improves capacity and reduces congestion.

3.2.1.7.3. Considerations

- Highly complex and expensive to design and construct.
- Requires substantial land acquisition and engineering.

3.3. BASICS OF TRAFFIC SIGNAL CONTROL: OPTIMIZATION AND DELAYS

In order to make the best and safest use of the available carriageway space at a highway intersection, traffic signals function by assigning distinct time periods to opposing traffic movements. Through the signals' cycle, priority can change over time. Signal linking can be utilized in urban areas, specifically when a road has several intersections throughout its entire length, to enable nearly continuous traffic flow along the route.

In the UK, the choice to install traffic lights at a particular intersection is made by weighing the estimated improvement in accident characteristics and the reduction in delay that will result from it economically against the computed capital and operating costs. Traffic signals in the UK typically flash red, red/amber, green, and amber. The order is red, green, and amber in Ireland.

The installation of traffic signals is justified by the need to:

- Reduce delay to motorists and pedestrians moving through the junction
- Reduce accidents at the junction
- Improve the control of traffic flow into and through the junction in particular and the area in general, thereby minimizing journey times
- Impose certain chosen traffic management policies.

Traffic signals can however have certain disadvantages:

- They must undergo frequent maintenance along with frequent monitoring to ensure their maximum effectiveness
- There can be inefficiencies during off-peak times leading to increases in delay and disruption during these periods
- Increases in rear-end collisions can result
- Signal breakdown due to mechanical/electrical failure can cause serious interruptions in traffic flow.

The demand-to-capacity ratio for each flow path serves as the foundation for design in this case, just like it does for priority junctions and roundabouts. Nevertheless, another important factor is the traffic signal's current setting. The maximum traffic flow that can cross the stop line under the assumption of 100% green time is known as the saturation flow, and it is used to express the capacity of a given flow path. While they can be installed, traffic signals typically function according to a predetermined schedule. However, they can also be activated by vehicles. It is possible to program the fixed-time sequence to change based on the time of day. For the morning and evening peaks, the daytime off peak, and the late night/early morning conditions, a different program may typically run.

3.3.1. Phasing at a Signalized Intersection

Conflicting traffic streams can be divided via phasing. A set of conditions applied to one or more traffic streams is called a phase. Every traffic in a phase will receive the same signal indications at the same time during a single cycle. Consider a crossroads intersection, where traffic moving from east to west will clash with that moving north-south. This is the number of necessary phases because there are two conflicts. More than two phases will be needed for other intersections that are more complicated. An illustration of this would be a crossroads on one of the entry roads where there is a significant volume of traffic turning right, requiring a separate phase for that movement. For this reason, a three-phase system is created. Because each phase of a traffic cycle has time delays, as will be demonstrated below, efficiency and safety require that the number be kept to a minimum.

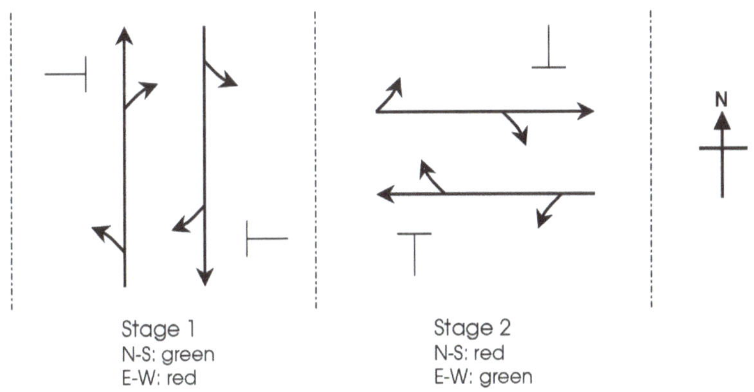

Figure 3.19. *Typical stages within a two-phase system.*

Source: Martin Rogers, Highway Engineering, ISBN 0-632-05993-1.

The sequential steps by which the control at the intersection is varied are frequently used to describe the control of traffic movement at a signalized junction. Stage control refers to this, where a stage typically starts at the beginning of the amber period and ends at the beginning of the subsequent stage. A traffic cycle has stages that are placed in a specific order. Typical two-phase and three-phase signalized intersections are depicted in Figures 3.19 and 3.20, respectively.

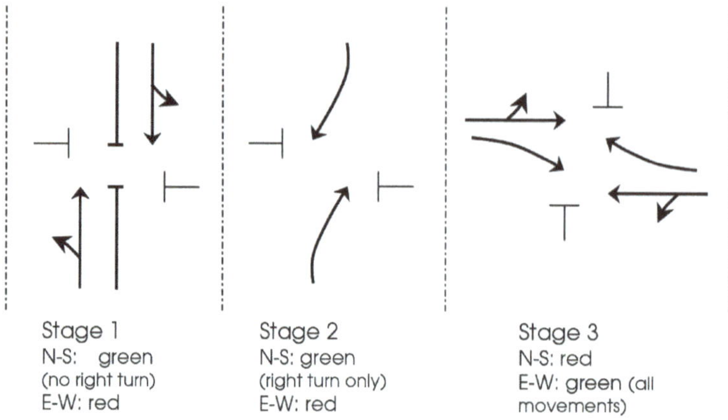

Figure 3.20. *Typical stages within a three-phase system (two lanes in each direction with a dedicated lane for right-turning vehicles in all cases).*

Source: Martin Rogers, Highway Engineering, ISBN 0-632-05993-1.

3.3.2. Saturation Flow

One way to estimate the capacity for each approach to an intersection is to add up the saturation flows of each lane within each pathway. The following are the primary variables that are pertinent to capacity estimation for each method:

3.3.2.1. Number, Width, and Location of Lanes

An approach road's total capacity is determined by adding the saturation flows of each of its lanes. Assuming a zero gradient, the average lane width approaching a signalized intersection is 3.25 m. At this value, the saturation flow for nearside lanes is set at 1940 pcu/hr, and for non-nearside lanes, it increases to 2080 pcu/hr.

3.3.2.2. Weather Conditions

In wet weather conditions, saturation flows decrease to 6% below their dry weather values.

3.3.2.3. Gradient

Over a distance of 60 meters measured back from the stop-line, an increase of 1% in the uphill gradient will result in a 2% decrease in the saturation flow value. Saturation values are invariant under descending gradients.

3.3.2.4. Turning movements

Saturation flows will decrease as the percentage of turning traffic increases in scenarios where there is no opposition. The saturation flow in situations where turning traffic is opposed is determined by the number of spaces in the opposing traffic flow as well as the storage capacity of the vehicles involved in this traffic movement.

The equations derived by Kimber et al., (1986) for predicting saturation flow are:

3.3.2.5. Unopposed Traffic Streams

The saturation flow rate is given by:

$$S_1 = (S_0 - 140d_n)/(1 + 1.5f/r) \text{ pcu/hr}$$

Where,

$$S_0 = 2080 - 42d_g \times G + 100(w - 3.25)$$

and

d_n = 1 (nearside lane) or 0 (non-nearside lane)

f = proportion of turning vehicles in the lane under scrutiny

r = radius of curvature of vehicle path, meters

d_g = 1 (uphill entry roads) or 0 (downhill entry roads)

G = percentage gradient of the entry road

w = entry road lane width, meters.

3.3.2.6. Opposed Traffic Streams

In this instance, the saturation flow in a given lane for right-turning opposed streams is given by:

$$S_2 = S_g + S_c \text{ (pcu/hr)}$$

The two components of this equation are computed as follows:

(1) S_g

S_g is the saturation flow occurring during the 'effective green' time within the lane of opposed mixed-turning traffic.

$$S_g = (S_0 - 230)/(1 + (T - 1)f)$$

$$T = 1 + 1.5/r + t_1/t_2$$

$$t_1 = 12X_0^2/(1 + 0.6(1 - f)N_s)$$

$$X_0 = q_0/\lambda n_1 s_0$$

$$t_2 = 1 - (fX_0)^2$$

where , X_0 is the degree of saturation on the opposing entry arm (ratio of flow to capacity);

q_0 is the actual flow on the opposing arm, measured in vehicles per hour of green time (excluding non-hooking right-turning vehicles);

l is the effective green time divided by the total cycle time, C;

n_1 is the number of lanes within the opposing entry arm;

s_0 is the saturation flow for each of the lanes on the opposing entry arm;

N_s is the number of storage spaces within the junction which the right-turning vehicles can use so as not to block the straight-ahead stream.

(2) S_c

Where, S_c is the saturation flow occurring after the 'effective green' time within the lane of opposed mixed-turning traffic.

(During a traffic phase, the effective green time is the actual green time plus the amber time minus a deduction for starting delays.)

$$S_c = P(1 + N_s)(fX_0)^{0.2} \times 3600/\lambda C$$

$$P = 1 + \Sigma_i(\alpha_i - 1)p_i$$

where, P is the conversion factor from vehicles to passenger car units;

α_i is the pcu value of vehicle type i;

p_i is the proportion of vehicles of type i.

The passenger car unit values used in connection with the design of a signalized junction are given in Table 3.1.

Table 3.1. *Passenger Car Unit Values*

Vehicle type	pcu equivalent
Car/light vehicle	1.0
Medium commercial vehicle	1.5
Heavy commercial vehicle	2.3
Bus/coach	2.0

Source: Martin Rogers, Highway Engineering, ISBN 0-632-05993-1.

3.3.3. Effective Green Time

The term "effective green time" refers to the amount of time needed to achieve a specific discharge rate over the stop line if the flow started and stopped simultaneously with an instantaneous color change as indicated by the signal head. It is possible to estimate the effective green time at an intersection by analyzing the movement of cars across the stop line. At the start of the green period and ending at the end of the amber period, cars are released over the stop line. Lost time is the period of time between the beginning of actual green time and the beginning of effective green time, as well as between the end of effective green time and the end of the amber period.

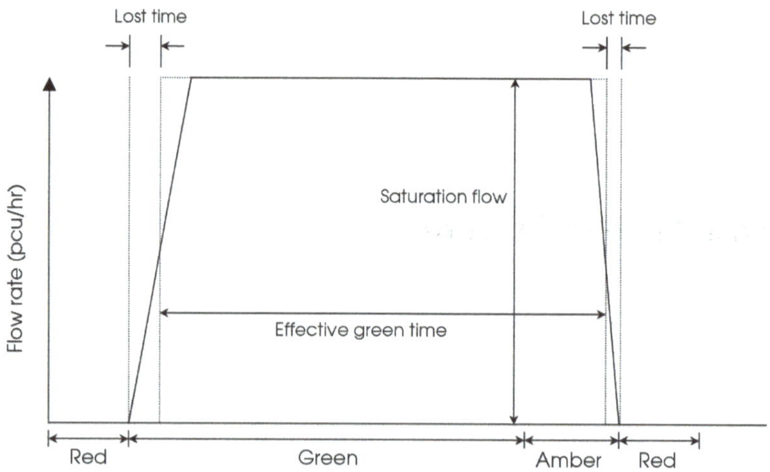

Figure 3.21. *Diagrammatic representation of actual green time, effective green time, and lost time.*

Source: Martin Rogers, Highway Engineering, ISBN 0-632-05993-1.

The flow across the stop line increases from zero at the beginning of each cycle when the light turns green and traffic starts to move off, and it keeps going up until saturation flow is reached. Until the light turns amber at the end of the phase, the flow level stays constant. While some cars will stop right away, others might take some time. The lights turn red, and the flow goes back to zero.

From Figure 3.20 it can been seen that the actual green time plus the amber period is equal to the effective green time plus the two periods of lost time at the beginning and end of the cycle. The effective green time is thus the length of time during which saturation flow would have to be sustained in order to obtain the same quantity of traffic through the lights as is achieved during an actual green period. It is denoted by a rectangle in Figure 3.20. This rectangle has exactly the same area as that under the actual flow curve (Figure 3.21).

Normally, the lost time is assumed to be taken as equal to 2 seconds, with the amber time set at 3 seconds. Effective green time is thus equal to actual green time plus 1 second.

3.3.4. Optimum Cycle Time

Assuming the signal system at an intersection is operating on fixed time control, the cycle length will directly affect the delay to vehicles passing through the junction. There is always an element of lost time every time the signal changes. If the cycle time is short, this lost time becomes a significant proportion of it, leading to inefficiencies in the working of the junction and consequent lengthy delays. Too long a cycle will result in all queuing vehicles being cleared in the early part of the green period, with the only vehicles crossing the stop line in the latter part being those that subsequently arrive, often quite widely spaced. This too is inefficient. (The discharge of traffic through a junction is at its most efficient when there is a waiting queue on the approach road.)

Webster (1958) developed a set of formulae for establishing the optimum signal settings in order to minimize the total delay to all streams on the approach roads. The optimum cycle time C_o is obtained from:

$$C_o = (1.5L + 5) \div (1 - Y)$$

where,

L = total lost time per cycle;

Y = the sum of the maximum y values for all of the phases which make up the cycle;

(y is the ratio of actual flow to saturation flow on each approach).

There is a minimum cycle time of 25 seconds based on safety considerations. A maximum cycle time of 120 seconds is considered good practice. Normally, the cycle time will lie within the range of 30 to 90 seconds.

3.3.4.1. Lost Time

Lost time per cycle consists of the time lost during the green period (generally taken as 2 seconds per phase) plus the time lost during what is known as the intergreen period. The intergreen period is defined as the period between one phase losing the right of way and the next phase gaining the right of way, or the time between the end of green on one phase and the start of green on the next, The intergreen period provides a suitable time during which vehicles making right turns can complete their maneuver safely having waited in the middle of the intersection.

If the amber time during the intergreen period is 3 seconds and the total intergreen period is 5 seconds, this gives a lost time of 2 seconds, as this is the period of time for which all lights show red or red/amber, a time during which no vehicle movement is permitted. The period of time lost to traffic flow is referred to as lost time during the intergreen period. It should not be confused with lost time due to starting delays at the commencement of each phase.

Table 3.2. shows an example of a two-phase system.

Table 3.2. *Typical Intergreen Period Within a Two-Phase Signal System*

Phase 1	G	A	A	A	R	R	Red
		Amber period (3 s)			Lost time (2 s)		←5 seconds intergreen period (3 s + 2 s)
Phase 2	R	R	R	R	R/A	R/A	Green

Source: Martin Rogers, Highway Engineering, ISBN 0-632-05993-1.

ROLE MODEL

GARRETT MORGAN: INVENTION OF THE THREE-POSITION TRAFFIC SIGNAL

Garrett Morgan (March 4, 1877–July 27, 1963) was an inventor and businessman from Cleveland who is best known for inventing a device called the Morgan Safety Hood and Smoke Protector in 1914. The invention was later dubbed the gas mask.

Early Life

The son of a formerly enslaved man and woman, Garrett Augustus Morgan was born in Claysville, Kentucky, on March 4, 1877. His mother was of Native American, Black, and white descent (her father was a minister named Rev. Garrett Reed), and his father, was half-Black and half-white, the son of the Confederate Colonel John Hunt Morgan, who led Morgan's Raiders in the Civil War. Garrett was the seventh of 11 children, and his early childhood was spent attending school and working on the family farm with his brothers and sisters. While still a teenager, he left Kentucky and moved north to Cincinnati, Ohio, in search of opportunities.

Although Morgan's formal education never took him beyond elementary school, he worked to educate himself, hiring a tutor while living in Cincinnati and continuing his studies in English grammar. In 1895, Morgan moved to Cleveland, Ohio, where he went to work as a sewing machine repairman for a clothing manufacturer, teaching himself as much as he could about sewing machinery and experimenting with the process. Word of his experiments and his proficiency for fixing things traveled fast, and he worked for numerous manufacturing firms in the Cleveland area. In 1907, the inventor opened his sewing equipment and repair shop. It was the first of several businesses he would establish. In 1909, he expanded the enterprise to include a tailoring shop that employed 32 people. The new company turned out coats, suits, and dresses, all sewn with equipment that Morgan himself had made.

Marriage and Family

Morgan married twice, first to Madge Nelson in 1896; they were divorced in 1898. In 1908 he married Mary Anna Hasek, a seamstress from Bohemia: It was one of the earliest interracial marriages in Cleveland. They had three children, John P., Garrett A., Jr., and Cosmo H. Morgan.

The Safety Hood (Early Gas Mask)

In 1914, Morgan was awarded two patents for the invention of an early gas mask, the Safety Hood and Smoke Protector. He manufactured the mask and sold it nationally and internationally through the National Safety Device Company, or Nadsco, using a marketing strategy to avoid Jim Crow discrimination—what historian Lisa Cook calls "anonymity by dissociation." At the time, entrepreneurs sold their inventions by conducting live demonstrations. Morgan appeared in these events to the public, with municipal fire departments, and city officials representing himself as his own assistant—a Native American man called "Big Chief Mason." In the South, Morgan hired whites, sometimes public safety professionals, to stage demonstrations for him. His newspaper advertisements featured smartly dressed white male models.

The gas mask proved very popular: New York City quickly adopted the mask, and, eventually, 500 cities followed suit. In 1916, a refined model of Morgan's gas mask was awarded a gold medal at the International Exposition of Sanitation and Safety and another gold medal from the International Association of Fire Chiefs.

The Lake Erie Crib Disaster

On July 25, 1916, Morgan made national news for using his gas mask to rescue men trapped during an explosion in an underground tunnel located 250 feet beneath Lake Erie. No one had been able to reach the men: Eleven of them had died as had ten others attempting to rescue them. Called in the middle of the night six hours after the incident, Morgan and a team of volunteers donned the new "gas masks" brought two workers out alive, and recovered the bodies of 17 others. He gave artificial respiration to one of the men he rescued.

Afterward, Morgan's company received many additional requests from fire departments around the country that wished to purchase the new masks. However, the national news contained photographs of him, and officials in several southern cities canceled their existing orders when they discovered he was Black.

In 1917, the Carnegie Hero Fund Commission reviewed the reports of heroism displayed during the disaster. Based on news reports that downplayed Morgan's role, the Carnegie board decided to give the prestigious "Hero" award to a minor figure in the rescue effort who was white, rather than to Morgan.

Morgan protested, but the Carnegie Institution said he had not risked as much as the other person had because he had safety equipment. Some reports say the Morgan gas mask was modified and used in World War I after the Germans unleashed chemical warfare at Ypres on April 22, 1915, although there is no strong evidence for it. Despite Morgan's popularity in the United States, there were dozens of other masks on the market by then, and most used in WWI were of English or French manufacture.

The Morgan Traffic Signal

In 1920, Morgan moved into the newspaper business when he established the "Cleveland Call." As the years went on, he became a prosperous and widely respected businessman and was able to purchase a home and an automobile, invented by Henry Ford in 1903. In fact, Morgan was the first African American to purchase an automobile in Cleveland, and it was Morgan's experience while driving along the streets of that city that inspired him to invent an improvement to traffic signals.

After witnessing a collision between an automobile and a horse-drawn carriage, Morgan took his turn at inventing a traffic signal. While other inventors had experimented with, marketed, and even patented traffic signals, Morgan was one of the first to apply for and acquire a U.S. patent for an inexpensive way to produce a traffic signal. The patent was granted on November 20, 1923. Morgan also had his invention patented in Great Britain and Canada.

Morgan stated in his patent for the traffic signal:

"This invention relates to traffic signals, and particularly to those which are adapted to be positioned adjacent the intersection of two or more streets and are manually operable for directing the flow of traffic...In addition, my invention contemplates the provision of a signal which may be readily and cheaply manufactured."

The Morgan traffic signal was a T-shaped pole unit that featured three positions: Stop, Go, and an all-directional stop position. This "third position" halted traffic in all directions to allow pedestrians to cross streets more safely.

Morgan's hand-cranked semaphore traffic management device was in use throughout North America until all manual traffic signals were replaced by the automatic red-, yellow-, and green-light traffic signals currently used around the world. The inventor sold the rights to his traffic signal to the General Electric Corporation for $40,000.

Other Inventions

Throughout his life, Morgan was always experimenting to develop new concepts. Though the traffic signal came at the height of his career and became one of his most famous inventions, it was just one of several innovations he developed, manufactured, and sold over the years.

Morgan invented a zig-zag stitching attachment for the manually operated sewing machine. He also founded a company that made personal grooming products such as hair dying ointments and the curved-tooth pressing comb.

As word of Morgan's life-saving inventions spread across North America and England, demand for these products grew. He was frequently invited to conventions and public exhibitions to demonstrate how his inventions worked.

Death

Along with many others, Morgan lost most of his wealth with the stock market crash, but it didn't stop his inventive nature. He developed glaucoma, but at the time of his death he was still working on a new invention: a self-extinguishing cigarette. Morgan died on August 27, 1963, at the age of 86. His life was long and full, and his creative energies were recognized both during and after his lifetime.

Legacy

Morgan's inventions have had a tremendous impact on the safety and well-being of people all over the world—from miners to soldiers to first responders to ordinary car owners and pedestrians. Another ongoing legacy is his weekly newspaper, originally named the "Cleveland Call" and now called the "Cleveland Call and Post." His achievements as a son of formerly enslaved people, against all odds, and in the face of Jim Crow era discrimination, are inspiring.

CLASS ACTIVITY

Highway Design Brainstorm Students will brainstorm in small groups to design a highway layout considering safety, traffic flow, environmental impact, and accessibility. They'll present their designs, highlighting key considerations and justifications for their choices.

SUMMARY

- A highway intersection is required to control conflicting and merging streams of traffic so that delay is minimized. This is achieved through choice of geometric parameters that control and regulate the vehicle paths through the intersection. These determine priority so that all movements take place with safety.
- At existing junctions, it will be possible to directly estimate peak hour and daily traffic flows together with all turning movements. In order for the measurements to be as representative as possible of general peak flow levels, it is desirable to take them during a normal weekday within a neutral month (April, May, June, September, or October).
- A priority intersection occurs between two roads, one termed the 'major' road and the other the 'minor' road. The major road is the one assigned a permanent priority of traffic movement over that of the minor road. The minor road must give priority to the major road with traffic from it only entering the major road when appropriate gaps appear. The principal advantage of this type of junction is that the traffic on the major route is not delayed.
- Mini roundabouts are small, compact circular intersections with a central island that is typically traversable by larger vehicles. The diameter of the central island is usually less than 4 meters, and the overall diameter of the roundabout is generally between 13 to 28 meters.
- Multi-lane roundabouts are larger and more complex than single-lane or mini roundabouts, designed to handle higher traffic volumes and multiple lanes of traffic.
- Turbo roundabouts, also known as spiral roundabouts or super roundabouts, are a variation of the traditional roundabout design that aims to improve traffic flow and capacity even further.
- Double roundabouts, or "twin roundabouts," consist of two closely spaced roundabouts connected by a short roadway segment. They are designed to manage traffic at intersections where a single large roundabout might not be feasible.
- Traffic signals work on the basis of allocating separate time periods to conflicting traffic movements at a highway intersection so that the available carriageway space is utilized as efficiently and safely as possible. Priority can be varied with time through the cycle of the signals. Within urban areas in particular, in situations where a road has a number of intersections along its entire length, signal linking can be used as a method for allowing almost continuous progression of traffic through the route.

REVIEW QUESTIONS

1. How do peak-hour traffic volumes influence the derivation of design reference flows from baseline traffic figures?
2. What are the key factors to consider when deriving design reference flows for roundabout intersections from baseline traffic figures?
3. How does the capacity of a roundabout compare to that of a signalized intersection?
4. What methods can be used to estimate traffic flow at a roundabout intersection?
5. Explain the impact of entry angle and roundabout diameter on the performance of a roundabout.
6. How do pedestrian and cyclist volumes affect the design and operation of a roundabout?
7. What is the primary objective of traffic signal optimization?
8. Describe the relationship between cycle length and delay in traffic signal control.
9. What are the common strategies used in traffic signal optimization?

REFERENCES

1. Binning, J. C. (1999). *OSCADY 4–User Guide*. Crowthorne, UK: Transport and Road Research Laboratory.
2. Burrow, I. J. (1987). *OSCADY–A Computer Program to Model Capacities, Queues and Delays at Isolated Traffic Signal Junctions* (TRRL Report RR105). Crowthorne, UK: Transport and Road Research Laboratory.
3. Department of Transport. (1981). Junctions and accesses: Determination of size of roundabouts and major/minor junctions (Departmental Advice Note TA 23/81). In *Design Manual for Roads and Bridges, Volume 6: Road Geometry*. London, UK: The Stationery Office.
4. Department of Transport. (1985). Traffic flows and carriageway width assessments (Departmental Standard TD 20/85). In *Design Manual for Roads and Bridges, Volume 5: Assessment and Preparation of Road Schemes*. London, UK: The Stationery Office.
5. Department of Transport. (1993). *The Geometric Design of Roundabouts* (Departmental Standard TD 16/93). London, UK: The Stationery Office.
6. Department of Transport. (1995). Geometric design of major/Minor priority junctions (Departmental Standard TD 42/95). In *Design Manual for Roads and Bridges, Volume 6: Road Geometry*. London, UK: The Stationery Office.
7. Department of Transport. (1996). Traffic appraisal manual. In *Design Manual for Roads and Bridges, Volume 12: Traffic Appraisal of Road Schemes*. London, UK: The Stationery Office.
8. Hunt, P. B., Robertson, D. I., & Bretherton, R. D. (1981). *SCOOT–A Traffic Responsive Method of Coordinating Signals* (Laboratory Report LR 1014). Crowthorne, UK: Transport and Road Research Laboratory.

9. Kimber, R. M. (1980). *The Traffic Capacity of Roundabouts* (TRRL Report LR942). Crowthorne, UK: Transport and Road Research Laboratory.

10. Kimber, R. M., & Hollis, E. M. (1979). *Traffic Queues and Delays at Road Junctions* (TRRL Report LR909). Crowthorne, UK: Transport and Road Research Laboratory.

11. Kimber, R. M., McDonald, M., & Hounsell, N. B. (1986). *The Prediction of Saturation Flows for Road Junctions Controlled by Traffic Signals* (TRRL Research Report 67). Crowthorne, UK: Transport and Road Research Laboratory.

12. O'Flaherty, C. A. (1997). *Transport Planning and Traffic Engineering*. Oxford, UK: Butterworth Heinemann.

13. Robertson, D. I. (1969). *TRANSYT–A Traffic Network Study Tool* (Laboratory Report LR 253). Crowthorne, UK: Transport Research Laboratory.

14. Semmens, M. C. (1985). *PICADY2–An Enhanced Programme for Modelling Traffic Capacities, Queues and Delays at Major/Minor Junctions* (TRRL Report RR36). Crowthorne, UK: Transport and Road Research Laboratory.

15. Webster, F. V. (1958). *Traffic Signal Settings* (Road Research Technical Paper No. 39). London, UK: The Stationery Office.

CHAPTER

4

Construction of Road Pavements, Drainage, and Materials

LEARNING OBJECTIVES

After studying this chapter, you will be able to:

- Describe highway construction materials and their relevant tests
- Define pavement
- Explain the drainage system for road construction

KEY TERMS FROM THIS CHAPTER

Asphalt	Concrete
Bituminous materials	Pavement
Aggregates	Emulsion
Portland cement	Admixtures
Epoxy	Permeability test
Dry density test	Aggregate gradation
Crushing test	Soundness test
Bitumen adhesion test	Bituminous macadam
Tack coat	Seal coat
Highway excavation	Geogrids

INTRODUCTION

In civil engineering, the building of road pavements, drainage systems, and material selection are essential elements that guarantee the robustness, security, and efficiency of road infrastructure. Every component is essential to managing water flow, supporting traffic loads, and preserving road integrity in a variety of environmental circumstances.

The process of building a pavement includes constructing a sturdy and stable surface for traffic. In order to create a stable subgrade, the process begins with site preparation, which entails clearing, excavating, and grading. Typically, the pavement structure is made up of three layers: the surface layer, base course, and sub-base. Compacted aggregates are frequently used to create the sub-base and base course, which distribute loads and offer structural support. The surface layer offers a smooth and long-lasting driving surface. It is primarily composed of two types: flexible (asphalt) and rigid (concrete). Concrete pavements are prized for their durability and strength, whereas asphalt pavements are preferred for their adaptability and simplicity of maintenance.

In order to prevent water accumulation, which can erode pavement structures and create dangerous driving conditions, effective drainage is essential for the longevity and safety of roads. Both surface and subsurface components are present in drainage systems. Water is directed away from the road surface by surface drainage elements like gutters, curbs, and side ditches. Water from beneath the pavement is removed by subsurface drainage systems, which include perforated pipes and drainage layers. This prevents water infiltration and possible erosion of the subgrade.

The longevity and performance of drainage systems and road pavements depend heavily on the materials chosen. Aggregates, asphalt, concrete, and geotextiles are examples of common materials. To guarantee stability and durability, aggregates have to fulfill certain requirements related to size, shape, and strength. Whereas concrete mixtures are selected for their strength and durability, asphalt mixtures are made to withstand traffic loads and weather conditions. Geotextiles are used to improve drainage, create layer separation, and strengthen soil stability.

4.1. HIGHWAY CONSTRUCTION MATERIALS AND THEIR RELEVANT TESTS

Large rights-of-way with significant length are a defining feature of highway construction. This means that materials required for highway construction are numerous and in great quantities. The material needed for highway construction is determined by the type of pavement—rigid or flexible. General information about the types of materials used in highway construction is provided in this post, along with a discussion of their characteristics.

These items of major importance which are used in normal highway construction are:

- Bituminous Materials
- Soil
- Aggregates
- Portland Cement Concrete
- Admixtures
- Pavement Marking Materials
- Structural Steel

4.1.1. Bituminous Materials

- Bituminous material, or bitumen in the form of asphalt, is one of the major highway construction materials used.
- Its use ranges from binders for the highest type pavements of complex engineering designs for the ultimate traffic, to direct spray applications.
- To protect an existing pavement or to provide a low-cost, all-weather road for a minimum of traffic.
- Asphalt materials, as used, range in consistency from semisolid to liquid products, which are processed and blended to conform to specification requirements for various types and grades.
- Asphalt is normally a residue product from the distillation of crude oil and, as refined, is a heavy, viscous material that is semisolid at normal atmospheric temperatures.

- By controlled processing or blending desired viscosity and hardness characteristics can be produced.
- This product is called asphalt cement and is also known as paving grade or penetration grade asphalt.
- Bituminous Materials, of the Highway Materials Manual contains the test procedures, sampling requirements, and test equipment.
- Necessary to characterize and control, within specification requirements, asphalt cement, cutback asphalts, emulsified asphalts, and asphalt-aggregate mixtures.

4.1.2. Emulsified Asphalt

- Emulsion, is made from asphalt cement.
- It is tiny particles of asphalt cement mixed with water and an emulsifying agent—usually a detergent.
- Emulsions are called liquid asphalts because, unlike asphalt cement, they are liquid at normal air temperatures and therefore do not require heat to liquefy.
- To produce emulsions, hot asphalt cement and water containing the emulsifying agent are pumped at high pressure through a colloid mill.
- The emulsifying agent coats the asphalt particles and puts an electric charge on their surfaces.
- This charge causes the asphalt droplets to repel one another so they don't combine.
- These charges are used to categorize emulsions as cationic (positive charge) or anionic (negative charge).
- Charges are important because they affect the compatibility of emulsion with mineral aggregates.
- An anionic emulsion should be used with limestone aggregate that usually bears a positive surface charge.
- A cationic emulsion should be used with sandstone because these aggregates usually bear a negative surface charge.
- After an emulsion is mixed with aggregates it sets or breaks.
- The asphalt droplets react with the aggregate and combine, squeezing out the water.
- The water then evaporates, leaving the asphalt droplets to set and produce a continuous film on the aggregates.

4.1.2.1. Advantages:

The following advantages have been reported for chip seals when using:

- An asphalt binder is less soft in hot weather and less brittle in cold weather.
- Improved asphalt binder adhesion and strength for increased cover aggregate/ chip retention.

- The ability for earlier sweeping to remove excess chips.
- The surface treatment of roads, streets, and highways with higher traffic volumes and speeds.
- Allows for the use of larger-size chips.

4.1.3. Soils

- Major works of man since the beginning of time have depended on the use of soils.
- Not only does soil form the foundation, or supporting surface for buildings, bridges, roadways, and culverts, but it is the most common constituent, in the form of aggregate, in the works themselves.
- Soils are the most basic abundant material in highway engineering and as such, are well characterized by the expression "common as dirt."
- The steps taken to prepare foundations manipulate and transport soils and prepare them for use as aggregates account for well over 70% of the total cost of construction for a normal highway project.
- It is easy to see then why knowledgeable consideration of every aspect of soils is very important to successful highway engineering.

4.1.4. Aggregates

- The surface, base, and sub base of pavements consist of aggregates or a mixture of aggregates with cement and water (Portland cement concrete) or a bituminous material (bituminous concrete).
- The strength properties of a layer are a function of both the quality and the proportions of ingredients.
- Aggregates are hard, durable, mineral materials obtained by mining or quarrying operations of deposits of sand, gravel, talus, or ledges, and may be defined as "granular material of mineral composition used either in combination with a binding medium to form bituminous concrete, Portland cement concrete, mortar, plaster, etc., or alone as in road bases, railroad ballast, filter beds, etc."

4.1.5. Portland Cement Concrete

- Portland cement concrete is a mixture of Portland cement, water, air, sand, and gravel or crushed stone.
- It is formed when the cement and water (paste) combine chemically and binds the entire mixture into a rock-like mass.
- It may be thought of as a two-component material; paste and aggregate.
- The paste is comprised of cement, water, and entrained air.
- The aggregate portion is generally composed of sand and gravel and comprises 70 to 80% of the volume of the concrete.

- Aggregates both fine and coarse, should be selected for their quality, strength, durability, and resistance to environmental effects.
- They should be smoothly graded. The quality of the concrete depends also on the quality of the paste.
- In properly made concrete each particle of aggregate is completely coated with paste and all of the space between particles is filled with paste.

4.1.6. Admixtures

- In recent years there have come into use a number of additives developed to improve various characteristics of concrete in both the plastic and hardened states.
- Other additives gaining in usage are known as water-reducing admixtures and set retarding admixtures.
- Water reducing additive is intended to impart more workability into plastic concrete with less water, thereby increasing its strength through a reduced water/cement ratio.
- Set retarding additive is intended to slow the setting characteristics of the mix and thereby allowing more time for placing and finishing.
- However, these additives are of diverse chemical compositions and can produce adverse reactions when used in a mix, either between the additives themselves when combined in a mix or with the cement, aggregates, or water making up the mix.
- Any time an additive is proposed for use in a mix, trial mixes should be made with the additives to evaluate their effectiveness and to check for any adverse reactions.

4.1.7. Pavement Marking Materials

- Pavement markings have important functions in providing guidance and information to drivers.
- Pavement markings convey important information to drivers without diverting their attention from the road.
- In some cases, they supplement regulations or warnings given by other devices.
- In other cases, they are used alone and produce results that cannot be obtained by any other device.
- Many materials can be used for pavement marking.
- The choice depends on cost, durability, reflectivity, pavement surface, and drying time.

4.1.8. Paint

Pavement markings made of water-based paint are most frequently applied, either hot or cold. It is inexpensive and dries quickly, but on rainy nights, it has only mediocre visibility.

4.1.9. Thermoplastics

The thermoplastic used to make these pavement-marking materials is heated to approximately 400°F before being applied to the pavement surface. To enhance reflections, scatter, blend, or use both types of glass beads. Thermoplastic materials exhibit extended durability, high reflectivity, and good visibility.

4.1.10. Epoxy

Epoxy is reasonably priced, has good visibility and reflectivity, and can also be applied as a liquid with the right tools. Marks last for three to four years.

4.1.11. Tapes

Pre-formed tape can be covered or inlaid into newly constructed areas. It is more expensive, but it lasts 7-8 years and has great visibility and reflectivity. With a lifespan of roughly four years, wet reflective tape offers outstanding wet night reflectivity. The durability is poor and the cost is medium.

4.1.12. Tests on Construction Materials

The safety and integrity of any construction project depend on the quality of its materials; just imagine building a building without rigorous testing on construction materials—a disaster waiting to happen, don't you think? In order to protect people's lives and financial investments, it is imperative that construction materials are tested before being used to build structures. Let's explore the field of testing construction materials and learn why testing construction materials is crucial to the building process.

Testing construction materials is essential to guarantee the safety and quality of materials used in different construction projects. Our roads, bridges, and buildings are primarily made of these materials, which range from asphalt to concrete and masonry. But how can we be certain that they adhere to the strict requirements required for durable, secure structures? The key tests carried out by materials testing laboratories hold the answer.

These tests include:

- Concrete testing
- Soil testing
- Steel testing
- Asphalt testing

Every test concentrates on particular characteristics that establish whether a material is appropriate for construction. By carrying out these tests, engineers and project managers can make sure that their projects comply with design specifications and legal requirements.

4.1.12.1. Concrete Testing

Since concrete is the most commonly used building material in the world, structural integrity testing of the material is crucial. Concrete satisfies design requirements when its characteristics like permeability, flexural strength, and compressive strength are assessed. Compressive strength is determined using standard techniques like ASTM C39 and the field-curing methods outlined in ASTM C31. Conversely, flexural strength testing evaluates an unreinforced concrete beam's or slab's ability to withstand bending failure. The coarse aggregate is one of the components that affect the permeability test, which measures the concrete's resistance to the passage of water and other fluids. Through these tests, engineers can identify possible problems or departures from the intended standards, enabling the necessary adjustments before the concrete is used in construction.

4.1.12.2. Soil Testing

The foundation of any construction project lies in the soil upon which it is built. Soil testing involves the following methods to assess the soil's capacity to sustain construction projects:

- Grain size analysis
- Compaction test
- Moisture test
- Moisture content test
- Specific gravity test
- Dry density test
- Atterberg limits test
- Proctor's compaction test

These methods are utilized to evaluate soil properties. One way to analyze grain size is by using sieve analysis, where particles are separated into various size ranges and the percentage of soil retained on each sieve is calculated based on the total mass of the soil sample. The compaction test, on the other hand, helps determine the soil's maximum dry density and optimum moisture content. These tests are essential in ensuring that the soil meets project requirements, which in turn enhances the stability and longevity of a structure.

4.1.12.3. Steel Testing

For many construction projects, steel is an essential component, but in order to guarantee its structural integrity, it must undergo extensive testing. Evaluation of properties using tests like ASTM E8 / E8M and ASTM A370 includes tensile strength testing, weld quality, and corrosion resistance. Corrosion resistance is assessed using various methods, including:

- Service History
- Field performance
- Accelerated corrosion tests
- Passivation testing
- Laboratory screening tests (e.g., ASTM F 746)

These techniques assist in evaluating the surface passivity of corrosion-resistant steel components and the atmospheric corrosion of coated specimens. Adherence to approved standards for steel corrosion resistance, such as ISO 12944 and ASTM standards for corrosion and wear, is crucial for achieving optimal corrosion resistance.

Construction projects can be assured of the structural integrity and longevity of their structural steel components thanks to steel testing.

4.1.12.4. Asphalt Testing

Testing is necessary to make sure that asphalt, a common material used in pavement construction, is of a high quality and complies with design specifications. Asphalt testing involves:

- Determining the asphalt content of samples of uncompacted asphalt mixtures
- Performing direct and indirect tests to measure asphalt properties
- Utilizing the Marshall method of asphalt mix design
- Executing a viscosity test to measure the resistance of the asphalt to flow.

Key indicators of quality in asphalt testing include:

- Aggregate gradation
- Binder content
- Air voids content
- Voids in mineral aggregate (VMA)
- Density measurement

Through the use of these tests, engineers can verify that pavement construction adheres to design guidelines, guaranteeing longevity and safety.

4.1.13. Aggregate Tests

In order to decide the suitability of the aggregate for use in pavement construction, the following tests are carried out:

- Crushing test
- Abrasion test
- Impact test
- Soundness test
- Shape test
- Specific gravity and water absorption test
- Bitumen adhesion test

4.1.13.1. Crushing Test

One way that pavement material can fail is through crushing under compressive stress. The crushing strength of aggregates is determined using a standardized test outlined in IS:2386 part-IV. The aggregate crushing value indicates the resistance to crushing under a gradually applied load. In the test, dry aggregates that pass through 12.5 mm sieves and are retained on 10 mm sieves are placed in a cylindrical measure in three layers, with each layer being tampered with 25 times. The specimen is then subjected to a compression test under standard load conditions. After being weighed and placed in the test cylinder in three layers, each layer is tampered with again. The specimen is then crushed under a gradually applied load of 40 tonnes at a rate of 4 tonnes per minute. The crushed aggregates are then sieved through a 2.36 mm sieve, and the weight of the passing material (W_2) is expressed as a percentage of the weight of the total sample (W_1), which is the aggregate crushing value.

$$Aggregate\ Crushing\ Value = \frac{W_1}{W_2} \times 100$$

An aggregate is considered exceptionally strong if its value is less than 10, and it is considered weak if its value is greater than 35 (Mathew, 2010) (see Figure 4.1).

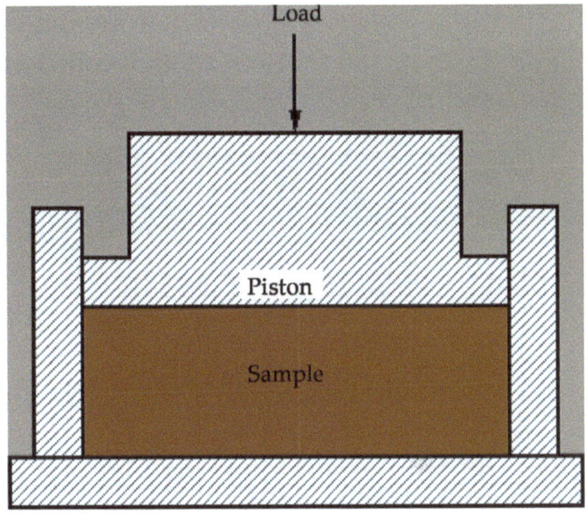

Figure 4.1. *Crushing test setup.*

Source. https://www.civil.iitb.ac.in/tvm/1100_LnTse/404_lnTse/plain/img4.gif.

4.1.13.2. Abrasion Test

The purpose of an abrasion test is to determine an aggregate's hardness and suitability for various pavement construction projects. The Los Angeles abrasion test, which is standardized in India (IS:2386 part-IV), is the recommended method for determining the hardness property. The Los Angeles abrasion test's basic goal is to determine the percentage of wear brought on by the aggregate and steel balls that are used as an abrasive charge rubbing against one another.

The Los Angeles machine is made up of a rotating circular drum with an internal diameter of 700 mm and a length of 520 mm that is mounted on a horizontal axis. Along with the aggregates, an abrasive charge made up of 48 mm diameter cast iron spherical balls weighing between 340 and 445 g is added to the cylinder. Depending on how the sample is graded, there are different numbers of abrasive spheres. Depending on the gradation, the amount of aggregates to be used typically varies between 5 and 10 kg. After that, depending on the aggregate gradation, the cylinder is locked and rotated for 500–1000 revolutions at a speed of 30-33 rpm.

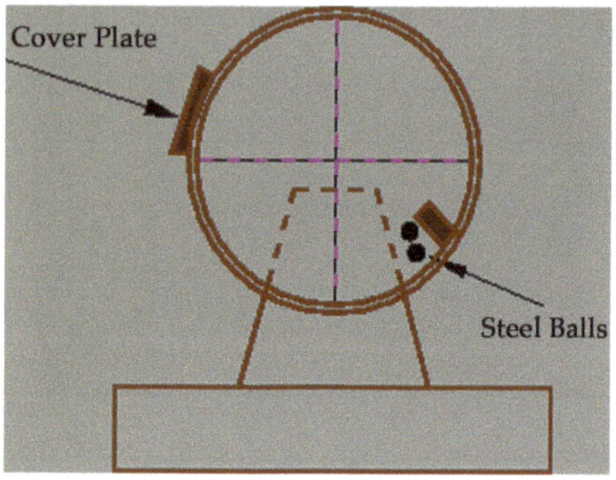

Figure 4.2. *Los Angeles abrasion test setup.*

Source. https://www.civil.iitb.ac.in/tvm/1100_LnTse/404_lnTse/plain/img5.gif.

The material is sieved through a 1.7 mm sieve after a certain number of revolutions, and the passed fraction is reported as a percentage of the sample's total weight. Los Angeles abrasion value is the name given to this value. In Indian conditions, a maximum of forty percent is permitted for the WBM base course. There is a maximum value of 35 for bituminous concrete. (Tom V. Mathew 2010) (see Figure 4.2).

4.1.13.3. Impact test

The purpose of the aggregate impact test is to determine how resilient aggregates are to impact. Aggregates that clear a 12.5 mm sieve and are held on a 10 mm sieve are placed into a cylindrical steel cup with an internal diameter of 10.2 mm and a depth of 5 cm. This cup is attached to the impact testing machine's metal base. Three layers of the material are filled, and each layer is tamped for a total of twenty-five blows. A metal hammer weighing between 13.5 and 14 kg is set up to fall with a free fall of 38.0 cm thanks to vertical guides, and the test object is struck 15 times. A 2.36 mm IS sieve is allowed to pass through the crushed aggregate. Additionally, the impact value is expressed as the proportion of aggregates that pass the sieve (W_2) relative to the sample's total weight (W_1). (Tom V. Mathew 2010) (see Figure 4.3).

$$\text{Aggregate impact value} = \frac{W_1}{W_2} \text{x } 100$$

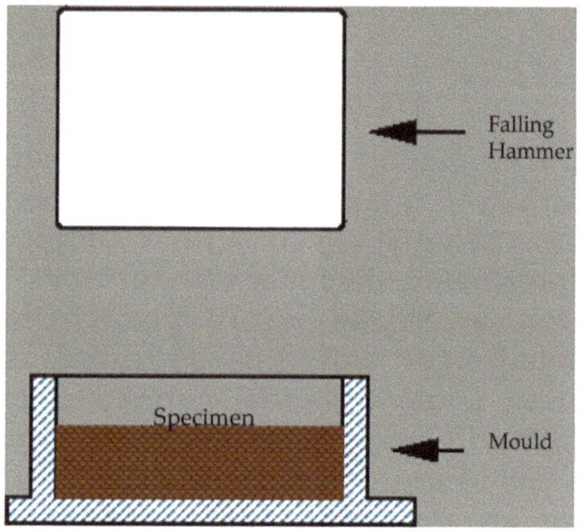

Figure 4.3. *Impact test setup.*

Source. https://www.civil.iitb.ac.in/tvm/1100_LnTse/404_lnTse/plain/img7.gif.

When using aggregates for a course, the impact value shouldn't be more than thirty percent. The highest percentage that can be applied to bituminous macadam is 35%. The maximum allowable amount for water-bound macadam base courses, as specified by the IRC, is 40%.

4.1.13.4. Soundness Test

By performing accelerated weathering test cycles, the purpose of the soundness test is to investigate the aggregates' resistance to weathering action. The porous aggregates that are exposed to freezing and thawing are prone to premature disintegration. Such aggregates are put through an accelerated soundness test, as outlined in IS:2386 part-V, to determine their durability. After 16 to 18 hours of alternating wetting in a saturated solution of either magnesium sulfate or sodium sulfate, aggregates of a given size are dried in an oven at 105 to 110°C to a constant weight. After five cycles, all undersized particles are sieved out and the weight loss of the aggregates is calculated. Additionally, the weight loss should not be more than 12% with sodium sulfate and 18% with magnesium sulfate solution.

4.1.13.5. Shape Tests

The proportion of flaky and elongated particles in the aggregate mass determines the shape of the particles in it. Higher workability and stability of mixes are harmed by elongated or flaky aggregates.

The percentage of aggregate particles by weight whose least dimension is less than 0 point 6 times their mean size is known as the flakiness index. India has standardized the test procedure (IS:2386 part-I). (Tom V. Mathew 2010) (see Figure 4.4).

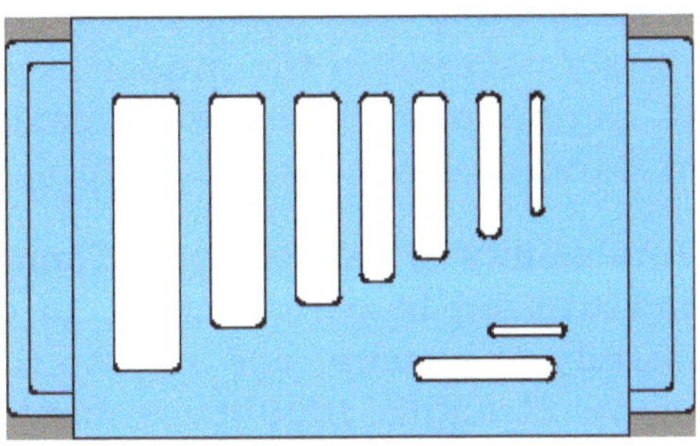

Figure 4.4. *Flakiness gauge.*

Source. https://www.civil.iitb.ac.in/tvm/1100_LnTse/404_lnTse/plain/img9.gif.

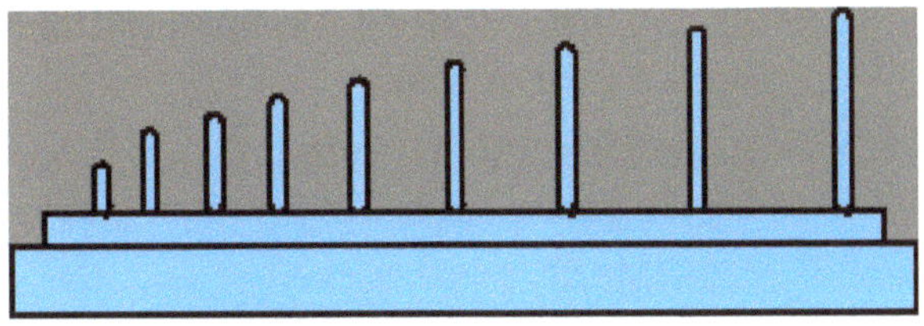

Figure 4.5. *Elongation gauge.*

Source. https://www.civil.iitb.ac.in/tvm/1100_LnTse/404_lnTse/plain/img10.gif.

The percentage of particles by weight whose largest dimension (length) is one and a half times their mean dimension is known as the elongation index of an aggregate. This test can be used on aggregates bigger than 6.3 mm. Additionally, this test is described in (IS:2386 Part-I). On the other hand, the elongation index has no established bounds. (Tom V. Mathew 2010) (see Figure 4.5).

4.1.13.6. Specific Gravity and Water Absorption

When designing concrete and bituminous mixes, aggregates' specific gravity and water absorption are crucial characteristics that must be considered. The mass of a solid divided by the mass of an equivalent volume of distilled water at a given temperature is known as its specific gravity. Two measures of the specific gravity of aggregates are used: apparent specific gravity and bulk specific gravity. This is because the aggregates may contain water-permeable voids.

- Apparent Specific Gravity, G_{app}, is computed based on the net volume of aggregates i.e., the volume excluding water-permeable voids. Thus,

$$G_{app} = \frac{M_D/V_N}{W} \qquad (1)$$

where, M_D is the dry mass of the aggregate, V_N is the net volume of the aggregates excluding the volume of the absorbed matter, and W is the density of water.

- Bulk Specific Gravity, G_{bulk}, is computed on the basis of the total volume of aggregates including water-permeable voids. Thus

$$G_{bulk} = \frac{M_D/V_B}{W} \qquad (2)$$

where, V_B is the total volume of the aggregates including the volume of absorbed water.

- Water absorption, the difference between the apparent and bulk-specific gravities is nothing but the water-permeable voids of the aggregates. We can measure the volume of such voids by weighing the aggregates dry and in a saturated, surface-dry condition, with all permeable voids filled with water. The difference between the above two is M_W. M_W is the weight of dry aggregates minus the weight of aggregates in saturated surface dry conditions. Thus,

$$\text{water absorption} = \frac{M_w}{M_D} \times 100 \qquad (3)$$

The specific gravity of aggregates normally used in road construction ranges from about 2.5 to 2.9. Water absorption values range from 0.1 to about 2.0% for aggregates normally used in road surfacing.

4.1.13.7. Bitumen Adhesion Test

All common types of road aggregates will adhere to bitumen well if they are dust-free and dry. There is essentially no adhesion issue with bituminous construction when there is no water present. An adhesion issue arises in cold, wet aggregate. By drying the aggregate and raising the mixing temperature, it is possible to solve this issue by eliminating moisture from the aggregate. Moreover, water presence results in the binder being stripped from the coated aggregates. When a bitumen mixture is water soluble, problems like these arise. The adhesion of bitumen binder to an aggregate in the presence of water is arbitrarily determined through a series of laboratory tests. The IRC specifies a static immersion test, which is a fairly basic one. The test's basic method involves immersing aggregate that has been completely covered in the binder in water that is kept at 40°C for 24 hours. According to IRC regulations, aggregates' maximum stripping value cannot be greater than five percent. (Tom V. Mathew 2010) (Table 4.1).

Aggregates have a significant impact on pavements' ability to transfer load. Therefore, they must undergo extensive testing prior to being utilized in construction. In order for the aggregates in the pavement to behave uniformly, they must not only be robust and long-lasting but also have the right size and shape. Strength, toughness, hardness, shape, and water absorption are all tested in aggregates.

Table 4.1. *Tests for Aggregates with IS Codes*

Property of Aggregate	Type of Test	Test Method
Crushing strength	Crushing test	IS : 2386 (part 4) -1963
Hardness	Los Angeles abrasion test	IS : 2386 (Part 5)-1963
Toughness	Aggregate impact test	IS : 2386 (Part 4)-1963
Durability	Soundness test-accelerated durability test	IS : 2386 (Part 5)-1963
Shape factors	Shape test	IS : 2386 (Part 1)-1963
Specific gravity and porosity	Specific gravity test and water absorption test	IS : 2386 (Part 3)-1963
Adhesion to bitumen	Stripping value of aggregate	IS : 6241-1971

Source. https://www.civil.iitb.ac.in/tvm/1100_LnTse/404_lnTse/plain/plain.html.

4.2. PAVEMENT

One kind of hard surface used for foot or vehicle traffic is a pavement, which is composed of a long-lasting surface material laid down on the area. Its primary job is to disperse the applied vehicle loads across several layers of the sub-grade. Sufficient skid resistance, good riding quality, good light-reflecting properties, and minimal noise pollution should all be provided by the road structure. Its objective is to lower the vehicle-transmitted load so that it doesn't surpass the subgrade's bearing capacity. The development of any construction is greatly influenced by the roads or pavements.

4.2.1. Requirements of a Pavement

An ideal pavement should meet the following requirements:

- Sufficient thickness to distribute the wheel load stresses to a safe value on the sub-grade soil,
- Structurally strong to withstand all types of stresses imposed upon it,
- Adequate coefficient of friction to prevent skidding of vehicles,
- Smooth surface to provide comfort to road users even at high speed,
- Produce the least noise from moving vehicles,
- Dust-proof surface so that traffic safety is not impaired by reducing visibility,
- Impervious surface, so that sub-grade soil is well protected, and
- Long design life with low maintenance cost.

4.2.2. Types of Pavement

There are major two types of pavement based on their structure: flexible and rigid.

4.2.2.1. Flexible Pavements

Wheel loads in the flexible pavement are transmitted from grain to grain via the granular structure's points of contact with one another. The stresses caused by wheel loads on the pavement are dispersed over a greater area and get less as you go deeper. It has numerous layers when taking into account the load distribution feature of flexible pavements.

Figure 4.6. *Flexible pavements.*

Source. https://civilengineeringtotalconceptcom.wordpress.com/wp-content/uploads/2021/02/image-25.jpeg.

Load is transmitted to the subgrade via a series of layers. Load is dispersed over a comparatively smaller area of the subgrade beneath the flexible pavement. Because flexible pavements are relatively inexpensive to install initially, they are more frequently found everywhere. But every few years, the flexible pavement needs upkeep and regular repairs. Furthermore, flexible pavement ages quickly; potholes and cracks are likely to occur as a result of inadequate drainage and high traffic volume (Figure 4.6).

1. Types of Flexible Pavements:

The following are major types of flexible pavement,

- *Conventional Flexible Pavements:* Layered pavement is what conventional flexible pavement uses. wherein inexpensive, low-quality materials are layered beneath expensive, high-quality materials to withstand the greatest amount of stress.
- *Full-depth Asphalt Pavements:* On top of the soil subgrade, bituminous layers are directly applied during construction. These pavement types work best in situations where traffic volume is high and local materials are scarce.
- *Contained Rock Asphalt Mats:* It is constructed by placing dense/open-graded aggregate layers in between two asphalt layers. Properly designed asphalt concrete is placed above the subgrade. This asphalt concrete will reduce the vertical compressive strain on the soil subgrade and protect it from surface water.

2. Pavement Layers or Road Construction Layers:

Road construction layers included (see Figure 4.7):

- *Compacted Subgrade (150 – 300mm):* Every pavement layer is based on the compacted subgrade. This layer receives the stress from all the pavement layers above it. Therefore, making sure that the soil subgrade is not overstressed is crucial. As a result, it needs to be appropriately compacted to the ideal density, close to the moisture content.

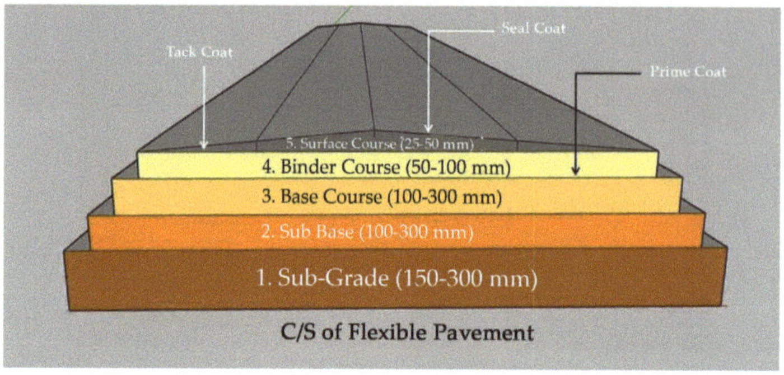

Figure 4.7. *Pavement layers.*

Source. https://civilengineeringtotalconceptcom.wordpress.com/wp-content/uploads/2021/02/image-15.png.

- *Sub-base Course (100 – 300 mm):* It is a substance layer situated beneath the base course. Providing structural support, enhancing drainage, and minimizing fines intrusion from the subgrade into the pavement structure are all important functions. When there is an open grade in the base course, a more fine-grained sub-base course can be used as a filler between the sub-grade and the base course.
- *Base Course (100 – 300 mm):* It is a layer of material that sits immediately beneath the binder course and aids in subsurface drainage as well as additional load distribution. A variety of materials, including crushed stone, crushed slag, and other untreated or stabilized materials, are used as base courses.
- *Prime Coat:* Low-viscosity cutback bitumen is spread onto an absorbent surface, such as granular bases, to create a prime coat before the binder layer is added. Its primary purpose is to create a connection between two layers. A prime coat creates a waterproof surface by filling in the gaps and penetrating the layers beneath it.
- *Binder Coat (50 -100 mm):* Most of the asphalt concrete structure is in the binder coat. Distributing the load to the base course is its primary purpose. The binder course doesn't need to be as high-quality as the surface course because it mostly consists of aggregate mixed with low asphalt. A more cost-effective design is produced when the binder course partially replaces the surface course.
- *Tack Coat:* A small amount of asphalt is applied to the surface for the tack coat. Usually, asphalt emulsion has been diluted with water. Its primary purpose is to ensure adequate bonding between two layers of binder course. To achieve this, it needs to be thin, evenly cover the whole surface, and set quickly.
- *Surface Course (25 – 50 mm):* It is the primary layer, which carries the weight of direct traffic and often has materials of the highest caliber. The material typically used to construct the surface course is graded asphalt concrete (AC).

The functions and requirements of this layer are:

- This course offers major characteristics like friction, smoothness, drainage, etc. Also, it prevents the entry of excess water into the underlying base, sub-base, and sub-grade,
- It should be provided with a hard surface to resist distortion under traffic and provide a smooth and skid-resistant riding surface,
  - *Seal Coat:* The seal coat is a thin layer with water-proof the surface and provides skid resistance.

4.2.2.2. Rigid Pavement

A cement concrete slab or cement reinforced concrete built directly on a subgrade or enhanced subgrade (with the use of a dry lean concrete layer) can be considered rigid pavement. Rigid pavements require adequate reinforcement or aggregate interlocking action to take care of the joints. The design is predicated on the slab's stiffness, which is backed by the subgrade soil's stiffness.

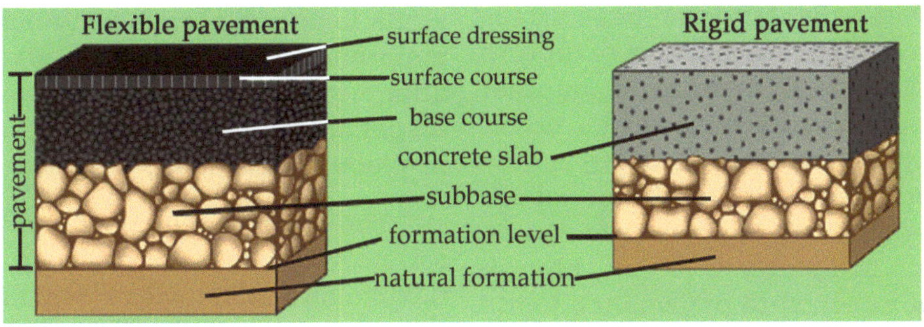

Figure 4.8. *Rigid pavements.*

Source. https://civilengineeringtotalconceptcom.wordpress.com/wp-content/uploads/2021/02/image-11.png.

Rigid pavement functions like an elastic plate resting on a viscous medium because load is distributed by the slab action. Since rigid pavements are made of Portland cement concrete (PCC), plate theory—which assumes an elastic plate resting on a viscous foundation—should be used to analyze them rather than layer theory. A condensed form of layer theory known as "plate theory" assumes that the concrete slab is a medium-thick plate that is plane both before and after loading (Figure 4.8).

1. Types of Rigid Pavements:

Rigid pavements can be classified into four types.

- *Jointed Plain Concrete Pavement:* Jointed Plain Concrete Pavements are made of plain cement concrete and have contraction joints placed closely apart. For load transfer across joints, dowel bars or aggregate interlocks are typically utilized. Their joint spacing is typically between 5 and 10 meters.

- *Jointed Reinforced Concrete Pavement:* Reinforcements can significantly increase the joint spacing to 10 to 30 m, but they do not significantly increase the structural capacity. For the transfer of load, dowel bars are necessary. Even after cracks, reinforcements aid in keeping the slab together.
- *Continuous Reinforced Concrete Pavement:* Complete elimination of joints is achieved by reinforcement.
- *Failure Criteria of Rigid Pavements:* For rigid pavement design, fatigue cracking has historically been regarded as the primary, or the only, criterion. The stress ratio between the flexural tensile stress and the concrete modulus of rupture determines the maximum number of load repetitions that can result in fatigue cracking. Pumping is now recognized as a significant failure criterion. Pumping is the process by which soil slurry is forced through cement concrete pavement's joints and fractures as the slab descends beneath large wheel loads. The three main causes of distress in rigid pavements are deterioration, faulting, and spalling.

4.2.3. Flexible Road Construction

The subgrade is covered with one or more layers of carefully chosen granular material that are compacted during the construction of flexible roads. Naturally occurring materials found in the earth are frequently used as-dug but manufactured, crushed aggregates work better. Crushed rock with an angular particle shape and a specific range and mix of particle sizes makes up the pavement on crushed aggregate (unpaved) roads, which are composed of one or more layers of manufactured, crushed aggregate. To help with compaction, the layers are wetted, which enables the particles to cling together and solidify. When more than one layer is used, the upper layer is made of granular material with a finer texture, and the lower layer has a coarser grading.

Pavements made of asphalt are pliable. Aggregate particles are bound together with bitumen to create a layer that is waterproof, strong, and flexible. Tarmac, also known as asphalt concrete, is a mixture of bitumen and aggregate. It is possible to apply a thin, flexible layer of bitumen-bound stone to aggregate roads, which will reduce maintenance, prevent dust, waterproof, and offer a safer driving surface. Stone chips are dispersed across the surface and secured in place by bitumen, which is sprayed on.

The most widely used building material for network and urban roads is layered flexible bituminous pavement. The subbase, an unbound granular base, or a base layer of road asphalt are placed first on the pavement before one or more unbound aggregate layers are added. The asphalt surface course comes last, followed by the asphalt binder course. The majority of the structural support is provided by the base course and binder course, and the surfacing offers a high skid resistance, smooth, waterproof riding surface. Bituminous flexible pavements typically have a 20-year design life. Failure can occasionally be caused by deep rutting brought on by subgrade failure, rutting, cracking, or aging of the asphalt layers. Pavement life can be significantly increased with routine surfacing layer maintenance (Figure 4.9).

Figure 4.9. *A highway constructed using flexible bituminous pavement*

Source. https://www.tensar.co.uk/getmedia/c414943c-935c-49a4-ae11-f0271163e0d2/EWs-rF0uThLCREeg4Gz8aFtSNK8Q8P9Yi7mjZtHNrCAZ18INMPsVr9jPxucDpdt2Z-XdfhRAJH1ScG-KPD-QJ3gPsycnzw4J5_Wi418BfSWqB-UHXIHCxzqxFLDGlTE9i-RCzd95.

4.2.4. Earthworks in Road Construction

Any task or operation carried out to make room for the construction of roads is referred to as "earthworks." Sufficient earthworks guarantee the durability of the road being constructed and aid in averting future maintenance and repair requirements.

When building a road, the "cut-and-fill" method is usually utilized for earthworks. The material is first removed from the site by excavation. After that, it is moved close by and used to build embankments. The cut-and-fill technique conserves extra material, lowering the need to dispose of soil, and uses less labor during construction.

4.2.4.1. Road Excavation Method

Excavation for roads or highways entails removing vegetation and topsoil in order to reveal the intended "formation level." In road construction, a bulldozer or tractor shovel are usually used for excavation. The subgrade's strength beneath the formation level will next be evaluated, and if it is found to be unstable, it will be stabilized.

In order to create embankments or a level surface, soil or rock that was removed from its natural location is moved to the fill locations (Figure 4.10).

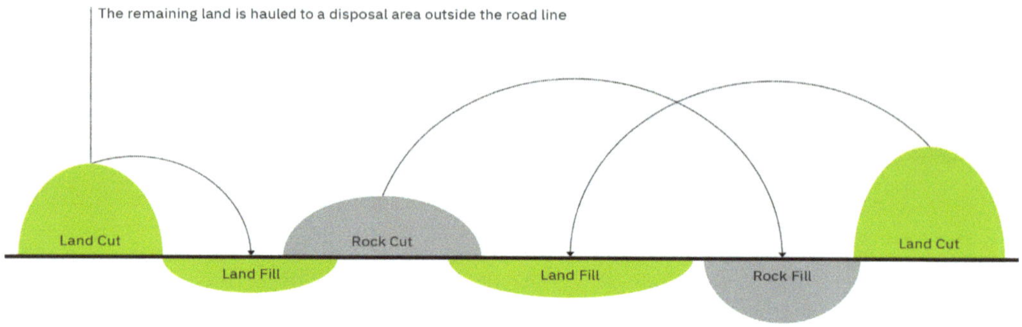

Figure 4.10. *Road or highway excavation.*

Source. https://www.tensar.co.uk/getattachment/ca7cbea1-9cc8-416a-a170-af1892c8931d/Land-Cut-Graphic.jpg.

4.2.5. Types of Road Surfaces

In order to extend the life of a road or pavement, various types of road surfaces are frequently layered on top of it. We have already discussed the various types of pavements used in road construction, their designs, and which ones are most prevalent in the UK. Road maintenance companies and councils use these road surface types across the UK:

- **Road/pavement renewal:** full replacement of the road surface type, where existing layers are removed and new ones installed.
- **Surface dressing:** liquid bitumen that is sprayed on existing road surfaces to extend life, prevent potholes, and reduce slipperiness.
- **Pavement slurry seal:** a mixture of bitumen emulsion, aggregate, and water that is sprayed over existing pavements to provide a more even surface.
- **Re-tread:** the existing surface is taken up and recycled by mixing it with bitumen and relaying it before a second surface treatment can be added.
- **Micro surfacing:** an additional asphalt protective seal coating that seals cracks and renews the surface.

4.2.6. The Future of Pavements

As in every other sector of the business, less expensive techniques for building roads are required. One way would be to have fewer maintenance requirements, which would minimize traffic disruption on crowded networks and save operating costs.

Increasing pavement life would be a second choice that would also improve sustainability. Reducing the amount of non-renewable materials used, the carbon emissions produced during construction and maintenance, and the reduction of traffic delays—which in turn raise carbon emissions and local air pollution—are some ways to accomplish this.

Furthermore, stronger pavements would make them less vulnerable to the effects of climate change, such as increased storm frequency and intensity in addition to higher and lower air temperatures. This represents a third, crucial potential path forward (Figure 4.11).

Figure 4.11. *Tensar InterAx helps to reduce the pavement depth at North Heybridge Estate Road with an estimated savings of £177,000.*

Source. https://www.tensar.co.uk/getattachment/8a1a2786-22e7-4f59-af1b-5464e736af23/road-580px.jpg.

4.2.7. The Power of Pavement Optimization in Road Construction

As previously stated, pavement optimization is the way of the future. The goal of pavement optimization is to build high-performing road pavements as quickly and cheaply as possible. Geogrids are frequently used for mechanically stabilized aggregate layers. This method, which is essential for flexible pavements, regulates aggregate displacement and postpones structural breakdown.

The phrase "pavement optimization" was first used by Tensar to describe its cost-effective design concepts. Pavement optimization requires finding a balance between improved traffic performance and decreased thickness, which typically results in material savings and a 20% cost reduction. In the long term, this saves money because it prolongs pavement life and reduces overall expenses. Furthermore, the carbon footprint of road construction projects is decreased by creating contemporary, optimized pavements, which is more environmentally friendly. Now that you are aware that pavement optimization can be a wise and economical approach to increase the lifespan, lower the cost, and lessen the environmental impact of roads, let's look at some real-world examples.

4.2.7.1. Road Stabilization

When building a road, unstable or weak ground may force the use of road stabilization as a substitute for digging up and replacing the subgrade. Using an in-situ mixture of stabilizing lime cement or another chemical agent, the properties of subgrade soil are modified. Check out this useful guide on road stabilization techniques and their applications for a thorough understanding of the various approaches.

Figure 4.12. *Confinement of aggregate particles with Tensar Interax geogrid.*

Source. https://www.tensar.co.uk/getattachment/6ee16a67-ed87-423a-ba4f-5fbbe23a05fa/interax-1-580-580-(1).jpg.

Placing a stronger soil layer—typically compacted aggregate material—over the weaker soil to distribute the load may be a more economical and environmentally friendly solution. This increases bearing capacity and frequently includes a subgrade-level geosynthetic. The utilization of geogrids in road and pavement stabilization enhances the road's effectiveness and provides an adaptable remedy for difficult terrain.

This method minimizes the need for large-scale excavation while bolstering road efficiency, making it a flexible fix for difficult ground situations (Figure 4.12).

4.2.7.2. Accelerated Pavement Testing (APT)

A technique called "accelerated pavement testing" (APT) evaluates the long-term performance of pavements in a shorter amount of time. APT effectively evaluates pavement response and performance by simulating extended in-service traffic conditions by applying realistic wheel loading to a layered pavement structure.

The aggregated wear and tear from numerous wheel passes on various layers affects the pavement's overall functionality and eventual breakdown. Accelerated pavement testing offers the only practical way to ascertain the long-term performance and lifespan of the entire pavement structure, as opposed to evaluating individual layers or materials separately. Explore our in-depth article for more information on accelerated pavement testing, which includes the methodology, results, and important takeaways from this vital pavement optimization aspect.

4.2.7.3. Geogrid Pavement Design

A strategic approach to building optimal road pavements that makes use of geogrids to improve stability and longevity is known as geogrid pavement design. Engineered materials called geogrids are incorporated into pavement layers to give aggregate materials mechanical stabilization. The objective of this design methodology is to optimize load distribution, reduce fatigue cracking and rutting, and shield the pavement structure from impurities and moisture. For a variety of flexible pavements, geogrid pavement design is crucial to assuring longevity and functionality. See this informative site on geogrid pavement design for a more thorough investigation (Figure 4.13).

Figure 4.13. *Site support facilities swiftly established by using Tensar InterAx geogrid.*

Source. https://www.tensar.co.uk/getattachment/88d9dbe1-d1e3-495a-8ebe-f1ede3df0fda/york-potash-(2).jpg.

4.2.8. Pavement Stabilization Technologies

Pavement stabilization can be simply defined as the process of changing the intrinsic properties of earthworks or pavement materials by adding a stabilization agent, such as a binder or granular material, to meet performance expectations during application. Recycling materials for pavement construction and maintenance and using the appropriate stabilization technology are the most economical ways to enhance the long-term performance of heavily trafficked pavements.

4.2.8.1. Materials for Soil Stabilization

Generally speaking, the stability of the underlying soils determines how well a construction structure will function over time. Unstable clay soils can lead to serious issues with structures or pavement. Soil stabilization is typically done to increase the bearing capacity of clay soil by using stabilizing agents (binder materials) to improve its geotechnical properties, such as compressibility, strength, permeability, and durability. Soft soils must be preconditioned for improvement of the mechanical strength before the construction of overlying structures. To improve a soil's geotechnical qualities, various stabilizers, such as lime and cement, are tested. Soil stabilization materials also included flocculants, ashes, wastewater sludge, and industrial wastes with pozzolanic value, such as fly ash and rice husk ash.

Furthermore, commercial road design is now utilizing the technology of nano polymer binder to stabilize soil and increase its strength. When dissolved in water, the fine particles known as nano polymers exhibit far greater efficacy in coating and binding aggregates, including soil particles.

The bond formed by the interlocking polymers becomes stronger when subjected to mechanical compaction. More compaction results in better binding. Using nano polymers to stabilize soft soils has several advantages, including a reduction in the carbon footprint, road crust, construction costs, maintenance, and the number of supporting layers typically used. It also uses less quarry aggregate and increases the strength and stability of the base and sub-base layers while also speeding up the construction process. Multiwall carbon nanotube, carbon nanofiber, and metal oxide nanoparticles, such as SiO_2, TiO_2, and Al_2O_3, are used to stabilize soil (Aroraa et al., 2019).

4.2.8.2. Stabilization of Aggregate Base Materials

Base stabilization is typically used by roadway engineers in the design and/or rehabilitation of pavement structures. In order to create a uniform base layer with improved strength, stability, and durability qualities that contribute to a long-term improvement in the performance of the supported pavement structures, the base aggregate materials are modified by the addition of a stabilizing additive. Nevertheless, additional stabilization techniques might involve drainage, decompaction, and enhanced grading through the addition of missing particle sizes. Among the stabilizers used are polymers, sludges, bitumen, cement, and other granular materials. Nevertheless, the most proprietary products being promoted for stabilizing base and sub base courses that gave promising results are the dry powders of some polymers and combination mixes of polymer additives with lime. Nepheline sludge was used in road pavements to increase the reliability of road structures during operation and prevent the early development of defects. Iron sludge was used in the sub-grade of the roller-compacted concrete pavement. The advantages of the slag addition are increasing the strength of the pavement by reducing the thickness and the cost of the pavement. Commercial dry powder polymers are produced by thermally bonding an inert fine carrier, like fly ash or silica fume, to a high-grade polymer. They work to increase the aggregate's hydrophobicity and waterproofness, which allows it to be applied to low-quality, clayey gravel, and sand-based aggregates with little to no plasticity. The reduced polymer stabilizer content tends to preserve the base course of the road pavement's increased strength and flexibility as well as its low water resistance (D. Cameron et al., 2016). The hydrated lime is combined with the dry powder polymers in certain applications. In order for the clay particles to adhere to the polymer, lime is added to help them flocculate. Here, the polymer is not applied to the lime.

4.2.8.3. Stabilization using Bitumen

Bitumen can be applied as an emulsion or foam to stabilize granular pavement materials, previously cement-stabilized materials, or reclaimed asphalt pavement. By applying bitumen to a mixture of recycled pavement materials and the underlying layer, a new base or sub base layer is created. Bitumen coats the aggregate particles, creating an adhesive bonding between the coating particles and increasing their strength. The aggregates with coatings become waterproof. To stabilize the aggregates without affecting the stability of the matrix, bitumen should be added within a certain level, though. It is anticipated that emulsified bitumen will not age the asphalt because it does not require energy during the stabilization process, in contrast to foamed asphalt, which requires heating for the foam to adhere to fine aggregate particles and create an asphalt-bound filler that can function as binding mortar. Insufficient fines can form asphalt-rich agglomerations of fines, which can lubricate the matrix and result in an unstable matrix with low strength (Wegman et al., 2017).

4.2.8.4. In-situ Recycling for Re-Condition Granular Roads

Recycled concrete aggregate and masonry could be used for in-situ recycling of road aggregate. Before being reinstalled, the marginally degraded in-situ pavement material needs to be improved with additives. The majority of the granular material that makes up the global road network is unbound and is covered in a thin layer of asphalt. Roads that are subject to trafficking and unfavorable environmental conditions may quickly lose their usability. Therefore, in order to meet the sustainability objectives of material reuse and greenhouse gas reduction through material transportation reduction, granular roads must be reconditioned through in-situ recycling. Bitumen may be utilized in the form of foamed bitumen or cold emulsion, but in certain cases, the road authority may find it to be too costly, or bitumen may not be appropriate if the pavement material contains plastic fines. When combined with lime, hydrophobic dry powder polymers can quickly and affordably create an improved road base (C. Hopkins et al., 2016).

4.2.9. Innovation in Road Construction

In addition, over the past 20 years, a lot of attention has been paid to the use of nanomaterials as asphalt mixture additives, such as nanosilica, nanoclay, and metallic nanoiron, which have reduced penetration values, increased viscosities and softening points, decreased phase angles, and increased complex moduli. In the case of concrete roads, several innovations have emerged in the production of developed types of concretes, such as fiber reinforced concrete (FRC), Ultra High-performance concrete (UHPC), self-compacting concrete (SCC), and High performance concrete (HPC). These materials and advanced technologies offer significant benefits to both the industry and the community (Yao, 2016).

4.2.9.1. Using polymers

Polymers are macromolecular, synthetic, or natural substances that are typically composed of millions of repeating, covalently bonded units, each of which is a relatively simple and light molecule. When used for sub-base, base, wearing course stabilization, and sealing, polymers are a very efficient way to improve the strength characteristics of well-engineered roads with appropriate drainage.

4.2.9.2. Using Fiber-reinforced Polymer (FRP) Composites

Fiber-reinforced polymer (FRP) composites are one technology that is demonstrating remarkable success in the building industry. Glass, aramid, and carbon fibers are mixed with polyester, epoxy, or vinyl ester resins to create fiber-reinforced polymers (FRP). They have the potential to be used in the construction of bridges and highway roads as bars in reinforced concrete pavement (instead of traditional steel bars) as well as in the retrofitting and repair of aging structures due to their high resistance to corrosion and fatigue. Composite fiberglass materials, which are used to manufacture construction elements like pipes (casings), have been developed and actively used in the construction of bridge structures on motorways. They are promising from an architectural and aesthetic perspective because, in comparison to metals, they have better durability and resistance to chemicals and wall pressure. Additionally, employing fiber-glass composite casings in bridge piers increases the accuracy of the bridge structure while it is in use. Additionally, tube-confined concrete technology helps to increase concrete loading and energy-absorbing capacities. On the other hand, self-mixing can be challenging and is perceived as a drawback due to the relatively high initial cost of FRP materials and products.

4.2.9.3. Asphalt Pavement using Polymers

For a very long time, aggregate binders in road construction has been made of asphalt, a petroleum-based binding agent. A typical asphalt road's composition consists of 95% sand, gravel, and stone and 5% bitumen. Nevertheless, temperature weakness in asphalt can cause rutting at high temperatures, fatigue at medium temperatures, and cracking damage at low temperatures. Polymeric modifiers can be added to bitumen and bituminous blends to improve their properties. In order to reduce the air void between the aggregates and bind them together to prevent bitumen from bleeding, modifiers are typically used to improve the properties of bituminous concrete mixes. Using various waste and virgin plastics could improve the asphalt's softening point, resistance to rutting, durability, viscoelastic property, and fatigue life. However, styrene-butadiene (SB) type polymers are used in approximately 80% of polymer-modified asphalt. Recycled waste tire rubber, or waste SB rubber, has also been utilized to reduce moisture absorption and porosity in road pavements, thereby enhancing performance, durability, and dependability.

The choice of base bitumen has less of an impact on the waste plastic binder's performance than the type of polymer that is incorporated and the mixing conditions. Additionally, adding recycled polystyrene up to 15% by volume to conventional asphalt mix decreased penetration and ductility while increasing flash and fire points, softening points, and viscosity of the modified asphalt. The addition of 4% recycled high density polyethylene to stone mastic asphalt produced promising wheel track and fatigue results. Lastly, adding low and high density polyethylenes and Crumb Rubber wastes to asphalt demonstrated a significant improvement in the binder's rheological properties (Khan et al., 2016).

On the other hand, polymer modification of coal bitumen has been used for many years to reduce carcinogenicity and enhance quality and technological aspects. It was mentioned that the addition of polyester resins resulted in distinct changes of rheological and viscoelastic properties. Household plastic waste and grinds could be recycled into pellets and flakes, then added to the asphalt mix as a complementary binder.

The addition of phenol-formaldehyde resin to coal-tar pitch was said to increase its softening temperature and viscosity as well as increase its thermal and mechanical resistance. In addition to decreasing the quantity of plastic dumped in landfills, the modified asphalt product is more affordable, long-lasting, and environmentally beneficial. While bitumen properties are somewhat improved by all of these polymers, there are still certain issues that will prevent bitumen modified with polymers from becoming widely used in the future. These issues include low aging resistance, poor storage stability, high cost, difficult dispersion, and separation issues.

4.2.9.4. Utilization of Geopolymers

More environmentally friendly construction materials are needed in the road construction sector. As of right now, geopolymer is a future sustainable building material that can be utilized in place of traditional Portland cement for modern road pavement construction and rehabilitation.

In contrast to conventional concrete's hydrated calcium silicate binder system, geopolymer concrete is a novel class of concrete based on an inorganic alumina silicate binder system. In order to protect natural resources, an increasing amount of industrial wastes and recycled materials, such as fly ash, slag, and silica fume, have been used in the pavement industry as alumina silicate precursors. Alkaline liquids (sodium or potassium hydroxide and sodium silicate or potassium silicate) activate the precursors of alumina silicate to form the binder. Strength increased to an appropriate level of alkaline solution concentration. The unconfined compressive strength of recycled asphalt pavement-fly ash geopolymer showed to increase with decreasing $NaOH/Na_2SiO_3$ ratios for both room temperature and 40 C curing of the geopolymer mixture. While high calcium content in standard road base material can rapidly react with alkaline activators, such that a self-cured geopolymer could be used. Geopolymers are used to stabilize soil, as pavement base materials and as successful stabilizers for pavement sub base materials, as well as very rapid-setting materials for rapid road repair of concrete infrastructures (Hoy, et al., 2016).

4.2.9.5. Using Bioasphalt

People are looking for alternative binders to petroleum asphalt that can be used in pavement construction because the cost of petroleum-based asphalt has been rising recently. Using alternative binders has the advantage of preserving and, in certain situations, improving asphalt performance while lowering energy and natural resource consumption. Bioasphalt is one type of substitute for asphalt. Although bio-bitumen hasn't been used in definitive protocols yet, bioasphalt is used for a number of reasons, including the possibility of replacing petroleum-based asphalt mixtures with mixtures made of plants and trees. Bio-asphalt binders also have advantages over petroleum-based asphalt binders in terms of economic, social, and environmental benefits.

Pavement uses biopolymers instead of petroleum-based polymers. Living organisms produce biopolymers. A portion of the biopolymers are soluble in water, and the majority are biodegradable. The greatest example of a biopolymer is natural rubber. To improve its beneficial qualities, it can be vulcanized or chemically cross-linked. Elastic or rubber joints spanning the connection are composed of three-dimensional molecular networks, where chemical bonds bind extended molecules together. Based on the asphalt rejuvenating mechanism, bio-modifiers made from leftover cooking oil have been used to recover the aged asphalt binder. It absorbs solvents and swells but does not melt. It was predicted that adding 6–8% of the bio-modifiers to the aged asphalt would effectively reduce its viscosity and stiffness, which would be advantageous for its resistance to fatigue and low-temperature cracking. The best percentage of bioasphalt to use as a rejuvenator is 23% coconut shell bioasphalt and 17% bitutechRAP bioasphalt.

On the other hand, used cooking oil is mostly recycled globally to produce biodiesel. About 10% of a black, viscous heavy oil by-product is coproduced during this process; this by-product is thought to have the potential to be a bio-oil modifier for creating bio-asphalt based on leftover cooking oil, which could eventually replace traditional petroleum-based asphalt. The results of the rheological performance tests showed that the addition of bio-oil reduced the control asphalt's stiffness and increased its fatigue resistance.

Additionally, biochar produced by pyrolysis of biofuel was used to modify asphalt, improving its performance as an asphalt binder against aging and deformation in warm tropical climates. Lignin is an organic polymer that gives plant cells their structure and form. It comes from the paper industry as a byproduct as well. Research has been done on lignin as a potential partial bitumen substitute.

A natural peat fiber and finely ground peat powder have been investigated as potential modifiers to improve bitumen's resistance to rutting. The use of lignin in combination with bitumen demonstrated polymer-like behavior, adding functionally to the asphalt binder and potentially delaying oxidative aging and extending the lifetime of asphalt pavements. The rheological data revealed the presence of a fiber network with strain-dependent elastic effects and stiffening effects from the powder fraction.

Despite the fact that bio-asphalt is one of the green technologies used in road construction and that it can partially replace petroleum-based asphalt binders with significant benefits to the economy, society, and environment, some research has shown that while most biomaterials can improve the rheological properties of modified asphalt at low temperatures, they also have an impact on the rheological properties of asphalt at high temperatures (Abdulnaser M. Al-Sabaeei, et al., 2020).

4.2.9.6. Plastic Road Technology

While the idea of using waste plastics for road construction is relatively new, various recycled plastics have been used globally as binder modifiers or as an alternative to aggregates. Nevertheless, plastic road technology—a road made completely of plastic—recycles plastic into roads and is a global innovation ecosystem led by the United States, the Netherlands, Scotland, India, and the Netherlands. The Plastic Road is a hollow, modular road structure that is prefabricated from recycled plastics. Its prefabricated nature, light weight, and modular design make it easier, faster, and more efficient to build and maintain than traditional road structures. Tar and polymer-based glue made from shredded plastic waste are the main ingredients of plastic roads. Plastic roads have several benefits over conventional roads, including a lower carbon footprint, a melting point of approximately 66°C, and the absence of potholes and cracks. Since waste is used in the fabrication process rather than virgin materials, plastic roads are made at a lower cost. The hollow area beneath the plastic road's upper surface can be used for a number of purposes, including the installation of electric vehicle charging stations and sensor devices, the passage of cables and pipes to prevent digging disasters, and the storage of water to reduce the risk of flooding. Despite the benefits of plastic road technology that have been mentioned, the long-term effects are still unknown.

4.2.9.7. Solar Roads

Globally distributed green thermal sources include solar energy. The largest solar thermal collectors are probably road pavements, which collect solar radiation all day long and store some of it for later use when a fluid is pumped through them. Road pavements are heated and cooled by this technology in many countries during the winter and summer, respectively.

However, a new technology called "solar roads" allows for the engineering of roads with solar panels—photovoltaic (PV) panels strengthened by silicon resin—that can be used for both driving and walking. Silicon possesses the ability to absorb solar radiation and transform it into electrical energy. The photovoltaic panels are typically set into the pavement and covered in a clear layer of concrete to keep cars out.

Light-emitting diode (LED) lights are used in the panels to create the lines and signage, eliminating the need for paint. Street lighting, video surveillance, and traffic light systems can all be powered by solar roads. To stop snow and ice from building up, they have built-in heating components. Additionally, solar roads could be constructed with paving that allows electric cars to be recharged through induction simply by traveling over it. This would encourage the use of environmentally friendly, electric vehicles.

The difficulties with solar roads stem from the fact that PV systems' performance is heavily influenced by both internal and external factors, such as shading. For example, a 25% increase in the area of shading on a photovoltaic module surface resulted in a 33.7% reduction in power. In addition, improper panel angling makes it difficult to capture solar energy effectively, and maintaining PV panels is expensive and labor-intensive. Furthermore, there is currently no manufacturing process that can be used to build solar roads that are both affordable and durable enough to withstand daily driving conditions while producing power at a commendable efficiency. On the other hand, solar roads could produce a significant portion of the nation's electricity needs if they completely replaced paved surfaces.

4.2.9.8. Sustainable Concrete using Supplementary Cementing Materials (SCMs)

Alternative materials, such as manufactured sand and aggregates of iron, zinc, steel, and copper slag, have been proposed as solutions to the problems of depleting natural resources and reducing greenhouse gases released into the environment. Efforts have been made to test and apply various combinations of blended cement or Portland cement as well as industrial by-products known as SCMs, such as fly ash, ground granulated blast-furnace slag (GGBS), silica fume, rice husk ash, and natural pozzolans (calcined shale and metakaolin). Using SCMs could result in the production of concrete that is more cost-effective. The concrete mixtures appeared to have better chemical stability, more strength, and less heat of hydration and permeability.

4.2.9.9. Self-Healing Pavement

Self-healing technology is a method that allows materials to repair themselves after being damaged, reducing the extent of the damage and prolonging the functionality and lifespan of the damaged area. Smart materials are designed to address issues such as cracks, stress, and environmental factors like temperature, pressure, and oxygen levels. For example, asphalt pavement is a self-healing material that can repair micro-cracks during rest periods, restoring its strength and stiffness. By adding modifiers and additives to the asphalt mix, the self-healing capabilities of the pavement can be enhanced. It is important for self-healing technology to be able to withstand the harsh conditions during construction and while in use.

There are three primary technologies used for self-healing in asphalt pavement design: nanoparticles (NPs), induction heating, and rejuvenation. Nanoparticles, such as nanoclays and nanorubbers, are added to enhance the properties of asphalt, while induction heating is used to improve self-healing by utilizing electrically conductive materials like carbon fibers, graphite, and steel fibers. This technology primarily relies on capillary flow and diffusion of asphalt binder at elevated temperatures.

However, aged asphalt binder undergoes oxidation during construction and maintenance, which turns the binder oils into resins and the resins into asphaltenes. This process results in agehardening and a higher viscosity than with fresh binder. A rejuvenating agent, like a cationic emulsion, can be added to the asphalt mix to restore its viscoelastic state. A rejuvenator's main function is to make the oxidized asphalt binder less stiff and to flux it, thereby changing the asphalt mix's properties and prolonging the life of the pavement. The development of self-healing polymers, which include a microencapsulated agent and a catalytic chemical trigger within a polymer matrix, has allowed them to automatically heal cracks. Through capillary action, the embedded microcapsules release the healing agent into the crack plane. Contact with the embedded catalyst causes the healing agent to polymerize, bonding the crack faces. It has been possible to design self-healing materials that include calcium lactate and live bacterial spores in water capsules and self-contained pods. The pods release calcite, or limestone when they come into contact with water. The concrete is then strengthened when this calcite fills in the damaged areas. It is estimated that self-healing concrete can reduce lifetime costs by up to 50% by obviating the need for repairs. The development and application of microbial self-healing technology are restricted by the type of microorganism, its survival rate, and the void left by microbial death. Smart concrete is currently undergoing testing in labs to see how long the bacteria can survive on its own, but researchers are optimistic that it will be used in the building sector very soon. Self-healing material is demonstrated in Figure 4.11 as a way to stop concrete cracks.

4.2.9.10. Shape Memory Alloys

One other intelligent material is the shape-memory alloy (SMA), a special type of adaptive material that can convert thermal energy into mechanical work. When properly treated with heat, various alloys can exhibit this effect, allowing them to be deformed at low temperatures and then revert to their original shape when heated. SMA materials have two distinct features: a shape-memory effect and super-elasticity. The shape-memory effect allows the alloys to return to their original shape when heated, up to their phase transformation temperature. Super-elasticity is the property of alloys that can recover a large amount of strain.

SME is caused by a shift in the crystalline structure of the material between the phases known as Austenite and Martensite, which is dependent on both temperature and stress. The three primary forms of SMA are alloys made of copper, zinc, and aluminum; the low-temperature phase, martensite, is relatively soft, and the high-temperature phase, austenite, is relatively hard. Commercially available and less expensive than NiTi ., iron- and copper-based SMAs like Fe-Mn-Si, Cu-Zn-Al, and Cu-Al-Ni are also available. However, research has shown that adding a fourth element to the ternary alloy of Cu-Al-Ni SMA can change the structure and/or morphology of the martensitic phase, improving the mechanical properties of the. SMA nanomaterials. They are also very helpful in addressing issues with uneven joints that lead to fatigue cracking and tensile damage in pavements (Cai et al., 2003). SMAs can function as semiactive, active, or passive parts to lessen the harm that earthquakes and other environmental factors can cause. One potential use for SMA materials is in self-repair. The phase transformation that creates compressive stress in the structures to prevent cracking may be triggered by the increased stress in the SMA when concrete structures begin to fracture. Moreover, phase transformation in ASM can cause shrinkage in response to temperature increases. It is possible to use this shrinkage to counteract the thermal expansion. As a result, SMA might be a very promising material to address issues with stress and cracking caused by temperature. On another hand, bars made from shape-memory alloy can significantly reduce the residual deformation under both cyclic loading and reversed cyclic loadings (Figures 4.14 and 4.15).

In order to support civil engineering applications, a novel iron-based shape-memory alloy known as memory steel is being produced on an industrial scale. Memory steel can be used as a pre-stress material for reinforced concrete when it is made into bars and strips that can be attached to buildings to increase their load-bearing capacity. Once installed, SMA steel bars can be prestressed and deformed, and they have been shown to be effective when used as external end fixed reinforcements for brand-new concrete structures or when retrofitting older ones without requiring the degree of invasive work that is typically needed (Hesami et al., 2015). Moreover, the SMA can be used to create smart strands, which are actuators that can be activated by changes in internal stress or external heating and that can be adjusted to handle various settlement, time, and temperature-dependent effects. Usually, the strands undergo mechanical deformation before being incorporated into concrete constructions like highways and bridges. Throughout the structure's life, the pre-stressing and self-repair effects can be triggered as needed. Shape-memory alloys (AMAs) have the primary benefit of lowering repair costs in concrete structures.

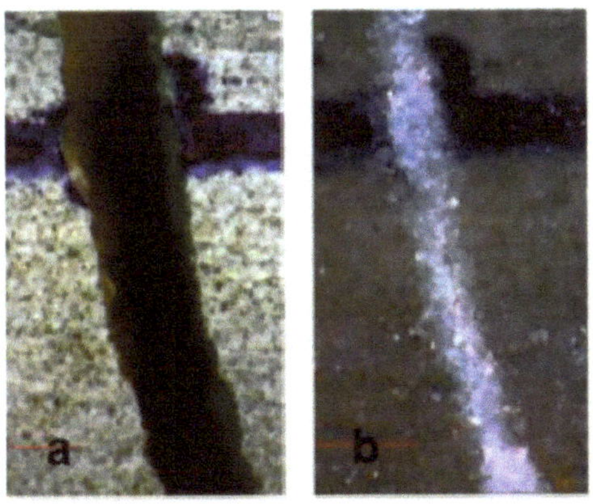

Figure 4.14. *(a) ~ 500 mm crack, (b) Self-healing in 28 days.*

Source. https://encrypted-tbn0.gstatic.com/images?q=tbn:ANd9GcT_vazbf60y5o9GUFKgh0tNIlfIKGOtV2UlTjuUX0CNCMWk6bd JCvONbCgyoM2yaQrJw9k&usqp=CAU

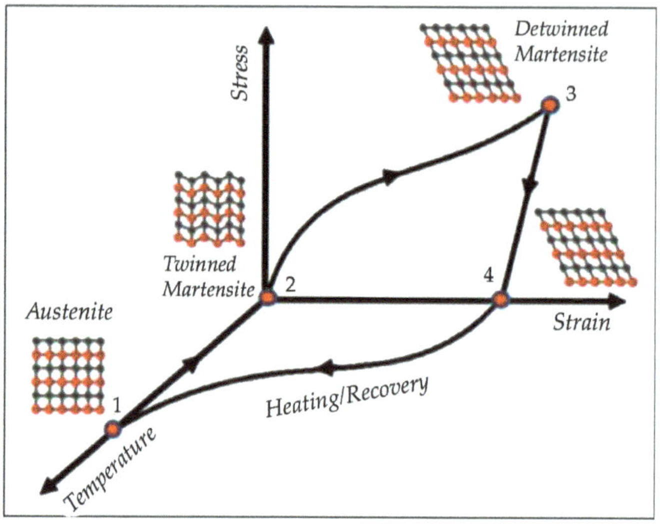

Figure 4.15. *Shape memory effect mechanism.*

Source. https://www.researchgate.net/profile/Yousif-Bas-2/publication/350979473 /figure/fig2/AS :1014442577907715@1618873366862/Shape-memory-effect-mechanism-87_Q320.jpg

4.2.10. Illuminating Cement (Glow-in-the-dark Concrete) Solar-Powered Cement

Luminous cement and concrete is an emerging technology with limited research available. However, the use of glow-in-the-dark concrete in coatings for buildings and highways has the potential to reduce the need for energy-intensive lighting. The credit for inventing illuminating cement goes to Mexican scientist Jose Carlos Rubio Avalos, who created a material that absorbs light during the day and emits it at night. This technology involves mixing phosphorescent materials with cement to create a glow that can illuminate walkways, buildings, and roads without the need for electricity. By harnessing and storing solar energy during the day, the cement is able to emit light at night without any additional power source. Thanks to this technology, drivers can see the road without depending on headlights from their own car or traffic lights. A significant technological advancement was the addition of photoactive materials to cement. This enables it to function even on cloudy days and gives it the propensity to absorb and emit light. The glowing surfaces have the potential to reduce street lighting expenses, and their production process is environmentally friendly since no harmful byproducts are created—just water vapor is released. In terms of price, it has been shown that a piece measuring one square meter and perhaps three millimeters thick would cost between $60 and $70. One of the few fundamental studies on illuminating cement involves the creation and testing of illuminating cement through the addition of rare-earth mineral phosphorescent strontium aluminate powder to a cementitious mixture. At room temperature, the phosphorescent strontium aluminate shows a lengthy afterglow and a wide spectral distribution of luminescence in the visible range. Typically, the mixture is applied as thin top layers of epoxies, sealants, or cement over an existing layer of asphalt or concrete. The study's findings demonstrated that samples with 30 and 50% of the cement's mass in rare-earth mineral phosphorescent powder were phosphoresced. However, the large amount of strontium aluminate powder required to produce a usable glow in the cement makes the production of the illuminating cement impractical and costly.

One of the other factors that restrict the use of luminescent concrete is that conventional concrete is opaque, meaning that light cannot pass through it. As a result, any photoactive material that is present below the top surface neither releases nor absorbs light energy. In an attempt to change the crystalline structures that form when concrete cures, researchers are attempting to address this issue by incorporating binding agents and clear aggregates. Phosphorescent sealants, paints, and coatings that would preserve existing concrete while adding phosphorescent utility are the focus of research and innovations surrounding illuminating concrete.

The materials used in road construction have a significant impact on technology, upkeep, and cost; therefore, research into new pavement materials is crucial to meeting use requirements. Future trends in the construction of smart roads are supported by the spread of nanomaterial technology, the use of sustainable binding materials with improved viscoelastic range, and the integration of concrete and asphalt materials into the category of smart materials that tend to have the functional characteristics that traditional pavement materials do not. On the other hand, advancements in road and traffic safety are made possible by the use of non-destructive measurements in road engineering along with technologies that allow for continuous monitoring of the state of road pavement condition, such as the installation of smart sensors into the pavement network.

4.3. DRAINAGE SYSTEM FOR ROAD CONSTRUCTION

Open concrete drains and pipe drains are two examples of the traditional drainage systems that are frequently used in road construction. A drain's design takes into account various factors, such as hydraulic, structural, environmental, sociological, and maintenance attributes. The importance of each factor is determined by engineering judgment and sustainability requirements.

4.3.1. Open Concrete Drain

Open drains can be installed on one or both sides of the road in various shapes such as trapezoidal, rectangular, or square. The channel can be lined with either mass concrete or reinforced concrete. To ensure the drain works effectively, the grade must be just right to allow for proper flow without causing damage to the lining. While open drains are easy to maintain, they can be unattractive and potentially dangerous for road users. In flat terrain where the invert slope is less than 2%, a trenchless drainage system may be a better option.

4.3.2. Pipe Drain

Pipe drains are typically buried underground, aligned, and have a cylindrical or rectangular shape. Drainage via pipes replicates the movement via conduits.

Pipe drains are best suited for residential developments with high values and densities. It is safer and has a nicer appearance than an open drain, but it is more difficult to maintain and far more expensive to build. Pipe drains need to mobilize allowable flow velocities between minimum and maximum in order to work properly. However, its conveyance becomes woefully inadequate when the terrain slopes less than 2% or the drain inverts. In that case, the trenchless drain system is a viable substitute.

4.3.3. Trenchless Drainage System

The trenchless drainage system offers a solution for removing runoff from road pavement located on flat terrain, presenting a viable option to traditional open or pipe drains in this type of environment.

4.3.3.1. Concept

The idea behind trenchless drainage systems is to apply grass cover and an engineered absorption field to allow runoff to penetrate into the subsurface. This is related to the general idea of a "French drain." The following highlights the trenchless system's components.

- *Absorption field:* An absorption field is an excavation or a trench, backfilled with relatively permeable material when compared with the native or in-situ soil. The excavation creates a wider surface area for the water to infiltrate and percolate in the underlying soil medium.
- *Grasses:* The benefits of grasses as an aid to effective drainage of an area have been highlighted (Coppin et al., 1990). The foliage intercepts raindrops that attempt to fall onto the ground to cause splash erosion. This, in consequence absorbs the terminal energy of the rain drop thereby reducing its erosivity. The foliage also provides an evapotranspiration medium for the system. The stem of the grass in conjunction with the leaves interrupts the free flow of the runoff, thereby increasing the flow roughness. This provides a suitable opportunity for infiltration to take place. The root system has numerous channels for water infiltration and also acts as a transition channel for evapotranspiration. Generally, grasses bind soil particles into a matrix and provide numerous infiltrations and evapotranspiration pathways.

A combination of absorption field and grass in an engineered fashion provides a good infiltration sump for road wash. It also restricts the migration of composite earth materials, such as silts and fine sand, in the runoff that could block infiltration channels.

4.3.3.2. Preliminary Studies

Trenchless drain design necessitates the study and consideration of a few basic parameters. These include the amount of rainfall in the area that is being studied, the permeability and hydraulic conductivity of the native or in-situ soil, the characteristics of the backfill material, and the width of the designed road.

- *Rainfall Intensity.* The rainfall intensity of the area can be obtained from rainfall studies of the nearest metrological station. For optimal design the value may be determined from maximum daily rainfall data of 10 years return period. Rainfall intensity (I) can be calculated from the expression,

  $$I = {}^{D_r}/_{T} \quad (1)$$

  where Dr = depth of rainfall in cm, T = duration of rainfall in min.
- *In-situ Soil Properties.* The permeability or hydraulic conductivity of the in-situ soil is required and this can be obtained through a number of simple field techniques. One of such technique is the Auger Hole method (Oswaldo PLS, et al., 2018), performed within a relatively shallow depth. A number of such a test is carried out along the length and breadth of the proposed road right-of-way. The soil permeability, K is estimated from the equation,

  $$K = {}^{C\Delta H}/_{\Delta T} \quad (2)$$

  where C = the geometry factor of the auger hole, ΔH = change in water level in the hole at time interval ΔT.

The ring infiltration method can also be used in estimating *K*. It uses the principle of falling head permeameter. The evaluation follows the following equation:

$$K = \left({}^{A_1 L}/_{A_2}\right)(T_1 - T_2)\ln\left({}^{H_1}/_{H_2}\right) \quad (3)$$

Where A_1 = cross-sectional area of the observation tube, A_2 = cross-sectional area of the soil tube, L = length of the soil tube, H_1 = head at the time T_1, H_2 = head at the time T_2.

In addition to the estimation of soil permeability, other basic properties of the in-situ soil would be studied. These include the grain size distribution, the Atterberg limits, specific gravity, specific weight; etc. They can be used for the classification of the soil.

- *Backfill materials.* The materials that could be used to backfill the dug trench must have passed through a series of cycles of abrasion and then become stabilized after the erosive components of the material may have been dissolved and removed. This ensures that the likely cementing agents and organic matters have been washed away. Sands and gravels fall into this category. They can be sourced from alluvial deposits or processed to obtain the required properties.

The backfill material should have greater permeability than the surrounding soil. Porosity can be determined through laboratory tests on samples of the backfill, including analysis of grain size distribution, specific gravity, and specific weight. Studies have indicated that backfill made of poorly graded materials with high values of diameter at 30% passing and uniformity coefficient are preferable (Owuama et al., 2020).

- *Road Width.* The width of the roadway is an input parameter in trenchless drainage design. It is a reference catchment area of the drain. The size may be obtained from preliminary design specifications or working drawing for the construction. Consideration would be given to the effective width of the road for drainage purposes. This may include areas outside the limit of asphalt coating. However, the actual width to be used in the sizing of the drain will, principally, be dependent on the judgment of the drainage engineer.

CASE STUDY

69-YEAR-OLD AIRPORT CONCRETE PAVEMENT HOLDS 95 PCI

The apron of the Brunswick-Golden Isles Airport (BQK) in Glynn County, GA, was constructed in June of 1953 and boasts a 95 pavement condition index (PCI; on a 0-100 scale) measured as recently as 2018. This regional airport is adjacent to a deep-water port and serves as a vital connection point for business travel, plane repair, and tourism. With exceptional performance metrics, this permanent pavement is only scheduled for preventative maintenance, and authorities plan to depend on the concrete for years to come.

The Georgia Department of Transportation (GDOT) estimated that the Brunswick-Golden Isles Airport generates nearly $155 million in economic activity annually, supports over 1,600 jobs, and serves as a vital regional connection for local business. BQK's apron was completed on June 2, 1953, when the U.S. Navy owned the airfield. The pavement has undergone no major maintenance or repair since its construction, lasting three and a half times longer than its design life. The 10-inch concrete pavement was built atop 6 inches of stabilized subbase with 12.5-by-15-foot panels, similar to modern specifications. Over the last 69 years, BQK has been steadfast in serving approximately 28,000 operations annually, including thrice daily flights to Atlanta Hartsfield-Jackson airport, two maintenance and repair operations (MROs), and a high volume of private flights from both small and commercial aircraft.

According to a 2019 Work History Report published by GDOT, the concrete apron was not touched between 1953 and 2006, then routine preventative maintenance of joint sealing and partial-depth patching was applied. The records indicate that in addition to the apron, a concrete runway and taxiway had also been constructed in 1953 and were in active service until they were replaced in 2014, meaning the concrete pavement had performed for 61 years. Going on seven decades, the concrete apron at BQK has seen an adjacent runway receive four asphalt overlays at an average of every 17 years. The investment made into concrete pavement for the Brunswick-Golden Isles Airport has paid dividends.

Durability is Key to Sustainability

The engineers who designed the BQK apron in 1953 were not concerned about the carbon intensity of the pavement, but the choice they made to use concrete delivered a sustainable result. Georgia has been hit by 66 hurricanes and many other extreme weather/inundation events while the concrete apron has been in service. By avoiding the environmental and economic costs associated with rebuilding the pavement, the concrete apron has proved to be a resilient and sustainable asset to Brunswick-Golden Isles Airport and will continue to do so well into the future.

CLASS ACTIVITY

Applying the pavement layers, which can include asphalt, concrete, or other materials, depending on the project specifications. This involves preparing the pavement mix, placing and compacting it, and ensuring proper thickness and smoothness. The activity in the scope of work for a road construction project can vary depending on the specific project requirements and conditions.

SUMMARY

- The construction of road pavements, and drainage systems, and the selection of materials are critical components in civil engineering, ensuring the durability, safety, and functionality of road infrastructure.
- Highway construction can be characterized by a large right-of-way having a length of a considerable amount. Due to this Highway construction requires materials that are needed in large quantities.
- The surface, base, and sub-base of pavements consist of aggregates or a mixture of aggregates with cement and water (Portland cement concrete) or a bituminous material (bituminous concrete).
- Concrete is the most widely used construction material worldwide, making its testing paramount for structural integrity.
- Properties such as tensile strength testing, weld quality, and corrosion resistance are evaluated through tests like ASTM E8 / E8M and ASTM A370.
- Asphalt, a common material used in pavement construction, requires testing to ensure its quality and adherence to design specifications.
- The aggregate impact test is carried out to evaluate the resistance to the impact of aggregates.
- A pavement is one type of hard surface made from durable surface material laid down on an area which is intended to carry vehicular or foot traffic.
- In Flexible pavement, wheel loads are transferred to subgrade by grain-to-grain transfer through the points of contact in the granular structure.

- Rigid pavement can either be a cement concrete slab or cement reinforced concrete constructed directly on subgrade or improved subgrade (using a dry lean concrete layer).
- Accelerated pavement testing (APT) is a method that assesses long-term pavement performance in a condensed time frame.
- Open drains can be provided on either or both sides of the road. It can be trapezoidal, rectangular, or square in shape.

REVIEW QUESTIONS

1. What are the Highway construction constitutes materials? Explain each of them.
2. What is the soil testing for a construction project?
3. Explain aggregate tests for use in pavement construction.
4. What are the types of pavements?
5. Describe the pavement stabilization technologies.

REFERENCES

1. Abd El-latief, R. A. E., Rivera-Armenta, J. L., & SalazarCruz, B. A. (2018). Modified Asphalt. In *IntechOpen*. DOI: 10.5772/intechopen.76832.
2. Al-Humairi, S. N. S. (2019). Recent advancements in the metallurgical engineering and electrodeposition. *IntechOpen*.
3. Al-Sabaeei, A. M., Napiah, M. B., Sutanto, M. H., Alaloul, W. S., & Usman, A. (2020). *Journal of Cleaner Production, 249*, 119357.
4. Arora, A., Singh, B., & Kaur, P. (2019). *Materials Today, 17*, 124–130.
5. Asmaela, N. M., & Waheed, M. Q. (2018). *Merican Scientific Research Journal for Engineering, Technology, and Sciences*. (Online), 38–43.
6. Baker, M. B., Abendeh, R., Abu-Salem, Z., & Khedaywi, T. (2016). *International Journal of Applied Environmental Sciences, 11(1)*, 183–192.
7. Basu, D., Puppala, A. J., & Chittoori, B. (2013). Proceedings of the 18th Conference on SMGE, Paris 2013, *Volume 1*, 1155–1162.
8. Bednar, R., & Merritt, S. (2018). ENGR-090 project. Swarthmore College, Department of Engineering.
9. Bella, F., Calvi, A., & D'Amico, F. (2012). *Procedia–Social and Behavioral Sciences, 53*, 943–952.
10. Brasileiro, L., Moreno-Navarro, F., Tauste-Martínez, R., Matos, J., & RubioGámez, M. C. (2019). *Sustainability, 11*, 646.
11. Callister Jr., W. D., & Rethwisch, D. G. (2009). *Materials Science and Engineering: An Introduction*. John Wiley and Sons, 8th ed.
12. Cameron, D., Hopkins, C., & Rahman, M. (2016). *Procedia Engineering, 143*, 26–33.

13. Casey, D., McNally, C., Gibney, A., & Gilchrist, M. D. (2008). *Resources, Conservation and Recycling, 52*, 1167–1174.
14. Chang, W. S., & Araki, Y. (2016). *Civil Engineering Proceedings, Institute of Civil Engineers, 169*, 87–95.
15. Coppin, N. J., & Richards, I. G. (1990). *The Use of Vegetation in Civil Engineering*. London: Butterworths.
16. Crucho, J., Picado-Santos, L., Neves, J., & Capitão, S. (2019). 9, 3657.
17. De Maeijer, P. K., Soenen, H., Van den Bergh, W., Blom, J., Jacobs, G., & Stoop, J. (2019). *Infrastructures, 4*(3).
18. Department of Irrigation and Drainage Malaysia. (2015). *Urban Storm Water Management, Volume I and II*. Kuala Lumpur: Reproduced by IES.
19. Djumari, S., DaimYami, M. A., Nasution, M. F., & Setyawan, A. (2017). *Procedia Engineering, 171*, 1413–1420.
20. Dolce, M., Cardone, D., & Marnetto, R. (2001). In *Smart Structures and Materials*, International Society for Optics and Photonics, 4330, 249.
21. Eijk, V. D., Olsen, J. S., & Zhang, Z. (2007). In *Proceedings of the First International Conference on Self-Healing Materials*, Netherlands.
22. Faruqi, M., Castillo, L., & Sai, J. (2015). *Journal of Civil Engineering Research, 5*, 21–27.
23. Firoozi, A. A., Olgun, C. G., & Baghini, M. S. (2017). *International Journal of Geo-Engineering, 8*, 26.
24. Gao, Y., He, B., Xiao, M., Fang, Z., & Dai, K. (2018). *Construction and Building Materials, 165*, 548–559.
25. Gargav, A., & Chauhan, J. S. (2016). *International Journal of Engineering Development Research, 4*, 473–476.
26. Gupta, P. K. (2013). *International Journal of Advanced Structural Engineering, 5*, 19.
27. Hameed, M. Z., & Taha, M. R. (2013). *Australian Journal of Basic and Applied Sciences, 7*, 576–581.
28. He, J., Chu, J., Tan, S. K., Vu, T. T., & Lam, K. P. (2017). *Journal of Marine Georesources and Geotechnology, 35*, 593–602.
29. Hesami, S., & Sadeghi, V. (2015). *International Journal of Pavement Research and Technology, 8*, 251–258.
30. Hopkins, C., Cameron, D., Rahman, M., & Rabbi, A. (2016). *Procedia Engineering, 143*, 26–33.
31. Hoy, M., Arulrajah, A., & Horpibulsuk, S. (2016). *Construction and Building Materials, 117*, 209–219.
32. Jameson, G., & Hennessy, G. (2019). *Guide to Pavement Technology Part 4D: Stabilized Materials*. Austroads. Sydney.
33. Ji, J., Yao, H., Suo, Z., You, Z., Li, H., Xu, S., & Sun, L. (2016). *Journal of Materials in Civil Engineering, 29*.

34. Khalifa, N. A., Zulkiple, A., & Ogab, M. (2020). *International Journal of Pavement Research and Technology, 13*, 240–246.

35. Khan, I. M., Kabir, S., Alhussain, M. A., & Almansoor, F. F. (2016). *International Conference on Sustainable Design, Engineering and Construction, 145*, 1557–1564.

36. Khan, S., & Deulkar, V. (2017). *International Journal of Latest Trends in Engineering and Technology, 7*, 130–136.

37. Kvitko, A., Shendrik, V., & Mukharryamov, I. (2018). *Transportation Research Procedia, 36*, 404–410.

38. Little, D., & Nair, S. (2009). National cooperative highway research program (NCHRP), *Transportation Research Board, 144*, 1–67.

39. Liu, R., Zhibin, Z., & Rui, Z., et al., (2007). *Nanotechnology Synthesis Study: Research Report*. Texas Department of Transportation.

40. Lucena, L. C. d. F. L., Juca, J. F. T., Soares, J. B., & Filho, P. G. T. M. (2014). *Transport and Environment, 33*, 210–219.

41. Lv, Q., Chang, C., & Zhao, B. (2018). *Soil Mechanics and Foundation Engineering, 54*, 409–413.

42. Ma, T., Zhao, W. D., & Li, R. (2019). *Synthetic Materials Aging and Applications, 48*, 132–143.

43. Malaiskiene, J., Skripkiunas, G., Vaiciene, M., & Karpova, E. (2017). *IOP Conference Series: Materials Science and Engineering, 251*, 012025.

44. Mathew, T. V. (2010). Pavement materials: Aggregates.

45. Meral, M. E., & Diner, F. (2011). *Renewable and Sustainable Energy Reviews, 15*, 2176–2184.

46. Mirzababaei, M., Arulrajah, A., & Ouston, M. (2017). *Procedia Engineering, 189*, 25–32.

47. Mishra, B., & Gupta, M. K. (2020). *Proceedings of the Institution of Civil Engineers–Municipal Engineer, 173*, 87–97.

48. Mustafa, R. J., Gomaa, M. R., Al-Dhaifallah, M., & Rezk, H. (2020). *Sustainability, 12*, 608.

49. Oad, P. K. (2016). *MSc thesis*, Queensland University of Technology.

50. Oswaldo, P. L. S., Stephanie, S. A., Gerlange, S. D. S., Aline, B. S., & Aivara, I. S. P. (2018). Determination of hydraulic conductivity by auger hole method. *Biofix Scientific Journal, 3*(1), 91–95.

51. Owuama, C. O. (2012). Conceptual design of a trenchless drain in a flat terrain. *Journal of Environmental Science and Engineering A, 1*(12), 1301–1307. David Publishing USA.

52. Owuama, C. O., & Owuama, K. C. (2020). A drainage system for road construction on flat terrain in New Owerri Nigeria. In *Proceedings of the 2nd International Conference, EGRWSE-2019*, University of Illinois, Chicago, IL, USA: Sprinkler Publications.

53. Pardeshi, B. M., & Raut, S. A. (2018). *International Journal of Chemical Physics Sciences, 7*, 37–41.

54. Pereira, T. C. (2014). *MSc Thesis in Civil Engineering, Faculty of Science and Technology University of Lisbon*.
55. Qureshi, T., & Al-Tabbaa, A. (2020). *Self-Healing Concrete and Cementitious Materials*. IntechOpen.
56. Raadama, S., & Schulze, F. E. (1980). Surface field-Drainage System, design and management system, Chapter IV. In *Drainage Principles and Applications*. IILRI. pp. 69–89.
57. Remišová, E., Decký, M., Mikolaš, M., Hájek, M., Koval⊠ík, L., & Mecár, M. (2016). *IOP Conference Series: Earth and Environmental Science, 44*, 022016.
58. Richard, S., & Macmillan, J. (1988). *International Higher Education*.
59. Rojas, N. X., & Mennis, J. (2019). *International Journal of Environmental Research and Public Health, 16*, 3704.
60. Saride, S., & Jallu, M. (2020). *Journal of Transportation Engineering Part B: Pavements, 146*.
61. Sarie, F., Bisri, M., Wicaksono, A., & Effendi, R. (2015). *IOSR Journal of Environmental Science, Toxicology and Food Technology, 9*, 53–59.
62. Sasidharan, M., Torbaghan, E., & Burrow, M. P. N. (2019). *K4D Helpdesk Report*, Institute of Development Studies, Brighton, UK.
63. Shafii, M. A., & Noh, N. M. (2018). *ARPN Journal of Engineering and Applied Sciences, 13*, 8280–8284.
64. Sihombing, A. V. R., Subagio, B. S., Susanto, E., & Yamin, A. (2019). *IOP Conference Series: Materials Science and Engineering, 508*.
65. Simatupang, M., Mangalla, L. K., Edwin, R. S., et al., (2020). *Geosciences, 10*, 132.
66. Singh, R., Srivastava, P., Singh, P., Sharma, A. K., Singh, H., & Raghubanshi, A. S. (2018). *Ecological Indicators, 105*, 505–515.
67. Takahiro, T., Fernando, C., Yoshii, T., & Shirayanagi, H. (2018). *Transportation Research Procedia, 34*, 211–218.
68. Teerawattanasuk, C., & Voottipruex, P. (2019). *International Journal of Pavement Engineering, 20*, 1264–1274.
69. Trenchlesspedia. (2020). Falling Head Test. Retrieved from www.trenchlesspedia.com
70. Van Dort, J. A., & Bos, M. G. (1980). Main drainage systems, Chap IV. In *Design and Management of Drainage Systems*. Wageningen: ILRI press. pp. 161–171.
71. Vinayakamurthy, M. (2017). *Master's thesis*, Arizona State University, USA.
72. Vliet, D., Slaghek, T., Giezen, C., & Haaksman, I. (2016). E&E Congress, Prague, Czech Republic.
73. Wang, C., Xie, T., & Cao, W. (2019). *Materials and Structures, 52*, 52–98.
74. Wang, J. Y., Soens, H., Verstraete, W., & Belie, N. (2014). *Cement and Concrete Research, 56*, 139–152.

75. Wattanachai, P., & Suwan, T. (2017). *IOP Conference Series: The 4th International Conference on Manufacturing and Industrial Technologies*, 27–29 May 2017, Lisbon, Portugal.

76. Wegman, D. E., Sabouri, M., Korzilius, J., Kuehl, R., & Intertec. (2017). LRRB Report 2017RIC02 December.

77. Wenyu, Q., Liu, X., & Kong, Z. (2018). *Open Journal of Social Sciences, 6*, 90–96.

78. White, G., & Reid, G. (2018). 8th Symposium on Pavement Surface Characteristics, Brisbane, Queensland.

79. Wong, B. Y. F., et al., (2019). *IOP Conference Series: Materials Science and Engineering, 495*.

80. Xiao, R., Polaczyk, P., & Zhang, M. (2020). *Journal of Transportation Research Board, 2674*, 22–32.

81. Zhang, Y., Liu, X., Apostolidis, P., Gard, W., Ven, M., Erkens, S., & Jing, R. (2019). *Materials, 12*, 4176.

CHAPTER 5

Traffic Engineering

LEARNING OBJECTIVES

After studying this chapter, you will be able to:

- Define the definition and scope of traffic engineering
- Explain the traffic characteristics
- Understand traffic flow
- Identify traffic control device
- Discuss traffic management
- Define road safety and accident studies

KEY TERMS FROM THIS CHAPTER

Traffic flow
Intersection design
Road capacity
Intelligent transportation systems
Safety analysis
Pedestrian facilities
Roundabout design
Signal optimization
Traffic modeling
Traffic congestion
Traffic surveys
Travel demand management
Traffic simulation
Traffic signal timing

INTRODUCTION

In order to ensure the safe and effective movement of people and goods, traffic engineering is an essential subdiscipline of civil engineering that focuses on the planning, design, and operation of traffic systems. It entails the investigation of traffic patterns, the evaluation of transportation systems, and the application of plans to enhance both the functionality and state of the roads. To comprehend traffic patterns and forecast future trends, traffic engineers combine simulation tools, data analysis, and mathematical models. The ability to predict the future is crucial for creating long-term transportation plans and for making the current infrastructure as efficient as possible in meeting changing traffic demands. Improving road safety through lower accident rates and efficient enforcement of traffic laws is one of the main goals of traffic engineering. This entails planning roads with the proper markings, signs, and signals to safely direct vehicles and pedestrians. In order to reduce traffic and travel time, engineers also pay attention to the capacity and effectiveness of transportation networks. This frequently entails the deployment of intelligent transportation systems (ITS), which make use of technology to dynamically manage traffic flow and give drivers access to real-time information.

Apart from resolving pressing traffic issues, traffic engineering is crucial for sustainable development and urban planning. Engineers can suggest changes to the infrastructure that lessen possible traffic congestion and environmental effects by examining how new developments affect the traffic conditions as they currently exist. This proactive strategy balances the demands of mobility with environmental stewardship, enabling cities to grow in a more planned and sustainable way. The multidisciplinary field of traffic engineering touches on many facets of environmental engineering, public policy, and urban planning. To create and execute efficient traffic management plans, stakeholders such as local governments and urban planners must work together. Traffic engineers work to develop safer, more effective, and environmentally friendly transportation systems that improve both urban and rural communities' quality of life. They do this by conducting ongoing research and innovation.

5.1. DEFINITION AND SCOPE OF TRAFFIC ENGINEERING

The planning, designing, and management of road networks and transportation systems all involve the application of engineering principles and technology, which is known as traffic engineering. It uses techniques like traffic signal control, road design, traffic management, and transportation planning to try to solve different problems with traffic flow, safety, and efficiency.

5.1.1. Scope

The scope of traffic engineering is broad and covers several key areas:

- **Traffic Flow Analysis**:
  - Studying and modeling traffic patterns and behaviors.
  - Analyzing traffic data to identify congestion points and develop solutions.
- **Traffic Control Systems**:
  - Designing and implementing traffic signal systems.
  - Developing intelligent transportation systems (ITS) for real-time traffic management.
  - Managing traffic signs, road markings, and signals to guide and control traffic.
- **Roadway Design**:
  - Planning and designing roadways, intersections, and interchanges.
  - Ensuring roads are designed to accommodate current and future traffic volumes safely.
  - Implementing geometric design principles for safer and more efficient roads.
- **Traffic Safety**:
  - Identifying and addressing safety issues through accident analysis and prevention strategies.
  - Designing safety features like guardrails, pedestrian crossings, and speed control measures.
  - Promoting and implementing measures for reducing traffic accidents and enhancing road safety.
- **Transportation Planning**:

- Developing short-term and long-term transportation plans.
- Conducting feasibility studies and impact assessments for new transportation projects.
- Coordinating with urban planning to ensure the integration of transportation systems with community development.

- **Public Transportation Systems**:
  - Designing and optimizing public transit routes and schedules.
  - Enhancing the efficiency and accessibility of buses, trains, and other public transportation modes.
  - Promoting the use of public transportation to reduce road congestion and environmental impact.
- **Non-Motorized Transport**:
  - Planning and implementing facilities for pedestrians and cyclists.
  - Enhancing the safety and convenience of walking and cycling infrastructure.
  - Promoting alternative modes of transport to reduce reliance on motor vehicles.
- **Environmental Impact**:
  - Assessing and mitigating the environmental impact of transportation systems.
  - Implementing sustainable practices in road design and traffic management.
  - Encouraging the use of eco-friendly transportation options.
- **Data Collection and Analysis**:
  - Utilizing traffic sensors, cameras, and other technologies for data collection.
  - Applying statistical and computational methods to analyze traffic data.
  - Using data-driven approaches to inform traffic engineering decisions and policy-making.
- **Policy and Regulation**:
  - Developing and enforcing traffic laws and regulations.
  - Setting standards for road design, signage, and traffic control devices.
  - Engaging with stakeholders to develop and implement transportation policies.

5.1.2. Importance of Traffic Engineering in Urban Planning and Transportation Systems

Because traffic engineering has a major impact on the sustainability, safety, and efficiency of urban environments, it is an essential component of urban planning and transportation systems. Here is a detailed look at its importance:

Enhancing Traffic Flow and Reducing Congestion

To maximize the flow of cars and pedestrians, traffic engineers design road networks, traffic signals, and control systems. Engineers can lessen traffic congestion and consequently commuters' frustration by efficiently managing traffic flow. This involves:

- **Signal Timing Optimization:** Adjusting traffic signals to minimize stops and delays.
- **Intersection Design:** Improving intersections to handle high traffic volumes more efficiently.
- **Traffic Modeling:** Using simulations to predict traffic patterns and plan accordingly.

Improving Safety

The primary goal of traffic engineering is safety. The number of accidents and fatalities can be considerably decreased by traffic engineers by creating safer roads and putting safety measures in place. This includes:

- **Roadway Design:** Creating safer road geometries, such as wider lanes, better lighting, and clearer signage.
- **Pedestrian and Cyclist Safety:** Designing crosswalks, bike lanes, and pedestrian bridges to protect non-motorized road users.
- **Traffic Calming Measures:** Implementing speed bumps, roundabouts, and other measures to reduce vehicle speeds in residential areas.

Supporting Sustainable Transportation

In order to lessen the negative environmental effects of urban transportation systems, traffic engineering is essential in promoting sustainable transportation options. This includes:

- **Public Transit Integration:** Designing road networks that support efficient public transit routes and prioritizing bus lanes.
- **Non-Motorized Transport:** Encouraging walking and cycling through the development of dedicated lanes and pathways.
- **Green Infrastructure:** Incorporating elements like permeable pavements and green spaces to manage stormwater and reduce urban heat islands.

Economic Benefits

Because efficient traffic management lowers the costs associated with congestion, like wasted fuel and lost productivity, it can have a major positive economic impact. Furthermore, by facilitating easier access to services and businesses, well-designed transportation networks can strengthen local economies.

Adaptation to Technological Advances

Autonomous vehicles and smart city technologies are two examples of how traffic engineering must evolve with technology. This involves:

- **Intelligent Transportation Systems (ITS):** Implementing technologies that allow for real-time traffic monitoring and management.
- **Vehicle-to-Infrastructure Communication:** Preparing roadways for future communication between vehicles and traffic control systems to enhance safety and efficiency.

Data-Driven Decision Making

Decision-making in modern traffic engineering is largely dependent on data. To make evidence-based changes, traffic engineers gather and examine data on traffic volumes, speeds, and accident rates. This strategy guarantees that modifications to the transportation network are focused and efficient.

Environmental and Social Equity

Traffic engineering ensures that transportation improvements benefit all members of the community, including those who live in underserved areas, which can help address issues of social equity and environmental justice. This involves:

- **Equitable Access:** Designing transportation systems that provide equal access to all neighborhoods.
- **Environmental Impact:** Reducing pollution and noise in residential areas, particularly those close to major roadways.

5.1.3. Historical Development and Evolution of Traffic Engineering

The fascinating journey of traffic engineering's historical development and evolution is reflected in the expansion of urban areas, technological advancements, and shifting societal needs. Here is a detailed overview:

5.1.3.1. Early Beginnings (Ancient and Medieval Periods)

- **Ancient Rome:** The Romans were among the first to develop a systematic approach to road construction and maintenance. They built extensive road networks with features like milestones and rest areas.
- **Medieval Cities:** In medieval Europe, cities grew organically with narrow, winding streets. There was little formal traffic management, but some cities began to impose regulations on market days to manage congestion.

5.1.3.2. 19th Century: The Industrial Revolution

- **Urbanization:** The Industrial Revolution brought rapid urbanization, leading to overcrowded streets filled with pedestrians, horse-drawn vehicles, and eventually bicycles.
- **Traffic Regulation:** The need for traffic regulation became evident. London, for instance, saw the introduction of the first traffic police in 1865 to manage the increasing congestion.

5.1.3.3. Early 20th Century: The Automobile Era

- **Rise of Automobiles:** The advent of the automobile in the early 1900s transformed urban transportation. Cities had to adapt quickly to the new mode of transport.

- **Traffic Control Devices:** The first traffic signals were installed in Cleveland, Ohio, in 1914. These manually operated signals laid the groundwork for automated traffic control.
- **One-Way Streets:** Cities began experimenting with one-way street systems to improve traffic flow. San Francisco introduced the first one-way streets in 1909.

5.1.3.4. Mid-20th Century: The Post-War Boom

- **Highway Development:** The post-World War II era saw significant investment in highway infrastructure, especially in the United States with the Federal-Aid Highway Act of 1956, which led to the creation of the Interstate Highway System.
- **Traffic Engineering Profession:** The field of traffic engineering became more formalized with the establishment of professional organizations such as the Institute of Transportation Engineers (ITE) in 1930.
- **Advancements in Traffic Control:** The development of the first electric traffic signal by Garrett Morgan in 1923 and the introduction of traffic circles (roundabouts) in the 1930s marked significant advances.

5.1.3.5. Late 20th Century: Technology and Innovation

- **Computers and Simulation:** The use of computers in the 1960s and 1970s revolutionized traffic engineering. Traffic simulations and models allowed for better planning and analysis.
- **Intelligent Transportation Systems (ITS):** The 1990s saw the emergence of ITS, integrating telecommunications and information technology with transportation engineering. This included developments in traffic signal coordination, electronic toll collection, and real-time traffic information systems.
- **Pedestrian and Cyclist Consideration:** There was a growing awareness of the need to accommodate pedestrians and cyclists, leading to the development of bike lanes, pedestrian zones, and improved crosswalk designs.

5.1.3.6. 21st Century: Sustainable and Smart Transportation

- **Smart City Concepts:** The integration of advanced technologies like the Internet of Things (IoT), big data analytics, and AI into traffic management systems. Smart traffic signals, adaptive traffic control systems, and connected vehicle technologies are becoming commonplace.
- **Sustainable Transportation:** Emphasis on reducing the environmental impact of transportation has led to innovations in public transit systems, promotion of electric vehicles, and development of green infrastructure.
- **Complete Streets:** A modern approach to urban street design that considers the needs of all users, including motorists, pedestrians, cyclists, and public transit users. This comprehensive approach aims to create more livable and inclusive urban environments.

5.1.3.7. Key Milestones in Traffic Engineering Evolution

- **First Traffic Light:** Installed in London in 1868 but operated manually.
- **First Electric Traffic Signal:** Installed in Cleveland, Ohio, in 1914.
- **First Modern Roundabout:** Introduced in the United Kingdom in the 1960s, replacing traditional traffic circles.
- **Federal-Aid Highway Act of 1956:** Paved the way for the Interstate Highway System in the United States.
- **Institute of Transportation Engineers (ITE):** Founded in 1930, contributing to the formalization of traffic engineering as a profession.
- **Advent of Intelligent Transportation Systems (ITS):** In the 1990s, marked a significant shift towards integrating technology with traffic management.

5.2. TRAFFIC CHARACTERISTICS

5.2.1. Road User Characteristics

The study of road users is crucial to the field of traffic engineering. They assume diverse roles such as those of drivers, cyclists, and pedestrians. and evaluate the circumstances in their own special manner. The primary characteristics of road users that dictate the safe operation of a vehicle on the road are physical, mental, and psychological factors.

The important road user characteristics are:

- Physical Characteristics

In traffic engineering, vital physical attributes like vision, hearing, strength, and road user reaction are examined.

- *Vision:* As seen in the Figure 5.1, the field of vision consists of three cones: peripheral vision, fairly clear vision, and acute or clear vision.

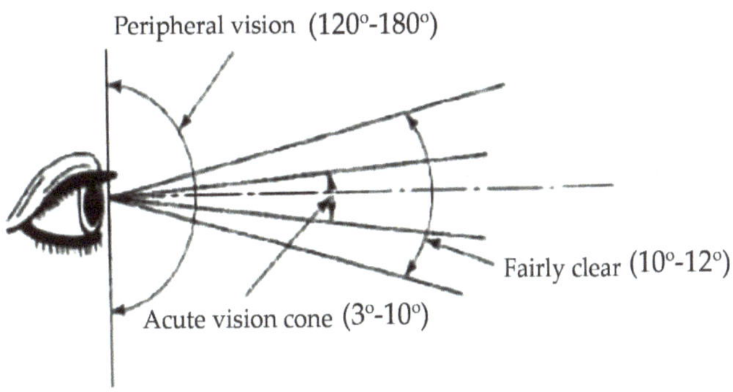

Figure 5.1. *Field of vision.*

Source: https://theconstructor.org/transportation/road-user-characteristics-traffic-engineering/34936/.

  - Acute or clear vision cone: This vision is around 3 to 10 degrees around the line of sight.
  - Fairly clear vision cone: This vision is at 10 to 12 degrees around the line of sight. A vision in this field can identify the color and shape.
  - Peripheral Vision: This field of vision extends up to 90 degrees to the right and left of the centerline of the pupil, as shown in Figure 5.1. It also extends above and below the line of sight by 60 and 70 degrees respectively.

A peripheral vision does not show stationary objects. It only detects the movement of objects within this field.

- *Hearing:* Hearing characteristics are more important for pedestrians and cycles compared to drivers.
- *Strength:* In comparison to other traits, this one is less significant. Parking heavy vehicles is challenging due to one's lack of strength.

Mental Characteristics

The road user's skill, intelligence, experience, knowledge, and literacy are among their key mental attributes. For safe traffic operations, one must possess a thorough understanding of vehicle characteristics, driving techniques, traffic laws, and traffic behavior.

Psychological Factors

An individual's response to traffic situations is influenced by psychological factors when driving. Anger, fear, superstition, impatience, anxiety, and so forth are examples of these reactions. This has an impact on traffic flow.

Environmental Factors

The different environmental factors that control the behavior of the road include the traffic facilities, the characteristics of the traffic stream, and the atmospheric conditions. It may be difficult to pass because of the heavy or mixed traffic in the stream. As a result, how a road user behaves varies depending on the flow of traffic.

5.2.2. Vehicular Characteristics

5.2.2.1. Dimension

- Length of vehicle affects turning radius, extra widening, parking design, etc. (Maximum length 18 meters)
- Width of vehicle affects lane width, shoulder width, and parking design. (Maximum width 2.44 meters)
- The height of vehicle affects the height of bridges, the height of the pole, etc. (Maximum height 4.75 meters)

5.2.2.2. Weight of Loaded Vehicle

Weight affects the design gradient & thickness of the pavement (Figure 5.2).

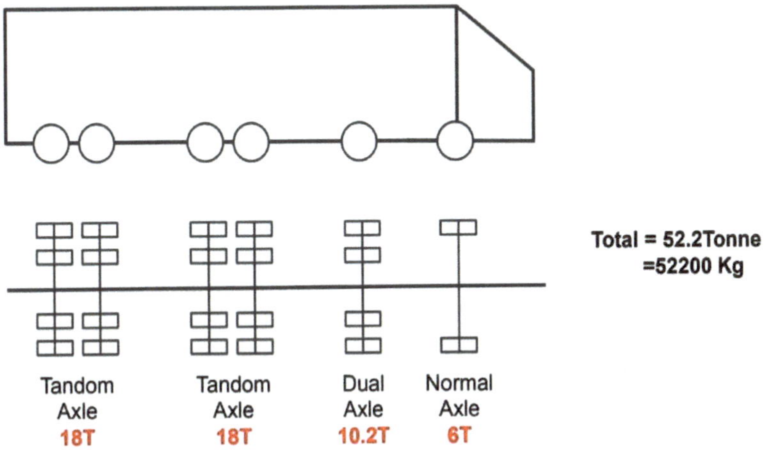

Figure 5.2. *Weight of Loaded Vehicle.*

Source: https://www.slideshare.net/priyamane3/traffic-engineering-highway-drainagepdf.

5.2.2.3. Power of Vehicle

It affects the limiting & design value of the gradient.

5.2.2.4. Speed of Vehicle

It affects all geometrical elements & traffic control devices.

5.2.3. Breaking Characteristics

To determine the skid resistance of pavement, the breaking test requires at least two of the three parameters listed below.

- initial speed (u m/s)
- length of skid distance (L meter)
- Time of break application (t sec)

⇒ Retardation

v-u = at

(v=o)

-u= at

-a= u/t

$$\text{Retardation} = \frac{\text{initial speed}}{\text{time}}$$

⇒ Skid Distance

$L = ut + \frac{1}{2}at^2.$

$a = \frac{-u}{t}.$

$L = ut + \frac{1}{2}\frac{-u \cdot t^2}{t}.$

$L = ut - \frac{1}{2}ut.$

$L = \frac{1}{2}ut.$

Skid distance $= \frac{1}{2} \times$ initial speed $\times$ time.

⇒ Skid Resistance

$v^2 = u^2 + 2aL$ (v=0)

$u^2 = -2aL$

$L = \frac{u^2}{-2a}$.....(i)

Δ KE = work done

$-\frac{1}{2}mu^2 = -m \cdot g \cdot f \cdot L.$

$L = \frac{u^2}{2gf}$.....(ii)

From equations *i* & *ii:*

$(-a) = gf$

$f = \frac{-a}{g}.$

Skid Resistance $= \frac{\text{Retardation}}{\text{accelaration due to gravity}}$

⇒ Breaking Efficiency:

$$\eta_B = (\frac{f_{\text{obtained}}}{f_{\text{max}}} \times 100).$$

$$f_{\text{obtained}} = \frac{u^2}{2gL}$$

5.3. TRAFFIC FLOW

The study of how cars and drivers travel between two locations and interact with one another is known as traffic flow. Traffic flow analysis is particularly challenging because driver behavior is unpredictable. However, traffic streams typically have some reasonable consistency and can be roughly represented mathematically because drivers tend to behave within a reasonably consistent range.

The dynamic shift is displayed by the traffic flow characteristic at various points during the day. Because drivers typically behave within the stream range, traffic congestion can be somewhat predicted and controlled despite the complexity of non-linear traffic flow.

The catch is that this holds true only if there is infinite precision known about the variables that determine their representation. The difficulty at hand is figuring out how to handle these erratic, various traffic issues. Determining and comprehending the traffic variable conditions is a crucial issue that forms the basis of various techniques.

It is crucial to comprehend the two primary streams of traffic movement in a network, interrupted and uninterrupted, as defined by the traffic flow theory.

- "All the flow regulated by vehicle-vehicle interactions and interactions between the vehicle and the roadway," such as the traffic on the highway, is known as uninterrupted flow. A few key elements are mentioned below:
  - There is no controlled access.
  - Can be seen in freeways and long sections of rural highways
  - There are no traffic signals (stop points)
  - Stream characteristic depends on geometric design, vehicles, and others
- Flow interruption: Every flow is controlled by an outside source, like a traffic signal. A few key elements are mentioned below:
  - Traffic signals can be applied
  - Stopping and restarting points
  - Platoons may happen

5.3.1. Traffic Flow Primary Variables

When describing, simulating, and researching the dynamics of traffic, numerous variables are essential. Three key variables in our problem are flow, density, and speed. These are crucial for planning, assessing, and managing transportation systems; more about them is provided below:

- Speed can be defined as "The rate of change of its position" expressed as the distance per unit of time. Generally, it is expressed by kilometers per hour (km/h) or mile per hour (m/h). The symbol of the speed is (υ), and the free flow speed is designated by (V_{max}). Speed is one of the most important variables which can measure the performance of different types of roads.

- Flow is "The amount of traffic passing a point or on a lane or roadway during a designated time interval." The symbol for flow is (q), and the maximum flow is (q_{max}). It is expressed by the number of vehicles using the road per specific time.

$$q = \frac{3600\,N}{t_{measured}}$$

- Density, the third main variable, it is "The number of vehicles occupying a length of a lane or roadway at a particular instant." It is expressed by the number of vehicles per kilometer. The symbol for density is (k).

$$K = \frac{N}{L}$$

K: Density

N: no. of vehicles

L: length of roadway

5.3.2. Relationships Between Speed, Flow, and Density

Density, flow, and speed are all connected to one another. It is not difficult to observe the relationships between density and speed in the real world.

5.3.2.1. Basic formula

The formula below shows the relation between the variables:

$$q = \rho * \upsilon$$

Where:

q: Flow

ρ: Density

υ: Speed

It is sufficient to choose two of them for collection in order to make gathering the primary traffic data along the road easier, and the third variable can be found using the diagram. As a result, the continuous curves and the majority of these key components are depicted in the diagrams under the influence of shifting environmental factors and the non-homogeneity of the flow of traffic. The Figure 5.3 shows the relationships between the main basic road variables.

- Speed-Density diagram
- Speed-Flow diagram
- Flow-Density diagram

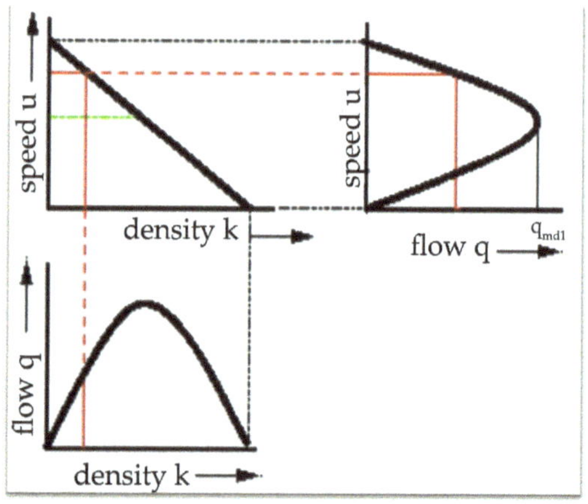

Figure 5.3. *Traditional fundamental diagram of traffic.*

Source: https://uomustansiriyah.edu.iq/media/lectures/5/5_2018_02_25!09_19_57_AM.pdf.

- Three crucial statutes can be identified in the above figure, which depicts the relationships between the primary basic road variables.

- The facility has zero density (k=0) and zero flow when there are no cars present. The speed under this condition is entirely theoretical and would be determined by the first driver to choose, most likely the highest possible number.
- The flow (q) is zero when the density rises to the point where all cars stop (k= 0). Cars are unable to pass a certain point on the road since there is no movement. It is known as the jam density when there is no more movement at all.
- Low and medium density and flow are represented by flow increasing (density as well) from zero while speed is beginning to decrease. It takes longer for density to reach capacity because the speed decreases gradually as density grows. One can find capacity in both low and high densities as well as high and low speeds. To further interpret the noteworthy points on these diagrams collectively, there is a speed-flow density diagram.

5.3.3. Time Mean Speed

Time mean speed ($\overline{V_t}$) = arithmetic mean of speeds of vehicles passing a point

$$\overline{V_t} = \frac{1}{N} \sum_{n=1}^{N} v_n$$

5.3.4. Space Mean Speed

The harmonic mean of the speeds that pass a point over a given amount of time is known as the space mean speed. It also equals the average speeds over a length of roadway

$$\overline{V_s} = \frac{N}{\sum_{n=1}^{N} \frac{1}{v_n}}$$

5.3.5. Headway

5.3.5.1. Time Headway

The time difference (in seconds) between the front of the car's arrival at a highway location and the front of the next car's arrival at the same location is known as the time headway ($\overline{h_t}$).

The average time headway ($\overline{h_t}$) = average travel time per unit distance * average space headway.

$$\overline{h_t} = \bar{t} * \overline{h_s}$$

5.3.5.2. Space Headway

Space headway ($\overline{h_s}$) is the difference in position between the front of the vehicle and the front of the next vehicle

$$\overline{h_s} = \overline{V_s} * \overline{h_t}$$

5.3.6. Modeling Traffic Flow

Three categories can be used to group the foundations of traffic flow theory and its applications:

5.3.6.1. Microscopic Traffic Modeling

Micro-simulation models can simulate the behavior of every single vehicle in a network. Theoretically, these models offer a more accurate and "purer" depiction of real-world driver behavior and network performance. These models address the movement of a single vehicle within the traffic system, including the driver's movement, speed, and travel time.

Microscopic models have a wide range of advantages. We summarized them in bullets below:

- Micro-simulation models are very useful where the increasing system complexity and uncertainty can be involved in the operation of the urban traffic networks.
- The microscopic models are based on some traffic factors which have the ability to manage the movement of individual vehicles in a transport network.
- Microscopic simulation models are able to evaluate various traffic management alternatives in order to determine the optimum solution for any traffic scenario and provision of visualization for the case under study.
- Traffic data is rarely constant and repeatable, so many reliable suggestions can be offered for different traffic problems such as congestions, incident management

5.3.6.2. Macroscopic Traffic Modeling

"Providing an aggregated representation of demand, typically expressed in terms of total flows per hour" is the definition of (also traditional transport models). Stated differently, they address the features specific to a given road segment.

5.3.6.3. Mesoscopic Traffic Modeling

Microscopic and macroscopic techniques are combined in mesoscopic models to produce highly efficient descriptions of all the important aspects of the road. For instance, changing lanes is a microscopic decision made by the drivers; however, this decision should be based on the densities and speeds determined by a macroscopic model.

5.4. TRAFFIC CONTROL DEVICE

Through the use of tactile or visual signs, traffic control systems inform, direct, and guide drivers. The main categories of traffic control devices are channelizers, signs, signals, barriers, and road design and marking.

Numerous traffic control devices can be adapted for use in smaller spaces and are commonly found on public highways. When driving on a public highway, drivers usually give signals and signs more attention. In smaller facilities, barriers and pavement markings might be more important components.

What are the Four Common Traffic Control Devices? The three main requirements for efficient traffic control systems are focus, reaction time, and regard for other drivers. There are numerous traffic control devices on the market designed to reroute cars away from property and pedestrians. To protect tenants and their properties, property managers can employ a range of traffic management techniques.

- Signs
- Traffic signals
- Road markings
- Barricades and bollards

5.4.1. Signs

Everybody has seen road signs. It's how we, as motorists, pedestrians, and other road users, obtain information. Changes in the speed limit are announced with signs. One way to indicate the direction of traffic is with a no-entry or one-way sign. Signs usually indicate where parking is available and where you can stop. Disobedience to regulation signals often leads to a movement offense.

In addition to being directional, warning, regulatory, and informative, signs also come in a range of sizes and shapes. Regulation and enforcement of traffic laws are accomplished through the use of signs like these.

- Stop signs
- Pedestrian crossing signs
- Speed limit signs

Warning signs, such as those alerting drivers to an approaching work zone, an animal crossing, or winding roads, inform drivers of potential hazards or construction. Drivers are guided by signs, like one that indicates how far it is to the next exit. To help cars get to their destination, route, and amenity information is displayed on directional or educational signs like these ones.

- Information on routes
- Exits
- Rest areas
- Gas stations
- Restaurants
- Tourist attractions

Vehicles are directed by signs when there is restricted or no access to a specific area. When an accident results in a lane closure, for example, signs should be posted sufficiently ahead of the scene to give cars enough time to adjust to the altered traffic patterns. Signs can also be used in these situations to notify drivers of proposed detours and expected delays.

Since road signs are composed of reflective material and come in vivid colors, they are easily visible at night when driven by headlights. The traffic regulation signs are the most important ones on the road. They declare that it is permissible to use the adjacent public route. The term "intersection control" refers to appropriate conduct at the intersection of two roads; yield and stop signs, in particular, are essential for preventing collisions.

5.4.1.1. Private Property Signage

The most common type of parking lot signage that is painted on lane separators and support pillars are striped caution signs. When used in conjunction with speed bumps, speed limit signs can be especially useful in alerting drivers to the maximum acceptable speed. Stop signs are essential in large parking lots and garages with crosswalks that could cause auto accidents. One-way signs can be useful in certain situations, particularly in garages and apartment buildings. Road markers can also serve as an indicator of traffic direction.

5.4.2. Traffic Signals

In addition to signage, properties employ traffic signals to direct traffic. Generally speaking, traffic signals ought to be able to draw in users, provide context, give them enough time to react, and minimize wasted time.

To regulate traffic flow, traffic control signals typically have distinct green, yellow, and red lights. When it is safe for pedestrians to cross at a designated crosswalk, pedestrian signals let them know. Typically, only areas with significant roads are suitable for using these kinds of traffic control systems.

5.4.2.1. Lights

Within the field of traffic management, red, yellow, and green traffic lights are frequently seen. The standard three lights can have lane indicators and other specific lane markers added to them. It's common to use flashing lights, either as a standard signal or hanging on their own. However, their meanings could vary depending on the context and color. Lights that never stop burning are commonly used as warnings or hazards.

5.4.3. Road Markings

A distinct kind of traffic safety device is the road itself. Road markings can be in the form of lines, patterns, words, reflectors, symbols, or other patterns on the pavement, curb, sides of islands, or adjacent permanent objects. These indicators consist of:

- **Pavement markings** – White paint is often used for these markings, with yellow paint for lanes or parking restrictions.
- **Curb markings** – These markings

are placed along roadside curbs to help drivers see them from a distance. They may include information on parking rules among other things.

- **Object markings** – To alert vehicles and pedestrians to certain physical impediments and dangers; labels may be attached.
- **Reflectors** – Reflective road markings may help cars navigate safely at night, and some hazard signs can reflect light up to 500 feet away.

Another crucial component of traffic control devices for managing traffic between buildings in a residential or commercial complex is road markers. To indicate expected behavior, any common road marker, such as shared lane markings (sharrows) or stop lines, may be used. Parking lots can also make use of road markers.

Stop lines and turn lanes can play a critical role in traffic regulation near the tight bends that may occur in restricted areas. Zebra stripes and crosswalks create demarcated pedestrian zones that help people get to their cars safely. Rumble strips are used on long stretches of road where vehicles may become distracted or fail to notice a sudden turn. Some driving behaviors are easier to adopt than others because of variations in environmental design and road construction. Additional elements that encourage safe driving habits are as follows:

- Speed bumps
- Curb extensions
- Roundabouts

5.4.3.1. Lanes

Road construction also includes traffic control. Rumble strips can be used to indicate lanes. By using sound and texture, they alert an approaching car that it is out of its lane. Road markings, such as lines and arrows, are used to mark the appropriate and lawful use of road surfaces. Among them are turn lane arrows, stop lines, and lane markings. Road markers such as "sharrows" warn drivers of unexpected traffic patterns by showing them that motorized vehicles and cyclists share the same space.

5.4.3.2. Parking

More often than one-way signage, parking lots use arrows to indicate the direction of traffic, much like one-way parking aisles. The most common road marking in a parking lot is, of course, the stall marker, which indicates the space allotted for each parked car. Additionally, markings for accessible areas greatly lessen user difficulties.

5.4.4. Barricades and Bollards

There are three primary types of traffic signs that serve as physical and visual barriers to impede drivers:

- **Removable posts** – These traffic posts can be swiftly taken down and rebuilt to offer temporary protection when there is a lot of traffic, whether it be from vehicles or pedestrians. Depending on its size and strength, the retractable traffic bollard can operate as a physical or visual deterrent.

- **Collapsible posts** – Some locations have permanent collapsible bollard posts, but they can also be raised or lowered to accommodate different clearance requirements. They are also ideal for providing brief protection. These come in handy in locations such as parks where driving is normally forbidden but passing utility or emergency vehicles may be permitted on a temporary basis.
- **Fixing posts** – Permanent constructions known as fixed traffic posts have the potential to provide physical protection from auto accidents. Depending on the type of surface, they can have embedded or surface-mounted installations. These are ideal for locations that require continuous security, like storefronts and building perimeters.

Flexible bollards are used to guide parking stops in back-in spots, acting as both a stop and placement guide. They also work great as lane markers on circular ramps where larger cars require more flexibility to use multiple lanes. For construction sites and transient hazards, cones and delimiters are great temporary channelizers. Later on, we will delve deeper into the topic of cones and temporary barriers.

In high-traffic work sites, barriers of all types, including fences, guardrails, and concrete barriers, are frequently used. Vehicles are kept apart from traffic and construction workers by barriers. Even opposing-moving traffic can have barriers separating it.

Barriers can be used to divert traffic away from makeshift construction sites. They can also be used to temporarily close off a lane or portion of the road when there are hazards that pose a risk to vehicles.

Another type of traffic control device that can shield people from harm and shield property from damage in the event of a collision is a traffic bollard. Pedestrians should always feel secure in the knowledge that they are safe from harm and that accidental car damage is always prevented.

5.4.4.1. Cones

The traffic cone is the most commonly used device for traffic direction. They can serve as a barrier to prevent vehicles from entering a highway or, conversely, be used in place of barriers to indicate the beginning or end of a work zone. There are many different colors and shapes of traffic cones to choose from, and some are even waterproof.

It can be difficult to see traffic cones when it's snowing or raining outside. Cones can be used in these situations to create a barrier that separates the travel lane edge from the shoulder, keeping the road dry and free of snow and ice.

It can also be used to temporarily block off one lane so that workers have access to the area without having to deal with cars stopping or veering past the barrier. Similar to how cars splattering water from puddles onto the pavement can lead to costly maintenance needs and pavement damage. Water can be kept off the pavement with the use of a barrier, saving costly repairs and preserving traffic flow.

5.4.5. Traffic Control System and Protection

Systems for controlling traffic do more than just manage and direct traffic. The importance of traffic control systems is highlighted by the following five factors:

- **Safety** – The protection of people near or in moving traffic is the main objective of traffic control systems. Road traffic accidents and injuries are more likely when there is uncontrolled traffic. Without traffic control systems, the road would become much more dangerous.
- **Organization** – For the purpose of directing cars at intersections with traffic, traffic control equipment is required. Anarchy in areas like intersections would result from a lack of traffic management. Ensuring efficiency and safety is the primary goal of traffic regulation. Notifying drivers when it is safe to drive and when they need to stop is the best way to improve safety and lower the number of accidents. Traffic direction safety and security benefits all users, including drivers, passengers, and pedestrians.
- **Reduce Accident Severity** – Traffic control systems make some accidents that happen at and around intersections less serious. At slower speeds, accidents have less of an impact as drivers learn to anticipate proper traffic flow.
- **Reduce Traffic Congestion** – Without order, there would be chaos in the traffic. Order is brought about by automated gates, toll booths, and traffic control systems at intersections. There will be much less opportunity for complete gridlock to arise as a result. Traffic jams increase the likelihood of crashes and make it more difficult for drivers to focus on the road.
- **Promote Fuel Efficiency** – In general, the longer a vehicle is driven, the higher its carbon emissions. The duration of time that cars are on the road is reduced through traffic management. This maintains traffic order and helps create a better future for society.

5.5. TRAFFIC MANAGEMENT

The process of prioritizing, reshaping, and routing network traffic to maintain the availability of vital applications is known as traffic management.

5.5.1. Arterial Road Traffic Management

Roads that serve as the main means of communication for traffic movement and that primarily transport traffic from one area to another are referred to as arterial roads. Major regional and interregional traffic movement must be accommodated by the arterial road system in a way that is both safe and effectively operates. The necessity and capacity of a specific arterial road to carry out this duty frequently depend on the conflicting tasks it must also complete. When creating appropriate traffic management plans, major arterial roads may need to take into account the needs for local public transportation or commercial or industrial access.

The goal of traffic management on arterial roads is to maintain a safe, convenient, and environmentally friendly environment. A few strategies for managing traffic on arterial roads include implementing the following:

- To give priority to the arterial road at intersections by means of stop or give way signs, traffic signals, etc.
- To control access to the arterial road by means of turn regulation, one-way sections, vehicle bans, channelization, median island, or guard rail.
- To eliminate hazardous objects or movements and increase capacity on arterial roads, by parking control, turn regulations, u-turn regulations, no lane-change, lane-use control, one-way, channelization, etc.
- To reduce the relative difference in speed in order to achieve compatibility among vehicles and reduce the severity of accidents by means of limiting the speed.
- To reduce noise, vibration, or other nuisance caused by vehicular traffic by means of limiting the speeds, and introducing vehicle bans or signals.

5.5.1.1. Different Types of Traffic Management Techniques Adopted in Arterial Roads

The basic techniques that should be given consideration when planning for good traffic management are:

- to simplify traffic flow in order to achieve similarity among components and stabilize the traffic flow

- to segregate road uses in space and time in order to reduce conflicts and simplify traffic flow
- to increase capacity in order to accommodate more vehicles
- to restrain traffic in order to reduce traffic volume

The basic elements that can be used to achieve effective traffic management practice are as follows:

- Speed Limits

  Speed limits are introduced with the following objectives:

  - To simplify traffic flow by reducing differences in speed (improves safety and mobility)
  - To reduce the severity of accidents by reducing the speed (improves safety)
  - Save fuel by avoiding excessive speeding (energy saving)
  - Helps to have enough reaction time for drivers and pedestrians (improves safety)

Disadvantages:

- Sometimes low speed limits may cause congestion and drivers may be inclined to disregard it.

Turn Regulations

Turn regulations are introduced with the objectives of

- to increase safety by reducing conflict points
- to increase capacity at intersections

Disadvantages:

- Additional travel and increased turning movements at other locations
- Delays due to additional turning movements

U-turn Regulations

U-turn regulations are introduced with the objectives of:

- To simplify traffic flow and increase capacity by eliminating U-turn movements in order to ensure safe and efficient flow

Disadvantages:

- Additional travel
- Increased turning movements at other locations
- Delays due to additional turning movements

Parking Controls

Parking controls are enforced with the objectives

- To increase capacity by freeing road space for the use of traffic in movement
- To provide drivers with a good vision field and to simplify traffic flow by eliminating parked vehicles
- To increase parking revenue by having a time control on parking
- To restrain traffic volume by prohibiting long-time parking
- To transfer people from private cars to public transport in order to reduce traffic congestion

Disadvantages:

- increased walking distance
- circulating vehicles searching for parking spaces
- effects on the environment by increasing traffic
- traders losing customers due to lack of on-street parking

No Standing

The no-standing situation is distinct from the no-parking situation, and can be described as follows:

Standing is the temporary stop of a car to allow for the loading or unloading of passengers. A vehicle is not permitted to stop, even for a short while, to pick up or drop off passengers when there is a sign prohibiting standing.

Other than stopping momentarily when the engine is off, parking is stopping. Therefore, a car may stand for a while to pick up or drop off passengers when a sign prohibits parking, but it is not permitted to stop with the engine off. As a result, prohibiting standing is a more severe driving restriction than prohibiting parking.

No standing is implemented with the objectives

- To increase capacity by freeing road space for the use of traffic mobility
- To secure a vision field for driving and avoid weaving maneuvers by eliminating standing vehicles on carriage way.

Disadvantages:

- Increased walking distance
- Reduced places for loading and unloading

Give Way

The give-way sign indicates observing the right-of-way rule.

Right-of-way Rule:

- When two vehicles enter an intersection from different approaches at approximately the same time, the driver of the vehicle on the left shall give way (yield) to the vehicle on the right.
- The driver of a vehicle intending to turn to the left shall give way (yield) to any approaching vehicle from the right.
- The driver of a vehicle approaching a roundabout should give way (yield) to any vehicle already in the roundabout
- At marked pedestrian crossings the drivers should always give way to the pedestrians on the crossing.

The give-way signs are introduced with the following objectives:

- To segregate major traffic from minor traffic and to give priority to the major flow by showing the right-of-way.
- To indicate the drivers who have priority at an intersection and avoid confusion.

Disadvantages:

- Sometimes a queue in the minor road flow
- May not be effective when traffic on both roads is heavy

Channelization of Traffic

Traffic is channelized with the objectives:

- To smoothen the traffic flow by restricting the driver choice, reducing the conflict points, and correcting the intersecting angles
- To guide the traffic to a suitable course
- To reduce the area of conflicts by controlling intersecting traffic streams by making them intersect at or near right angles.
- To ensure low relative speeds between merging traffic by making the streams merge at small angles.
- To discourage prohibited or undesirable movements
- To protect pedestrians from turning or crossing vehicles by providing refuge.

Disadvantages:

- When drivers do not observe, it is difficult to correct them.
- Continuous lane marking and regular maintenance may be costly.

No Lane-Change

No lane change at certain sections of roads is implemented with the following main objectives:

- to simplify traffic flow by eliminating weaving movements.
- to make the road section safer.

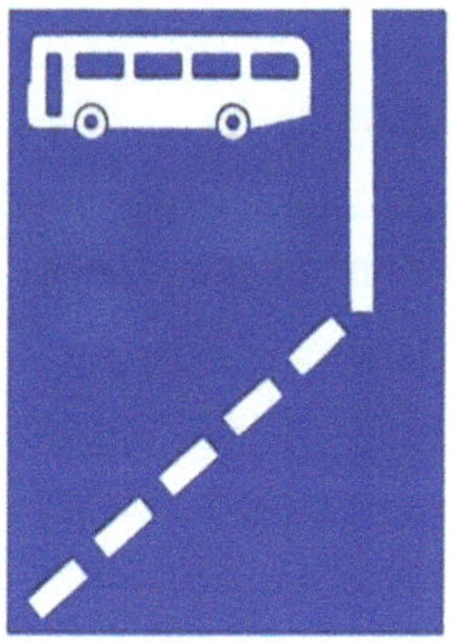

Disadvantages:

- Delay in certain lanes when the distribution of traffic over the whole lane changes.
- Unless the drivers maintain strict lane discipline this will be very hard to implement.

No overtaking

No overtaking sections of roads are implemented with the objectives:

- To segregate opposing flows
- To prevent accidents due to overtaking maneuvers
- To create a safe and stable flow

Disadvantages:

- Delays may be caused due to slow-moving vehicles.

One-way

One-way sections are implemented with the objectives:

- To increase capacity and travel speed by eliminating opposing vehicles performing right turns
- Traffic signals could be linked in a simple progressive pattern
- To reduce conflict points by segregating traffic by its direction
- May discourage undesirable through traffic in residential areas.

Disadvantages:

- Diverting traffic or additional travel distance
- effects on public transport routes and increased walking distance for passengers
- Effects on traders
- Hazards at transition areas between one-way and two-way operation
- Confusion for stranger

Bus Lanes

There are lanes reserved only for buses. The bus lanes may introduced with the following objectives:

- To reduce the travel time of buses by reserving lanes for the exclusive use of buses
- To restrain vehicle traffic by encouraging the use of buses and discouraging the use of private cars

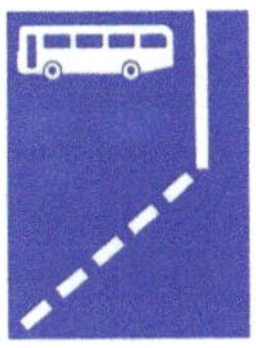

Disadvantages:

- To be effective strict lane discipline should be maintained
- Problems at intersections because of turning movements
- Reduced space available for other vehicles

Vehicle Ban

Certain vehicles can be banned from entering particular areas if required. These vehicle bans can be introduced with the objective:

- To eliminate particular categories of vehicles or all of them due to:
  - security reasons
  - ease congestion
  - improve traffic flow
  - discourage private vehicles

Disadvantages:

- Increased flow of diverted traffic and increased travel distance
- Effects on a particular business
- Diversion of public transport

5.5.1.2. Urban Arterial Roads

Traffic management should be implemented on urban arterial roads to guarantee safety, mobility, and a healthy environment—all of which are essential for the efficient operation of metropolitan areas. The following actions can be performed to meet these expectations:

- Re-allocate traffic over the existing road network by means of re-allocation measures.
- Restrain vehicular traffic to adjust it to the existing road network, by means of restraint measures.
- Integrate available measures in order to reinforce each other and achieve optimum pattern of traffic flow as a whole bb means of integration measures.

(a) Re-Allocation Measures

This involves redistributing traffic demand throughout the current road network in accordance with the purposes of the roads while taking the adjacent land's characteristics into consideration. The purposes of the various kinds of roads.

1) Arterial Roads

These roads serve for carrying heavy traffic, controlled by arterial control or special routing.

2) Collector roads.

Collector roads serve for collecting local traffic and distributing it to access streets.

3) Local streets.

Local streets are designed to provide convenient access to nearby properties while discouraging traffic.

Characteristics of abutting land:

- Residential area
- School area
- Shopping area
- Industrial area
- Business area (commercial area)

(b) Restraint measures

In order to promote the use of public transportation and discourage the use of private vehicles, this is being done. The system improves overall mobility, energy conservation, safety, and the environment despite the limitations placed on private vehicles. The following are some ways to discourage using private vehicles:

- Parking control
- Providing bus lanes
- Vehicle ban
- Pedestrian precinct

(c) Integration Measures

This is done to blend components so they counteract negative effects and strengthen one another without contradicting one another. It usually encompasses the entire region and attempts to coordinate the efforts of all relevant parties and agencies.

5.5.1.3. Rural Arterial Roads

To maintain rural areas' safety and good environmental conditions, traffic management should be implemented on rural arterial roads. The following actions can be performed to surpass expectations:

- To divert through traffic from rural residential or shopping areas. This may be achieved through, local bypasses, turn regulations, one-way sections, vehicle bans, speed limits, parking control, pedestrian precinct, and special routing.
- To restrict pedestrian crossing points and make them visible to the approaching vehicles. This may be achieved through pedestrian crossings and warning signs.
- To bring vehicles to low speeds where crossing movements are inevitable to improve safety and reduce the severity of accidents. The speed limit will help to achieve the objectives.
- To prevent head-on collisions due to overtaking maneuvers through no-overtaking zones.
- Give advance warning of hazardous conditions such as bends, crests, railway crossings, slopes, and slippery road stretches through warning signs or warning devices.

5.6. ROAD SAFETY AND ACCIDENT STUDIES

One of the main factors considered in all traffic management plans is traffic safety. Accidents happen and that is a loss to the community. There are two possible categories for these losses: (1) direct losses and (2) indirect losses.

- Direct losses are the loss of life, injuries, damage to vehicles and other property, etc.
- Indirect losses are the loss of productivity, repair costs, insurance costs, rehabilitation costs, and losses resulting from induced congestion, etc.

The economy is eventually impacted by all of these losses. As a result, it is imperative to lower the annual accident rate. Studies on traffic safety are detailed investigations conducted to identify the causes of accidents and devise strategies for averting them. To prevent accidents from happening or lessen their effects, contributing factors must be correctly identified.

5.6.1. Causes for Accidents

Let's take a general look at the causes of traffic accidents before moving on to accident analysis and accident prevention strategies. On our roads, five primary factors lead to collisions and injuries.

When trying to increase road safety, these issues must be carefully addressed.

- Speeding
- Negligence of road rules
- Drunk driving
- Fatigue
- Poor condition of vehicles and roads

These five are the most important, though there are many more that are less important.

5.6.1.1. Speeding

Most fatal and local traffic accidents involve speed as a contributing factor. Speeding is defined as exceeding both the posted speed limit and the speed that is appropriate given the current state of the road. Crashing may occur from this. Drivers must slowdown in certain situations, like heavy traffic, rain, drizzle, night driving, pedestrians on the road, and rough sections of the road.

5.6.1.2. Negligence of road rules

It is a well-known fact that many local drivers break road rules, whether intentionally or unintentionally. One potential solution to this issue is implementing ongoing driver education programs through various media outlets and driving schools. Encouraging participation in defensive driving, advanced driving, and safe driving programs for both fleet operators and the public can also help improve road safety. These educational programs are crucial for enhancing knowledge of road rules and promoting safe driving practices. Safe driving is largely dependent on awareness of potential hazards and maintaining safe driving behaviors. Priority should be given to defensive driving courses that focus on vehicle control skills and road rule knowledge. Additionally, implementing policies to limit the importation of incompatible vehicles, such as three-wheelers, can help reduce the number of road rule violations.

5.6.1.3. Drunk Driving

Some drivers have a widespread misconception that consuming a small amount of alcohol enhances their driving performance. Alcohol has been shown to have an adverse effect on one's ability to drive. It is a fact that people cannot drive as usual, even if they appear and behave as though they are not impaired by alcohol. Alcohol depresses you; it causes your brain to work more slowly. It lessens your capacity for situational response, decision-making, and action. More frequent roadside testing and law enforcement should fall under the purview of the police.

5.6.1.4. Driver Fatigue

Driver fatigue is another important factor that leads to fatal accidents. It is recommended that drivers take reasonable breaks before embarking on a lengthy drive, plan their schedules realistically, and stop for appropriate rest periods to prevent driver fatigue. Don't drive when you should be sleeping. The majority of fatigue-related accidents happen between 11 p.m. and 7 a.m. when the body naturally sleeps. The hours of early to mid-afternoon are also very risky. This is not to say that driver fatigue exclusively occurs during those times; rather, it is the most likely time to experience its consequences. Watch out for the symptoms of fatigue, restlessness, body aches, sluggish steering, and sore eyes to prevent accidents caused by exhaustion. When you notice any of these symptoms, stop at the closest safe location and take some time to relax until you feel fully awake.

5.6.1.5. Condition of Vehicle and Condition of Roads

Two other crucial elements that affect road safety are the state of the vehicles and the road itself. The state of brakes, lights and reflectors, tires, windscreen wipers, and rearview mirrors should all be in good condition to reduce the risk of an accident. The driver and the owner of the vehicle should oversee having them fixed if they are defective and maintained in good working order at all times. This is a severely underutilized area in the nation since there are no required vehicle fitness tests or road worthiness tests carried out because of financial constraints. The traffic police should examine their records to determine what proportion of accidents are caused by vehicles in poor condition considering the rising number of accidents and make recommendations accordingly. Additionally, the highway authorities have to investigate high-accident areas and take appropriate action to eradicate them.

5.6.1.6. Other Factors

There are a lot of additional variables that lead to traffic accidents that need to be carefully considered. On city roadways, large cars pose a challenge. Fleet operators are quickly realizing how important proper driver training is. To minimize heavy vehicles on national highways, every effort should be made to shift container movement from road to railway as much as possible. The careless bus drivers are another issue. They pose the biggest risk to road safety. Tight guidelines, like not allowing a bus to pass another without a good reason and requiring a minimum headway between buses, etc. might be presented.

5.6.2. Traffic Safety Study Measurement Methods

Traffic safety studies are conducted based on three broad measures.

- Measures based on reported accidents
- Measures based on traffic conflicts
- Measures based on public assessment of traffic safety

5.6.2.1. Measures Based on Reported Accidents

Research on traffic safety can be done using accident reports as a basis. The majority of nations classify different types of accidents into the following categories:

- **Fatal accidents** – These are accidents where at least one person is killed in the accident.
- **Minor injury accidents** – At least one person is seriously injured (i.e., admitted to hospital) but no one is killed.
- **Major injury accidents** – At least one person is injured but no one is killed or admitted to hospital.
- **Property damage type accidents** – No one was injured but the vehicle or road property was damaged.

The traffic accidents can be subdivided according to the following important factors.

- **Accident type** – depending on single or multiple accidents, transport modes involved in the accident, maneuvers before the accident, etc.
- **Persons involved** – according to age, sex, and other social and physical indications.
- **Location** – according to road category, and surface condition, whether inside or outside built-up areas, at intersections or mid-blocks, curves, or straight sections, etc.
- **Time** – by season, day of the weak, hour of the day.

- **Circumstances** – daylight or in dark, dry, or rainy.
- **Traffic characteristics** – hourly volume, speed distribution, etc.

This kind of information enables the establishment of various traffic safety "indicators.." The most direct measure of safety is the total number of accidents, which is why "before and after studies" (which will be covered later in this chapter) are of such fundamental importance. Organizing accidents based on their level of severity provides more details.

5.6.2.1.1. Level of Safety

One useful metric for assessing safety is the ratio of accidents to population. This metric can be used to compare the accident rates across nations or between cities within nations.

5.6.2.1.2. Accident Rate

The accident rate refers to the number of traffic-related accidents. When analyzing road sections, the accident rate is typically reported as accidents per million vehicle-kilometers, while for intersections, it is measured as accidents per million vehicles passing through. This metric is frequently utilized to assess the safety of different road segments and intersections due to the correlation between accident frequency and traffic volume. When comparing various modes of transportation, the most appropriate indicators include:

- **For passenger transport** – the number of accidents (whether fatal, injury, or property damage) per million passenger-kilometers.
- **For goods transport** – the number of accidents (whether fatal, injury, or property damage) per million tonne-kilometers.

5.6.2.2. Measures Based on Traffic Conflicts

Like reported accidents, traffic conflicts can be categorized, tallied, and used as a gauge of traffic safety. While there are several approaches that can be used to recommend which kinds of conflicts should be minimized or prevented, it is difficult to forecast with precision when these conflicts will arise. As a result, it is challenging to conduct traffic conflict studies and determine whether they have a greater potential to measure traffic safety than published accident studies.

5.6.2.3. Measures Based on the Public Assessment of Traffic Safety

It is theoretically feasible to prioritize government actions by ranking public evaluations of traffic safety. However, in order to be effective tools for the quantification of traffic safety, analyzes like conflict analyzes—which rely on a certain degree of subjective assessment—still require a lot of work. At this level, public assessments that are nearly solely based on subjective criteria are not covered in depth.

5.6.3. Data Collection

5.6.3.1. Data Sources

Road accident data can be gathered from a variety of sources, each with specific goals that affect the scope and type of data collection. The primary sources are:

- Police Department accident reports
- Coronial inquest reports (where available)
- Hospitals (regarding the severity of injuries)
- Road Authority reports on fatal and other road crashes
- Insurance company records (if obtainable)

Data from sources like hospitals, coronial inquests, and insurance company records are generally used sparingly. Many road authorities gather specific data on fatal and other traffic accidents, especially in cases where the crash's circumstances could give rise to legal action or where a claim could be made regarding property damage to the road authorities. In general, data from this source is not very useful for analyzing traffic crashes.

5.6.3.2. Police Department Road Crash Reports

Any traffic crash's initial investigation is under the exclusive authority of the police department. Data on traffic crashes is most frequently obtained from their reports. This consists of information gathered from all "reported" traffic collision sites and stored in a police department database. When it comes to "property damage type" accidents, information must be obtained from the various insurance companies involved as these incidents are typically not reported to the police.

Kumarag et al., (2003) have demonstrated that the number of traffic accidents on Sri Lankan roads has been rising over time. It has also shown that, aside from the country's rapidly growing car fleet, the following contributing factors have primarily contributed to this high accident rate.

- The rapid increase in the amount of travel undertaken by the population
- Shift from relatively safe forms of transport such as public transport to extremely unsafe modes such as the motorcycle
- Combine the effect of ineffective enforcement and blatant violation of road rules
- Poor road design
- Lack of safety interventions and poor maintenance programs
- High percentage of dangerous vehicles such as; recklessly driven private buses and heavy vehicles, three-wheelers and motorcycles, etc.

5.6.3.3. Reporting and Recording Criteria

The criteria for reporting the data should be maintained in the following format.

NAME OF TOWN:

STREET NAME:

TIME:

* Date of Accident: (day/month/year)

* Time of Day of Accident (hour: minute)

* Accident Report No:

LOCATION:

* Name & Type of Identifying Object:

* Distance from the Identifying Object:

* Direction from the Identifying Object:

WEATHER:

* Weather Type: (sunny, rainy, cloudy)

* Road Surface Condition: (wet, dry, ice)

VEHICLES INVOLVED:

* Type of Vehicles/Roadside Objects Involved:

* Sex and Age of Driver:

* Direction of Travel:

* Street of Travel:

* Stated Speed of Vehicle:

SEVERITY OF ACCIDENT:

* Severity of Accident: (Fatal, Major Injury, Minor Injury, Property Damage)

* Number of Persons Killed:

* Number of Persons Seriously Injured:

* Number of Persons with Minor Injuries:

* Number of Tow away Accidents:

DESCRIPTION OF THE ACCIDENT:

5.6.4. Accident Analysis

Upon gathering accident data, the analysis can be conducted as either a short-term or long-term traffic management plan, contingent on the depth of the investigation. In order to perform an exhaustive accident analysis, it is recommended to gather all accident data over a minimum of five years, if not longer.

5.6.4.1. Long-Term Traffic Management Schemes

This thorough investigation aims to identify the variables influencing the likelihood of accidents. For the purpose of analyzing accidents in long-term traffic management schemes, there are generally two approaches available.

5.6.4.1.1. The Clinical Method

Using this approach, an accident is examined to ascertain its cause and circumstances and to derive lessons from it regarding how future incidents might be prevented. This approach is employed when the kind of mishap is exceedingly uncommon and expensive (e.g. aircraft crashes, significant maritime mishaps).

5.6.4.1.2. Statistical Method

This approach looks for information on conditions or elements that are thought to be involved in accidents. The circumstances may have to do with the vehicle's features, the state of the road, or the driver's age, sex, personality, and level of alcohol use, among many other things. For every traffic unit involved in an accident, an effort is made to obtain exactly the same kind of information. The factors that are present in accidents under specific circumstances are then identified through statistical analysis of these data. Large amounts of standardized data are needed for the statistical approach, and these data are analyzed using mathematical models to look for relationships. The clinical approach differs in that it looks for relationships by interpreting data and only looks for unique, pertinent data in a smaller number of cases.

5.6.4.2. Short-Term Traffic Management Schemes

In short-term traffic management schemes, accident analysis is done using one of two methods.

5.6.4.2.1. Before and After Studies

Studies with before and after data are becoming more and more popular for assessing traffic safety. The same methodology can be used for traffic conflict studies as it can for reported accidents.

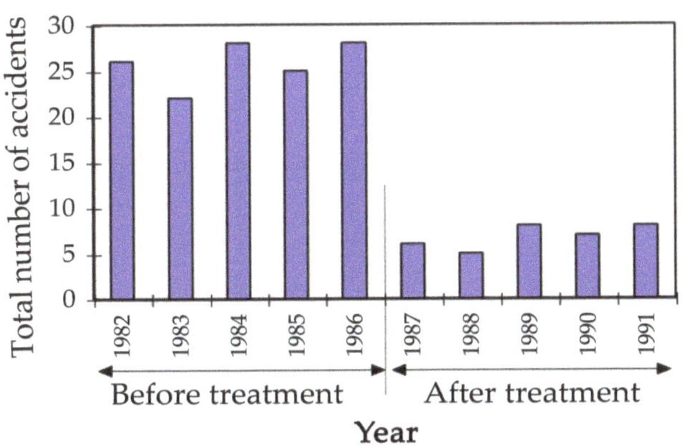

Figure 5.4. *Road accident statistics before and after the road surface treatment on High Street.*

Source: Kolita S. Weerasekera; An Introduction to Traffic Engineering; ISBN 978-955-514-691-3.

A comparison is made between the number of accidents that occurred prior to and during the same period of time following the implementation of the traffic management scheme. This provides a reasonable indication of how well the plan is working from a safety standpoint. Methodological errors resulting from before and after studies must be addressed. The issue of control factors needs special attention in order to account for the impact of traffic growth, variations in weather from one period to the next, and systematic changes in the number of accidents as a result of changing certain factors. The statistical methods required to determine the significance of observed changes should also receive a great deal of attention (Figure 5.4).

5.6.4.2.2. Detection of Hazardous Road Locations

One of the primary goals of some traffic management plans is to remove dangerous road locations, or "blackspots.." The identification of these locations is a significant issue. By taking into account the method of control, traffic volumes, type of road or junction, surface texture, weather conditions, etc., it is possible to determine which road sections or junctions are particularly dangerous. To determine which neighborhoods, streets, and intersections have higher accident rates than anticipated, carefully gathered empirical data on accident frequencies and rates for various types of junctions or road sections is required. Such data are also necessary to assess which corrective actions to take and how much of an impact they will have on traffic safety.

5.6.4.2.3. Collision Diagrams

The collision diagram, a schematic representation of all crashes occurring at a given location over a specified period, usually 1–5 years, is the fundamental tool used in site-specific crash diagnosis.

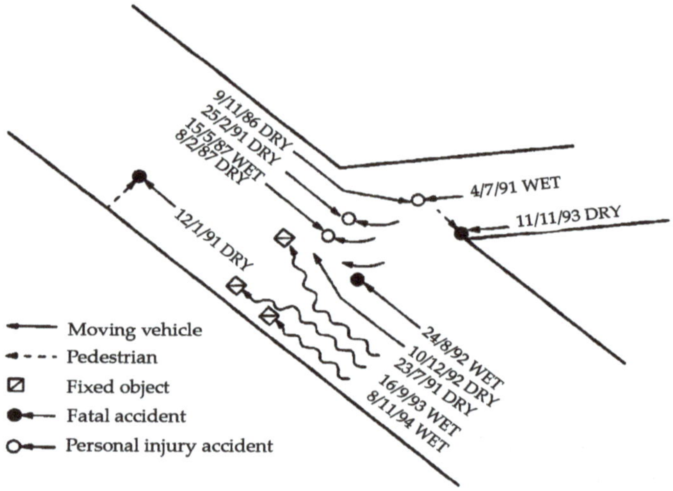

Figure 5.5. *A typical collision diagram at a three-way junction.*

Source: Kolita S. Weerasekera; An Introduction to Traffic Engineering; ISBN 978-955-514-691-3.

As seen in Figure 5.5., each collision at the scene is represented by a set of arrows, one for each vehicle or pedestrian involved, that indicate the kind of crash and the directions in which it occurred. Arrows can be labeled with codes that indicate the day, time, temperature, type of vehicle, and so on. A tabular summary of the crash details is presented after the data for each crash, which may be displayed on the collision diagram. The data may include crash type, crash severity, date and time of crashes, road condition, lighting condition, site geometry, and locational information (Ogden, 1994).

5.6.4.2.4. Blackspots

The dangerous places on the road (road sections or junctions) that are particularly risky due to factors like location, excessive speed limits, the volume of traffic, type of road or junction, surface roughness, weather, etc. are known as blackspots. However, dangerous locations deserving of corrective action may also be deemed to be road locations that have a high crash potential and are amenable to remedial treatment.

Police accident reports are a great resource for pinpointing the hotspots. To improve safety at those dangerous locations, the number of blackspots should be reduced gradually through the implementation of blackspot treatment programs, with priority given to the worst locations. Locations of hazardous roads are typically determined by utilizing one or more of the following criteria:

- Number of crashes greater than a set cut-off value
- Crash rate greater than a set cut-off value (where the rate is defined as the number of crashes per exposure or risk unit, e.g. crashes/10^8 vehicle kilometers)

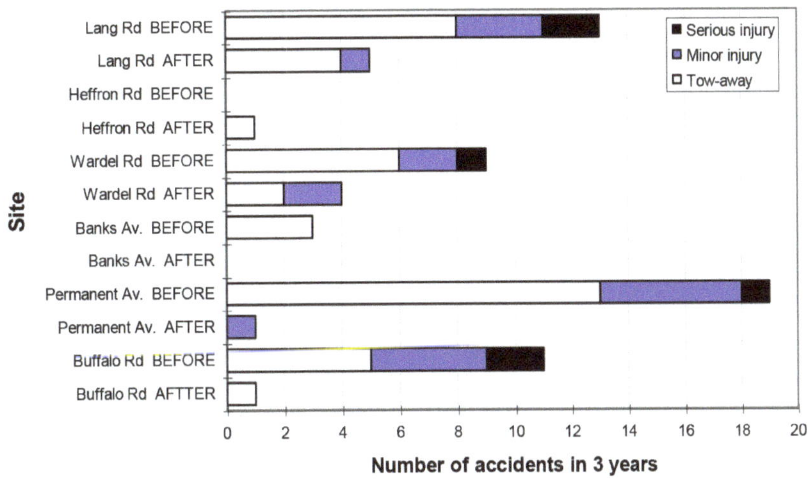

Figure 5.6. *Before and after accident results of a blackspot treatment program.*

Source: Kolita S. Weerasekera; An Introduction to Traffic Engineering; ISBN 978-955-514-691-3.

Regarding which of the aforementioned criteria is better suited for identifying dangerous locations, there isn't much consensus. The approach has the best chance of lowering the number of crashes because it concentrates attention on the areas where crashes happen most frequently. The second criterion finds the locations where the crashes have occurred because of something genuinely unusual (i.e., crash statistics are not just achievable in situations with heavy traffic) (Figure 5.6).

The most widely used performance indicators for benchmarking when identifying fatality hotspots worldwide are as follows (Ogden, 1994):

- Fatalities per 100,000 persons,
- Fatalities per 10,000 vehicles (registered), and
- Fatalities per 100 million vehicle kilometers of travel

While each of these has a place, none of them will give a complete picture of road deaths on their own. For instance, Australia has a very low road death rate based on vehicles, but India has a high road death rate based on population, while India has a low road death rate based on vehicles. South Africa has a population and vehicle rate that are slightly higher than those of most developed nations, but an exceptionally high percentage of those deaths involve pedestrians.

A comprehensive and concise set of definitions of hazardous road locations are given below:

- The highest risk locations can be determined separately based on accident history and the clusters of accidents that have occurred at:
  - **blackspots** – specific locations usually related to particular features of road geometry such as junctions, bends, or hills
  - **black sites** – specific lengths of road with high accident frequencies

- **black areas** – areas within which clusters of accidents occur; used primarily in urban areas where methods of reporting of accident locations may not be precise enough to identify, individual roads in a closely spaced network.

- Locations of intermediate risk may have too few accidents to identify specific sites from accident records alone, but observational studies combined with accident data or a combination of data from groups of related sites may reveal potentially dangerous features. These may be designated grey spots, sites, or areas in similar terms to those above
- Locations where common circumstances or traits predominate in accidents; in other words, there may be a high frequency of a specific type of accident that is not always connected to clusters of accidents.

To identify and address potentially dangerous roadside features that do not currently have a history of accidents but may in the future have a high frequency or severity of accidents, the two latter definitions mentioned above are particularly crucial. Any object, condition, or measurement that can be recorded or measured can be considered a roadway feature. These elements can be found on or off the road and include traffic control devices, pavement friction, junctions, railroad grade crossings, geometric features (lane width, shoulder width, grades, curvature, etc.), roadside obstacles (bridge parapets, guardrail, poles, drainage structures, etc.), and traffic control devices.

While reading accident study reports, it can be helpful to understand the definitions of the following frequently used terms.

- **Crash type** – the classification used to describe a crash in terms of the vehicle movements involved, e.g. run-off road, right-turn opposing, or the type of road user involved, e.g. vehicle/pedestrian, vehicle/bicycle.
- **Crash rate** – the number of crashes per unit of exposure.
- **Exposure** – the measure of opportunity for a crash to occur, e.g. amount of travel in vehicle/km, length of road, number of vehicles, or vehicle/vehicle, vehicle/pedestrian, etc. conflict.
- **Crash severity** – a measure of the consequences of a crash in terms of the most severe casualty class sustained by one of the persons involved, e.g. fatal, serious injury requiring medical treatment, injury not requiring medical treatment, no injury (i.e., property damage only).
- **Severity weighing** – the use of factors to give additional weight to the more severe and more costly crashes.
- **Casualty class** – a measure of the consequences of a crash in terms of the number of persons killed, injured, and not injured.

CASE STUDY

REAL-TIME TRAFFIC MONITORING FOR CONGESTION PREVENTION

Founded in 2017, Telelink City has already become a leading innovator in designing solutions for the smart cities of the future. Their main goal is to help city managers use technology to create cities where everyone thrives on innovation. Telelink City's flexible, fully customizable solutions can be deployed quickly, without specialized support or a large initial investment. They tailor their solutions to the unique challenges of each environment, addressing real problems with efficient, effective solutions. Telelink City partnered with GoodVision to implement real-time traffic monitoring technology that increased accuracy and decreased costs. Their main concern was the low accuracy of capturing traffic events with their existing solution, which had a significant impact on traffic control and congestion prevention.

Traditional Traffic Monitoring Lacked Accuracy

Prior to engaging GoodVision, Telelink City used conventional traffic data collection methods and traditional video detectors that simply didn't deliver the high-quality data they needed:

- **Environmental conditions** like sun, shade, and other obstacles reduced the accuracy of detection using traditional background subtraction video analytics methods.
- **Low accuracy** also meant that real-time traffic-control strategies couldn't be implemented to their full potential as the controller's reaction is only as accurate as the input data to it.
- **Micro and macro modeling** automation was impossible because the historical data was in a static format and of such low quality.

Source: https://casestudies.goodvisionlive.com/case-study-real-time-traffic-monitoring-for-congestion-prevention.

To compensate for these limitations, Telelink City adapted their algorithm by adding a scaling factor. This was based on some manual counting and calibration for each detector since accuracy varied in each situation. This provisional solution was far from scalable since it was impossible to automate.

Smart Cities Call for Future-Forward Technology

Telelink City sought a new solution that increased accuracy without increasing cost. Mladen Zamanov, Smart Solutions Engineer for Telelink City, said, "The field of traffic monitoring has seen many technological advancements, most notably in computer vision and the application of deep-learning models to detect and track objects in images. It was time for us to pursue a state-of-the-art technology solution."

A transition to real-time traffic monitoring would give Telelink City the improved data they needed, without increasing costs. But real-time technology also offers a host of other benefits that are appealing for smart cities:

- Decreased commute times
- Improved public safety
- Reduced environmental and public health costs
- Potential to optimize routes and traffic in real-time
- More efficient parking

Indeed, the potential benefits of real-time traffic monitoring were incredibly appealing to Telelink City, given their focus on supporting the traffic management needs of smart cities.

Stara Zagora Gets a Cutting-Edge Solution

Given these requirements, Telelink City chose GoodVision as their real-time traffic monitoring partner. Initially, the Telelink City team was concerned about the potential unknowns of implementing any new technology and whether GoodVision's platform would provide the robust data and real-time visibility they needed.

However, they were quickly impressed with the flexibility during setup and the additional data they could collect. Furthermore, they found that they could also leverage the GoodVision platform to collect additional data, such as vehicle speed, queue indicators, saturation flows, stopped or delayed vehicles, or traffic gaps. On top of that, GoodVision's solution provided full multimodal detection of vehicular traffic, bicycles, and pedestrians at once, with the possibility to even extend the set of vehicle types if required.

As a pilot project after technology evaluation in the lab, Telelink City and GoodVision launched a deployment in Stara Zagora, Bulgaria. Among the largest cities in Bulgaria, Stara Zagora has 150,000 residents, 70 buses for public transit, and a city management team of 40. The ultimate goal of the project was to provide an all-in-one mobility system that supported the seamless operation and integration of multiple transit options in the city.

Examples of traffic scene setup in GoodVision Live Traffic in Stara Zagora

Source: https://casestudies.goodvisionlive.com/case-study-real-time-traffic-monitoring-for-congestion-prevention

Real-time traffic monitoring played an important role in fulfilling that goal. GoodVision's solution was connected with the existing city IP cameras (2.0 megapixels, 25FPS). Traffic video analytics are based on GoodVision's proprietary artificial intelligence, which performs real-time detection, classification, and tracking of traffic attendants based on their appearance and their full trajectories are retrieved for further analysis. Video analytics run on off-the-shelf EDGE computing units equipped with NVIDIA Jetson Xavier processors. Each unit is capable of processing up to 8 live camera streams simultaneously. The edge unit was integrated with Telelink's traffic control system via API, providing alerts with a latency below 1 second.

Mladen Zamanov worked closely with the GoodVision team throughout the project. "Even though we weren't able to work together in person because of the pandemic, the great communication and collaboration between our teams led to a successful realization of this pilot project." Seamless remote configuration, API implementation, and shipping of pre-installed devices to the team at Telelink City is perfect proof of the modularity and simplicity of GoodVision's solution.

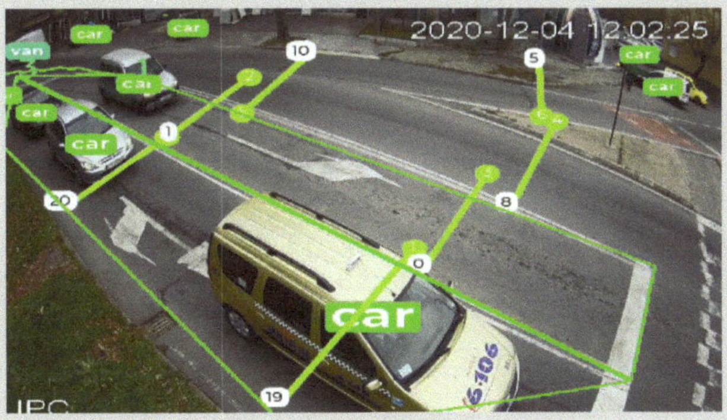

Source: https://casestudies.goodvisionlive.com/case-study-real-time-traffic-monitoring-for-congestion-prevention

The Partnership Yields Immediate Benefits

For Telelink City, the outcomes of the partnership with GoodVision have been stellar. Accuracy rates of traffic-event detection quickly jumped from 70-80% to a whopping 97% and beyond. Meanwhile, GoodVision's state-of-the-art technology also gave Telelink City the ability to capture more data and traffic events needed for their control, and in higher resolution than ever before. This deeper set of high-quality data meant that Telelink City could integrate more data into new traffic-control algorithms.

One of the biggest benefits for Telelink installations is the versatility of the solution, offering deployment on the off-the-shelf EDGE devices or on the servers in the data center, which covers pretty much all the various needs of Telelink customers.

	Initial Costs	Operational Costs	Accuracy	Speed	Classification
Inductive Loops	---	---	+++	+*	+*
Magnetic Sensors	---	---	+++	+*	+*
Video Detectors	+	+	+	-	-
Video Analytics (GoodVision)	+	+	+++	+++	+++

*Speed measurement and classification are possible with a specific setup that further increases the amount of installation work and cost.

Source: https://casestudies.goodvisionlive.com/case-study-real-time-traffic-monitoring-for-congestion-prevention

Zamanov noted that there's still plenty of potential for this partnership. "I do believe that this object detection and tracking technology has countless applications," he said. Both companies are working together to address Telelink's customers in Eastern and Central Europe. GoodVision Live Traffic is evolving into a fully integrated traffic monitoring solution in the central Video Insights platform.

Telelink City and GoodVision learned plenty of lessons in Stara Zagora that would benefit other cities. To learn more about how they implemented real-time traffic monitoring, please join us for our upcoming webinar! We'll explore this project and host a Q&A with industry experts from both companies.

CLASS ACTIVITY

Students will work in small groups to design a new traffic signal system for a busy intersection, considering factors such as traffic flow, pedestrian safety, and environmental impact. They will present their designs to the class, explaining their reasoning behind each element.

SUMMARY

- Traffic engineering involves the application of engineering principles and technology to the planning, design, and operation of road networks and transportation systems. It seeks to address various issues related to traffic flow, safety, and efficiency by employing strategies such as traffic signal control, road design, traffic management, and transportation planning.
- Road users play an important role in studying traffic engineering. They take up various roles like drivers, pedestrians, cycles, etc., and assess the conditions in their own unique way. The physical, mental, and psychological factors are the main road user characteristics that govern the safe operation of the vehicle on the road.
- Traffic flow is the study of the movements of individual drivers and vehicles between two points and the interactions they make with one another. Studying traffic flow is very difficult because a driver behavior is something that cannot be predicted. However, drivers tend to behave within a reasonably consistent range and, thus, traffic streams tend to have some reasonable consistency and can be roughly represented mathematically.
- Micro-simulation models can model each vehicle's behavior within a network. In theory, such models provide a better, and „purer,' representation of actual driver behavior and network performance.
- Many traffic control devices are used on public highways and can be modified for use in smaller facilities. On public highways, drivers frequently pay more attention to signals and signs. Pavement markings and barriers may carry a greater part of the load in smaller facilities.
- Traffic control signals frequently include unique green, yellow, and red lights to manage traffic flow. Pedestrian signals let pedestrians know when it is safe to cross at a designated crosswalk. These kinds of traffic control systems are often only used in locations with major roadways.
- The most widely used tool for directing traffic is the traffic cone. They can be used instead of barriers to mark the start or end of a work zone, act as a barrier to stop cars from accessing a highway, or vice versa. Traffic cones are available in a range of shapes and colors, and some can even be used underwater.

- Arterial roads can be defined as those roads that predominantly carry through traffic from one region to another region, and which form the principal avenues of communication for traffic movement. The prime function of the arterial road system is to provide for major regional and inter-regional traffic movement in a safe and operationally efficient manner.

REVIEW QUESTIONS

1. What are the main traffic characteristics studied in traffic engineering?
2. How do traffic volume and speed relate to each other in traffic flow analysis?
3. Define the concept of traffic flow in traffic engineering.
4. What are the three primary variables of traffic flow?
5. What is the purpose of traffic control devices?
6. What are the key components of an effective traffic management system?
7. What is the main goal of road safety studies?

REFERENCES

1. Bliemer, M. (2007). Dynamic queuing and spillback in analytical multiclass dynamic network loading model. *Transportation Research Record: Journal of the Transportation Research Board,* (2029), 14–21.
2. Carlson, R. C., Papamichail, I., Papageorgiou, M., & Messmer, A. (2010). Optimal motorway traffic flow control involving variable speed limits and ramp metering. *Transportation Science, 44*(2), 238–253.
3. Kadiyali, L. R. (1997). *Traffic Engineering and Transport Planning*. New Delhi: Khanna Publishers.
4. Kumarage, A. S., Wickremasinghe, S. M., & Jayaratne, M. D. R. P. (2003). Analysis of Road Accidents in Sri Lanka. *Transportation Engineering Division, Department of Civil Engineering, University of Moratuwa, Sri Lanka*.
5. Ogden, K. W. (1994). *Traffic Engineering Road Safety: A Practitioner's Guide*. Report No. CR 145. Canberra: Federal Office of Road Safety.
6. Parker, G. B. (1987). Cars must Go! Evicting the City Centre Squatter, Developing World Land Transport. Grosvenor Press International Ltd.
7. Slinn, M., Matthews, P., & Guest, P. (1998). Traffic Engineering Design–Principles and Practice. London: Arnold, Hodder Headline Group.
8. Weerasekera, K. S. (2000). The Need of Traffic Impact Assessments and Some Guidelines. *Sri Lanka Engineering News, Institution of Engineers Sri Lanka, 1*(42), March 2000.

CHAPTER 6

Hill Roads

LEARNING OBJECTIVES

After studying this chapter, you will be able to:

- Discuss on overview of hill roads
- Describe the special consideration in hill road design
- Focus on special structures in hill roads

KEY TERMS FROM THIS CHAPTER

Spurs	Groynes
Hill road	Fragile ecosystems
Rainfall	Geographic information systems (GIS)
computer-aided design (CAD)	River route
Ridge route	Gradient
landslide	groundwater seepage
Bench type	Box cutting
Semi bridge	Semi tunnel

INTRODUCTION

Hill roads are essential for connecting mountainous regions to the rest of a country, facilitating not only transportation but also economic development and social integration. These roads are often feats of engineering, designed to navigate steep gradients, unstable terrain, and extreme weather conditions. The construction and maintenance of hill roads pose significant challenges due to the natural topography and geological conditions. Engineers must consider factors such as slope stability, erosion control, and drainage to prevent landslides and ensure the longevity of the road. The alignment of hill roads is crucial; it involves careful planning to balance the need for a feasible gradient and minimizing the environmental impact. Switchbacks or hairpin bends are commonly used to manage steep ascents and descents, although they require skilled driving and careful design to ensure safety.

Moreover, the materials and methods used in constructing hill roads differ from those in plains. The choice of materials must account for local availability, durability, and the ability to withstand harsh weather conditions, including heavy rainfall and snow. Reinforced concrete, gabion structures, and retaining walls are frequently employed to provide additional support to the road structure and surrounding terrain. Proper drainage systems are critical in hill road design to manage surface water runoff and prevent water accumulation, which can weaken the road foundation and cause erosion.

Environmental considerations are also paramount. Construction activities must minimize ecological disruption, as hill regions often host unique and fragile ecosystems. Measures such as reforestation, controlled blasting, and soil conservation techniques help mitigate environmental impact. Additionally, hill roads must be designed with the local community in mind, ensuring accessibility and safety for residents and promoting sustainable development. The integration of local knowledge and practices can enhance the effectiveness of road projects.

Maintenance of hill roads is an ongoing process that demands regular inspection and timely repairs. Seasonal variations, such as monsoons and winters, bring specific challenges like landslides and snow blockages, requiring proactive and reactive measures. Technological advancements, including geotechnical monitoring systems and remote sensing, aid in the early detection of potential hazards and efficient management of maintenance activities.

Hill roads are vital lifelines for mountainous regions, enabling connectivity and fostering development. Their construction and upkeep require a multidisciplinary approach, combining engineering expertise, environmental sensitivity, and community involvement. Through innovative design, sustainable practices, and continuous maintenance, hill roads can effectively serve their purpose while preserving the natural beauty and ecological balance of the regions they traverse.

6.1. OVERVIEW OF HILL ROADS

A hill road is any route that traverses land with a cross slope of twenty-five percent or more. Particularly when the road follows a river route, there may be places along hill roads where the cross slope is less than 25%. Even so, hill roads are another term used to describe these sections. Therefore, it is necessary to consider the entire terrain when establishing a hill road. The topography, challenging and dangerous terrain, vast high-altitude areas, and extremes in climate are typical features of hilly regions. The area has a low population density and lacks the basic infrastructure that would be expected in a plain setting. Therefore, for the general development of other sectors as well, a strong, stable, and practical road must exist in hilly areas.

Design and Construction Problems: Road design and construction in hilly and mountainous areas are more complicated than in flat areas. It is caused by a number of regionally related factors. They are:

- A hilly or mountainous area is characterized by highly broken relief with vastly differing elevations and steep slopes, deep gorges, etc. which may unnecessarily increase road length.
- The geological condition varies from place to place.
- Hill slopes stable before construction may not be as stable due to increased human activities.
- There may be variations in hydro-geological conditions which may easily be overlooked during design and construction
- Due to highly broken relief construction of special structures should be done at different places. This increases the cost of the construction.
- Variation in the climatic conditions such as the change in temperature due to altitude difference, pressure variation, precipitation increases at greater height, etc.
- High-speed runoff occurs due to the presence of high cross slopes.
- Filling may overload the weak soil underneath which may trigger new slides.
- The need for the design of hairpin bends to attain heights.

6.2. SPECIAL CONSIDERATION IN HILL ROAD DESIGN

Because hilly terrain presents its own set of challenges, hill road design requires special considerations. Ensuring stability and safety while minimizing the impact on the environment is the main priority. These factors cover a wide range of topics, including materials, drainage, gradient, alignment, and ecological preservation (Sharma & Sharma, 2014).

6.2.1. Alignment of Hill Roads

In the hilly area, choosing an alignment is a difficult task. The designer ought to make an effort to select a quick, simple, affordable, and secure route. When creating hill roads, the path is chosen to follow valley bottoms, hillside crests, and, if necessary, over mountain passes. The route's length naturally increases due to its complex topography. The harsh geological conditions necessitate the provision of special structures. Climate and geological factors are significant in addition to the severely fractured relief, which plays a fixed role in determining the alignment and location of unique structures.

In locating the alignment special consideration should be made with respect to the variations in:

- Temperature
- Rainfall
- Atmospheric pressure and winds
- Geological conditions

6.2.1.1. Temperature

- Air temperature in the hills is lower than in the valley. The temperature drop is approximately 0.5° per 100 m of rising.
- On slopes facing south and southwest snow disappears rapidly and rain water evaporates quickly while on slopes facing north and northeast rain water or snow may remain for the longer time.
- Unequal warming of slopes, sharp temperature variations, and erosion by water are the causes of slopes facing south and southwest.

6.2.1.2. Rainfall

- Rainfall increases with an increase in sea level.
- The maximum rainfall is in the zone of intensive cloud formation at 1500-2500 m above sea level. Generally, the increase of rainfall for every 100 m of elevation averages 40 to 60 mm.
- In summer very heavy storms may occur in the hills and about 15 to 25% of the annual may occur in a single rainfall. The effects of these types of rainfall are serious and should be considered well.

6.2.1.3. Atmospheric Pressure and Winds

- It decreases with an increase in elevation.
- At high altitudes, the wind velocities may reach up to 25-30 m/s, and the depth of frost penetration is also 1.5 to 2 m.
- Intensive weathering of rocks because of sharp temperature variations which cause high winds.

6.2.1.4. Geological Conditions

- The inclination of folds may vary from horizontal to vertical stratification of rock. These folds often have faults. Limestone or sandstone folds may be interleaved with layers of clay which when wetted may cause fracturing along their surface. This may result in shear or slip fold.
- The degree of stability of hill slopes depends on the types of rock, degree of strata inclination or dip, occurrence of clay seams, the hardness of the rocks, and presence of groundwater.
- When locating the route an engineer must study the details of the geological conditions of that area and follow stable hill slopes where no ground water, landslides, and unstable folds occur.

6.2.2. Route Location in Hills

Route location in hills is a complex and critical aspect of road design, requiring a meticulous balance between engineering feasibility, environmental sustainability, and socio-economic benefits. The process begins with a thorough analysis of the topography, geology, and climatic conditions of the region. Engineers use topographical surveys, geological studies, and hydrological assessments to understand the natural landscape and identify potential challenges such as steep gradients, unstable slopes, and water flow patterns. The goal is to select a route that minimizes the need for extensive excavation and construction while ensuring the road's safety and durability. Environmental impact is a major consideration, as hill roads traverse ecologically sensitive areas. Planners strive to minimize deforestation, preserve natural habitats, and avoid disrupting watercourses. The route should also enhance connectivity and accessibility for local communities, promoting economic development and social integration. Technological tools such as Geographic Information Systems (GIS) and computer-aided design (CAD) are employed to simulate different route options and evaluate their impacts. The selected route must offer a balance between technical viability, cost-effectiveness, and minimal environmental footprint. By integrating these factors, the route location process aims to create roads that are safe, sustainable, and beneficial for both people and the environment. Hill roads may follow different paths according to the feasibility of the road. However, a hill road alignment varies for the sections along the valley bottom and along the mountain pass. The first is called the river route and the second is called the ridge route.

6.2.2.1. River Route

The location of a route along a river valley is the most frequent case of hill alignment as there is a great advantage of running a road at a gentle gradient. Also, there is a benefit of low construction cost and operation costs. However, a river valley may run through numerous horizontal curves. Requirements for the construction of large bridges over tributaries also may occur. It may also be necessary to construct special retaining structures and protection walls on hill side for safe guarding the road against avalanches.

Some important considerations:

- The road bed should be located sufficiently above and away from the maximum water level.
- When the roadbed is near the waste water course embankment slope should be well protected and stabilized.
- More care should be given to geological and hydrological structures.
- Best alternatives should be selected for crossing water sources.

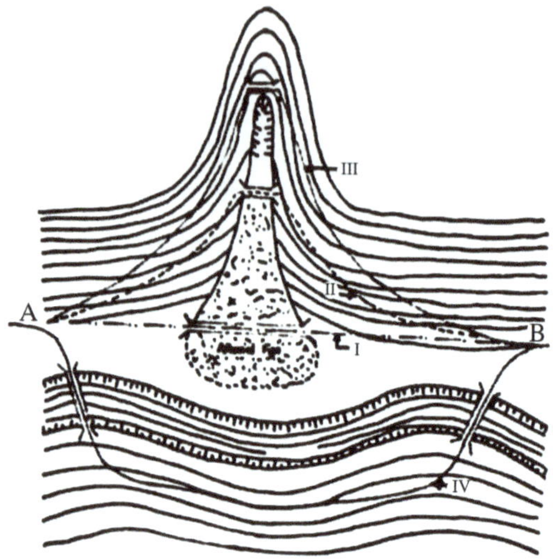

Figure 6.1. *River route.*

Source: https://abhashacharya.com.np/wp-content/uploads/2017/04/Hill-Roads.pdf.

For example, as shown in the Figure 6.1 a road is to be connected from A to B.

The first option creates a bridge by passing through the alluvial. That might not be possible, though, without solid foundations, which could significantly raise the cost of building.

The second option is situated above the alluvial fan that significantly shortens the bridge's length.

In a similar vein, alternative options such as routes III or IV might also be selected in light of the comparative analysis.

6.2.2.2. Ridge Route

A ridge route is a strategic choice in hill road design, involving the alignment of a road along the ridges of hills rather than through valleys or across slopes. This method offers several advantages in terms of stability, drainage, and scenic value, although it also presents unique challenges.

One of the primary benefits of a ridge route is its inherent stability. Roads located along ridges are less prone to landslides and soil erosion compared to those built on slopes or through valleys. The natural drainage of ridges ensures that water flows away from the road, reducing the risk of waterlogging and the consequent weakening of the road foundation. This effective drainage is crucial in preventing damage caused by heavy rainfall, which is common in hilly regions. Ridge routes also tend to offer better visibility and reduced gradient, making them safer for drivers. The alignment along the high points of the terrain can minimize sharp curves and steep climbs, which are often hazardous in mountainous areas. Moreover, ridge routes provide panoramic views, enhancing the travel experience and potentially boosting tourism in scenic regions.

However, constructing a ridge route involves specific challenges. The initial surveying and alignment require precise engineering to ensure the road follows the natural contours of the ridge. This can involve significant cutting and filling operations, which must be carefully managed to avoid destabilizing the terrain. Additionally, ridge routes may require more extensive infrastructure, such as retaining walls and protective barriers, to ensure the safety and longevity of the road. Environmental considerations are paramount when designing ridge routes. The construction process must aim to minimize deforestation and preserve the natural landscape. Mitigating the ecological impact involves implementing soil conservation techniques, reforestation projects, and careful management of construction debris. Moreover, the route must be planned to avoid disturbing local wildlife and natural watercourses (Figure 6.2).

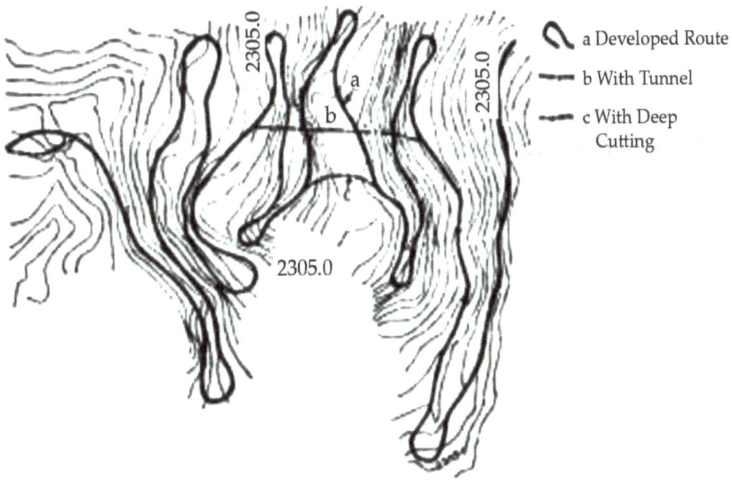

Figure 6.2. *Ridge route.*

Source: https://abhashacharya.com.np/wp-content/uploads/2017/04/Hill-Roads.pdf.

Some important considerations:

- It is characterized by the very steep gradient.
- A large number of sharp curves occurs on the road with hairpin bends.
- Extensive earthwork is required.

- The requirement for the construction of special structures.
- The necessity of long length away from the air route.

6.2.3. Gradient

A significant amount of earthwork is necessary for hill roads. To minimize earthwork and lower construction costs, gradients that are nearly at maximum are chosen. Steep gradients reduce the amount of earthwork and road length needed, but they also increase fuel consumption and decrease operating speed because the vehicles must shift into lower gears, which requires more energy. Thus, both of these considerations need to be made, and an appropriate solution needs to be selected. In hilly terrain, the total elevation change should not be greater than 100 meters, and in steep terrain, it shouldn't be more than 120 meters. The square parabola is the design for vertical curves.

The curves should be provided at all grade changes exceeding those indicated in the table below: (see Table 6.1).

Table 6.1. *Design for Grade Change*

Design speed	Maximum grade change not requiring a vertical curve	Minimum length of vertical curve
Up to 35kmph	1.5%	15m
40kmph	1.2%	20m
50kmph	1.0%	30m

Source: https://abhashacharya.com.np/wp-content/uploads/2017/04/Hill-Roads.pdf.

6.2.4. Design of Hair Pin Bends

Steep gradients arise from the route's length being significantly shortened when a curve is inserted inside a turning angle. It is preferable in these situations to circumscribe the curve instead of inscribing the curve around the turning point in order to round off the route. We refer to these compound curves as reverse loops or hairpin bends. The portion of the hill with the least cross slope and highest stability has a hairpin bend. It needs to be secure from groundwater seepage and landslides (Figure 6.3).

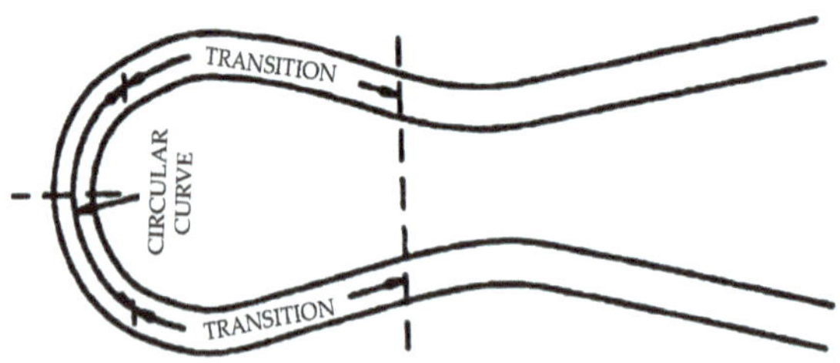

Figure 6.3. *Hairpin bends, where unavoidable, may be designed either as a circular curve with transition at each end or as a compound circular curve.*

Source: https://abhashacharya.com.np/wp-content/uploads/2017/04/Hill-Roads.pdf.

The diagram illustrates two types of symmetrical hairpin bends featuring a main curve ‘C,’ reverse curve ‘Cr,’ and tangents ‘m.’ The acute angle of the bend is denoted as ⊠, with the main curve having a radius of R and a total length of C, subtending an angle ⊠ at the center. Points A and B mark the apices of the reverse curves. Tangents are necessary between the ends of the reverse curves and the main curve for smooth transitions of super-elevation and extra width. When designing hairpin bends, factors such as the radii of the main and reverse curves (R and r) and the length of tangents (m) are taken into account. The design process involves determining the turning angles β at points A and B that meet the predetermined parameters of the bend (Figure 6.4).

According to NRS-2070, the design criteria are as follows:

Minimum design speed	20 kmph
Minimum radius of curve	15 m
Minimum length of transition curve	15 m
Minimum longitudinal gradient	4%
Maximum superelevation	10%

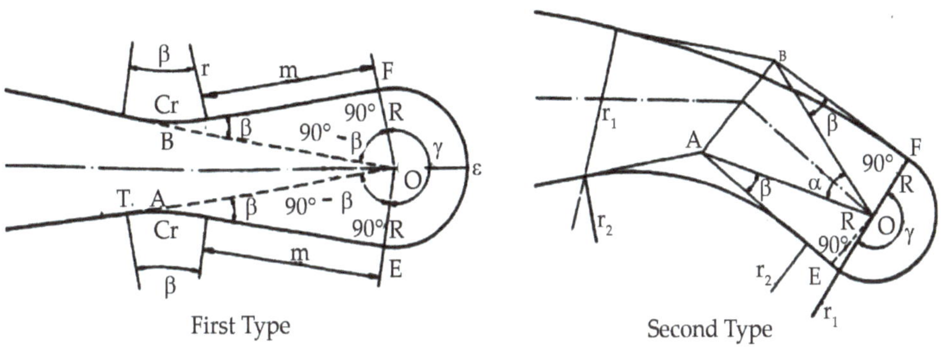

Figure 6.4. *Two different kinds of symmetrical hairpin bends.*

Source: https://abhashacharya.com.np/wp-content/uploads/2017/04/Hill-Roads.pdf.

6.2.4.1. Tangent Length of Reverse Curve

$T = r \tan \beta/2$

where,

T – Length of the tangent

r – Radius of the Reverse Curve

β – Deflection Angle

The separation between the starting point of the main curve and the apex of the reverse curve angle.

Plotting the results of these parameter calculations for hairpin bend can now be done on a contour map or paper. Hairpin bends of the first type are the bends that have been described thus far and have a reverse curve located on their convexities in the opposite direction. The convexities of the curve face on the same side in the bends of the second type can also be symmetrical or asymmetrical. These bends are introduced where the contours indicate a flat hill nose or shallow drainage basin.

6.2.5. Typical Cross Sections of Hill Road

In hilly terrain, the cross-section of a road is influenced by various factors such as the site's initial ground slope, the road formation's slope, the roadway's width, the size, and shape of the side drain, and more. Various types of road cross-sections are:

- Cut and fill
- Bench type
- Box cutting
- Embankment with retaining walls
- Semi bridge
- Semi tunnel
- Platforms

6.2.5.1. Cut and Fill

Cut-and-fill roads are less expensive and more ecologically friendly when the slope is two percent or higher. Generally, the cut mass balances the fill mass. Benches are constructed on the hillside's surface for sufficient stability. Their lengths range from 1.5 to 3.0 meters, depending on the slope, and their height is 0.5 meters (Figure 6.5).

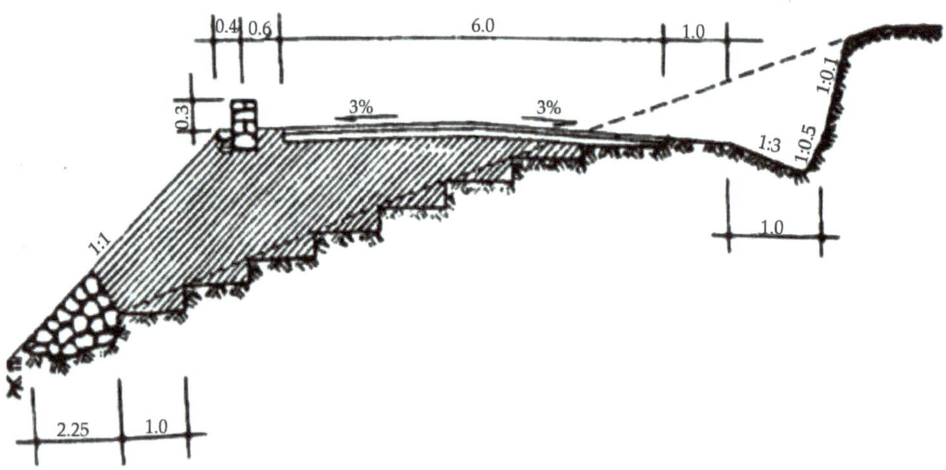

Figure 6.5. *Cut and fill the roadbed.*

Source: https://abhashacharya.com.np/wp-content/uploads/2017/04/Hill-Roads.pdf.

6.2.5.2. Bench Type

If the hillside is stable enough, a bench-type cross-section guarantees the total stability of the roadbed even though it requires a small amount more earthwork. The bench-type method is a widely used technique in hill road construction, particularly on steep terrains. This approach involves creating a series of horizontal steps, or "benches," cut into the hillside to provide a stable foundation for the road. Each bench serves as a platform for the roadbed, reducing the gradient of the slope and helping to manage issues related to stability, erosion, and drainage.

One of the main advantages of the bench-type method is its ability to enhance slope stability. By cutting the hillside into a series of flat or gently sloping steps, the natural tendency for the slope to slide is significantly reduced. Each bench acts as a stabilizing platform that supports the road above and minimizes the downward movement of soil and rock. This is especially important in areas prone to landslides or where the soil is loose and unconsolidated (Hill, J.R., & Liu, J.Y. 2008).

Drainage is another critical benefit of the bench-type method. Each bench can be equipped with drainage channels and culverts to direct water away from the road surface and into designated drainage paths. This helps to prevent water accumulation on the road, which can lead to erosion, potholes, and other forms of damage. Effective drainage also reduces the risk of water infiltrating the soil beneath the road, which can compromise the integrity of the slope and the road itself.

The bench-type method also allows for more controlled and manageable construction. By working on one bench at a time, construction crews can more easily manage excavation, material placement, and compaction. This sequential approach can improve the quality and safety of the construction process, as it allows for continuous monitoring and adjustment. However, the bench-type method requires meticulous planning and precise execution. The design of each bench must consider the specific geological and hydrological conditions of the site to ensure stability and effectiveness. The width and height of each bench are determined based on the slope's steepness, soil type, and expected load from the road traffic. Proper engineering analysis is essential to determine the optimal dimensions and placement of each bench (Figure 6.6).

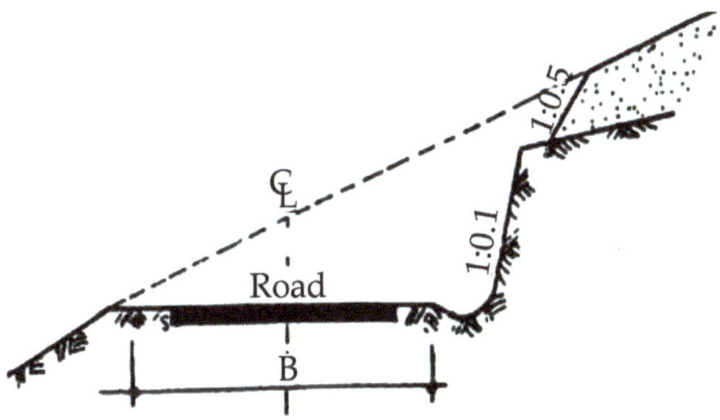

Figure 6.6. *Bench type.*

Source: https://abhashacharya.com.np/wp-content/uploads/2017/04/Hill-Roads.pdf.

6.2.5.3. Box Cutting

The road bed is designed with a trench-style cross-section when its placement along a hillside is unstable or inappropriate for various reasons. It greatly increases the amount of earthwork. It is presented to satisfy the geometric design requirements for a certain class of road (Figure 6.7).

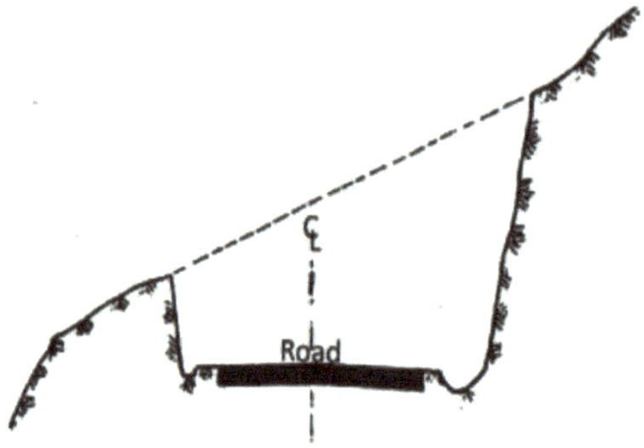

Figure 6.7. *Box cutting roadbed.*

Source: https://abhashacharya.com.np/wp-content/uploads/2017/04/Hill-Roads.pdf.

6.2.5.4. Embankment with Retaining Walls

The amount of earthwork required to build the embankment increases significantly on steep slopes between 30 and 35 degrees. Sometimes the retaining wall is included to improve stability and lower the cost of the earthwork. Furthermore, when embankment soil on steep terrain needs support, retaining walls are installed. To improve the stability of the roadbed, they could also be built on a less steep incline (Figure 6.8).

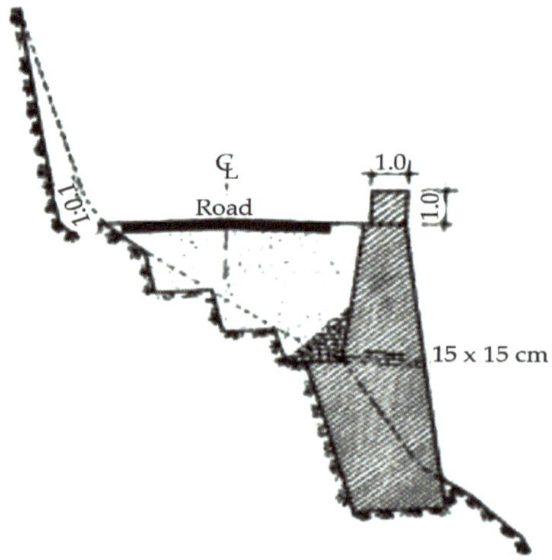

Figure 6.8. *Embankment with retaining walls.*

Source: https://abhashacharya.com.np/wp-content/uploads/2017/04/Hill-Roads.pdf.

6.2.5.5. Semi Bridge

A substantial retaining wall is required if the road is situated on a hilly slope. In these situations, a road bed with a semi-bridge-style structure may be built in order to minimize the amount of labor required.

The semi-bridge method is an innovative approach used in hill road construction to overcome the challenges posed by steep and unstable terrains. This technique involves constructing sections of the road on partially elevated structures or bridges, while the remaining sections are built on the natural terrain. The semi-bridge method combines the benefits of traditional road construction and full bridge construction, providing stability and reducing environmental impact.

The primary advantage of the semi-bridge method is its ability to navigate difficult terrain without extensive excavation or alteration of the landscape. By elevating portions of the road, engineers can avoid areas that are prone to landslides, erosion, or have poor soil stability. This reduces the risk of road failure and ensures a safer and more reliable transportation route.

Additionally, the semi-bridge method can significantly improve drainage. Elevated sections allow water to flow naturally beneath the road, preventing water accumulation and reducing the risk of erosion and road damage. Proper drainage is crucial in hilly areas, where heavy rainfall can lead to significant water runoff and associated problems.

The construction of semi-bridges involves using various materials and techniques to ensure stability and durability. Common materials include reinforced concrete, steel, and pre-stressed concrete. The design of the semi-bridge must account for load-bearing capacity, seismic activity, and other environmental factors. Piers and abutments are strategically placed to support the elevated sections, ensuring they can withstand the weight and forces exerted by traffic and natural elements (Figure 6.9).

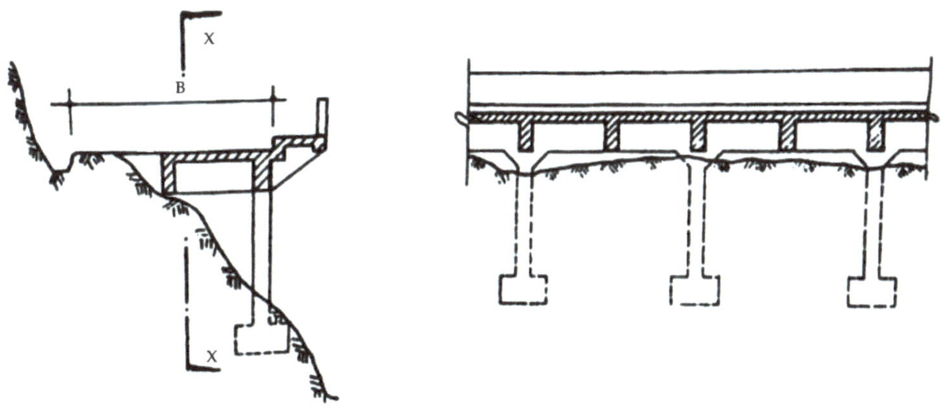

Figure 6.9. *Cross-section and L-section (X-X) of a Semi-bridge.*

Source: https://abhashacharya.com.np/wp-content/uploads/2017/04/Hill-Roads.pdf.

6.2.5.6. Semi Tunnel

To minimize rock works, the rock may be allowed to overhang the road when inscribing is to be cut into steep hills in stable rock faces. A semi-tunnel is a cross-section of that type.

The semi-tunnel method is a specialized approach in hill road construction that combines elements of both open-cut and full tunnel construction. This technique involves excavating a section of the hillside to create a partially enclosed roadway, providing the benefits of tunnel construction while minimizing some of the complexities and costs associated with full tunnels. The semi-tunnel method is particularly useful in areas with steep and unstable slopes, where traditional road construction methods would be impractical or unsafe.

One of the main advantages of the semi-tunnel method is enhanced slope stability. By partially enclosing the roadway, this method protects it from rockfalls, landslides, and other forms of slope failure that are common in hilly terrains. The semi-tunnel structure supports the surrounding hillside, reducing the likelihood of erosion and collapse. This approach also shields the road from extreme weather conditions, such as heavy rainfall and snow, which can cause significant damage to open-cut roads.

Drainage is another critical benefit of the semi-tunnel method. The design typically includes built-in drainage systems to manage water runoff effectively. These systems channel water away from the roadway and prevent water accumulation that could undermine the tunnel structure and lead to erosion or flooding. Proper drainage is essential in maintaining the integrity of the road and ensuring safe driving conditions.

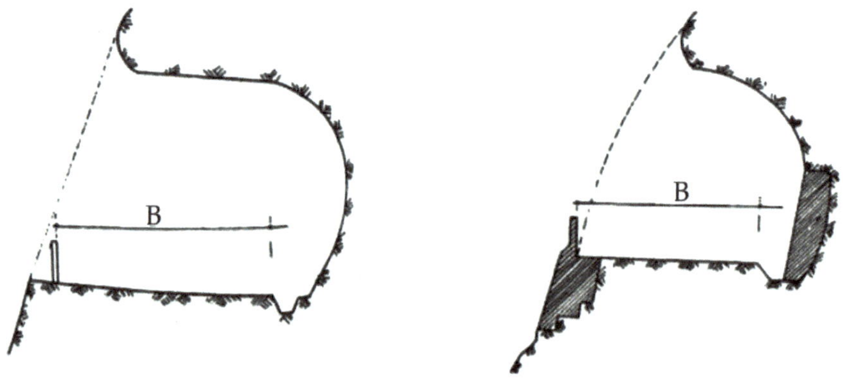

Figure 6.10. *With Accommodating Road-Way Only and With Retaining and Breast Walls.*

Source: https://abhashacharya.com.np/wp-content/uploads/2017/04/Hill-Roads.pdf.

Construction of semi-tunnels involves the careful excavation of the hillside and the installation of supporting structures such as retaining walls, arches, and reinforced concrete linings. The method requires precise engineering to ensure that the excavated section is stable and that the tunnel structure can bear the weight of the overlying rock and soil. Modern construction techniques, such as rock bolting and shotcrete (sprayed concrete), are often used to reinforce the tunnel walls and ceiling (Figure 6.10).

6.2.5.7. Platform

Platforms are typically cantilevered out of the rock on which the road is partially located on steep slopes, where moving the route into the hillside will require massive rock works, ultimately increasing the cost. Semi-tunnels cannot be built on these slopes (Figure 6.11).

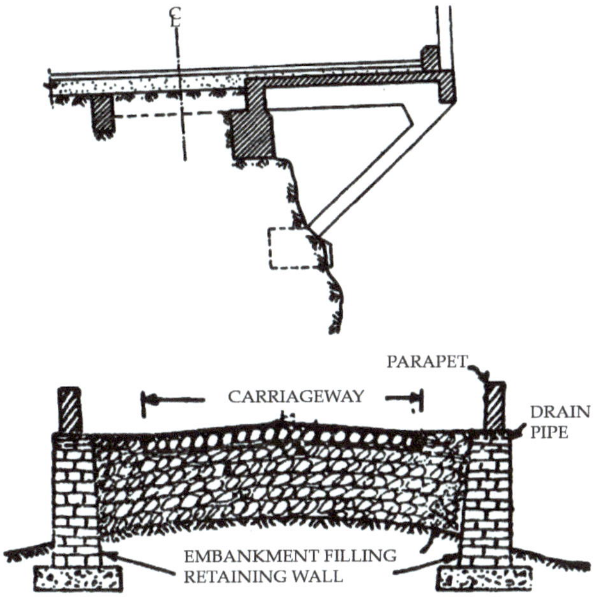

Figure 6.11. *Fully in Embankment.*

Source: https://abhashacharya.com.np/wp-content/uploads/2017/04/Hill-Roads.pdf.

6.3. SPECIAL STRUCTURES IN HILL ROADS

To handle the particular difficulties presented by mountainous terrains, such as steep slopes, unstable soil, and severe weather, special structures in hill roads are imperative. These constructions improve the durability, stability, and safety of hill roads, providing dependable mobility in these difficult conditions. A lot of specialized structures are needed when building hill roads because of the difficult hydrological and geological conditions as well as the severely fractured relief.

The following types of structures are mostly used on hill roads for strength durability and stability:

- Retaining structures
- Drainage structures
- Slope protection structures

6.3.1. Retaining Structures

A wall built to support an earth bank that is vertical or nearly vertical is called a retaining structure. To stop the hillside from sliding toward the road, retaining walls are built on the valley side. Situations where construction of retaining walls is required:

- Places where the valley side surface gets saturated in the monsoons and is likely to result in slip taking a part of the road with it.
- Places where undercutting by a stream or other watercourse causes damage to the valley side and the road. In the valley point where water flows over the road
- To achieve roadway width, where cutting into the hill is not economical or has to be restricted due to other reasons.

6.3.1.1. Types of Retaining Walls

A retaining wall is a structure made to withstand soil pressure from the side or to retain soil particles. The earth filling, liquid pressure, sand, and other granular materials behind the retaining wall structure could also be the cause of the lateral pressure. Retaining walls come in a variety of forms and are employed for many purposes.

1. Gravity Retaining Wall (see Figure 6.12).

- Gravity retaining wall depends on its self-weight only to resist lateral earth pressure.
- Commonly, a gravity retaining wall is massive because it requires a significant gravity load to counteract soil pressure.

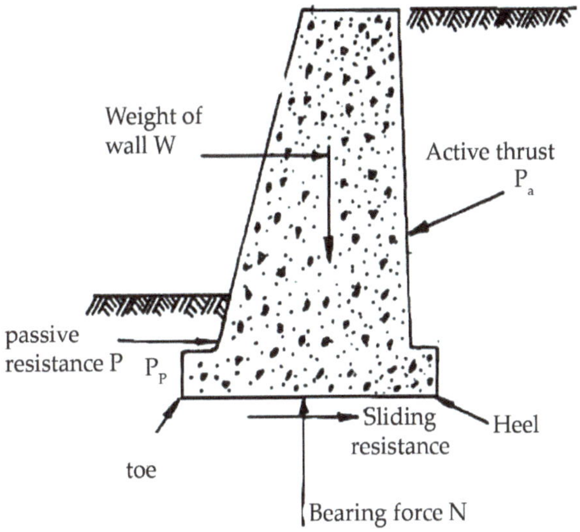

Figure 6.12. *Gravity retaining wall.*

Source: https://theconstructor.org/geotechnical/retaining-wall-types-use/24566/.

- Sliding, overturning, and bearing forces shall be taken into consideration while this type of retaining wall structure is designed.
- It can be constructed from different materials such as concrete, stone, and masonry units.
- It is economical for a height up to 3m.
- Crib retaining walls, gabions, and bin retaining walls are also types of gravity retaining walls

2. Crib Retaining Wall

- Crib retaining walls are a form of gravity wall.
- They are constructed of interlocking individual boxes made from timber or pre-cast concrete.
- Then, the boxes are filled with crushed stone or other coarse granular materials to create a free-draining structure.
- Basic types of crib retaining walls include reinforced precast and timber retaining walls.
- It is suited to support planter areas, but it is not recommended for support of slopes or structures.

3. Gabion Retaining Walls

- Gabion retaining wall walls are multi-celled, rectangular wire mesh boxes, which are filled with rocks or other suitable materials.
- It is employed for the construction of erosion control structures.
- It is also used to stabilize steep slopes.

4. Cantilever Retaining Wall

- Cantilever retaining wall composed of stem and base slab
- It is constructed from reinforced concrete, precast concrete, or prestressed concrete.
- Cantilever retaining wall is the most common type used as retaining walls.
- Cantilever retaining wall is either constructed on-site or prefabricated offsite i.e., precast.
- The portion of the base slab beneath the backfill material is termed as heel, and the other part is called the toe.
- Cantilever retaining wall is economical up to a height of 10m.
- It requires a smaller quantity of concrete compared with a gravity wall but its design and construction shall be executed carefully.
- Similar to gravity wall, sliding, overturning, and bearing pressure shall be taken into consideration during its design.

5. Counter-fort / Buttressed Retaining Wall

- It is a cantilever retaining wall but strengthened with counter forts monolithic with the back of the wall slab and base slab.
- Counter-fort spacing is equal to or slightly larger than half of the counter-fort height.
- Counter-fort wall height ranges from 8-12m.

6. Anchored Retaining Wall

- This type of retaining wall is employed when the space is limited or a thin retaining wall is required.
- Anchored retaining wall is suitable for loose soil over rocks.
- Considerably high retaining wall can be constructed using this type of retaining wall structure system.
- Deep cable rods or wires are driven deep sideways into the earth, then the ends are filled with concrete to provide anchor.
- Anchors (tiebacks) acts against overturning and sliding pressure.

7. Piled Retaining Wall

- Pile retaining wall are constructed by driving reinforced concrete piles adjacent to each other.
- Piles are forced into a depth that is sufficient to counter the force that tries to push over the wall.
- It is employed in both temporary and permanent work.
- Piled walls offer high-stiffness retaining elements that are able to hold lateral pressure in large excavation depths with almost no disturbance to surrounding structures or properties.
- Sheet pile walls are built using steel sheets into a slope or excavations

up to a required depth, but it cannot withstand very high pressure

- Sheet pile retaining wall economical till height of 6m

8. Mechanically Stabilized Earth (MSE) Retaining Wall

- It is among the most economical and most commonly constructed retaining walls.
- Mechanically stabilized earth retaining wall is supported by selected fills (granular) and held together by reinforcements, which can be either metallic strips or plastic meshes
- Types of MSE retaining walls include panel, concrete block, and temporary earth retaining walls.

6.3.1.2. Hybrid Systems

Retaining wall systems referred to as hybrid or composite use both mass and reinforcement for stability.

6.3.2. Drainage Structures

The detrimental effects of water are the primary issues that hill roads face. Water may reach the various sections of the road from various sources. Any methods required must be used to drain this water. The following subtopics can be used to study hill road drainage:

6.3.2.1. Drainage of Water from Hill Slope

Drainage of hill roads is a major issue when surface water runs off the hill and into the road. During strong storms, a catch drain is typically installed to collect the water in the middle of the slope because a lot of water and debris flow down the hill slopes. Water that has been caught in catchment drains is then directed by sloping drains to a culvert that crosses the road or to the closest watercourse. An arrangement for drainage from hill slopes is depicted in Figure 6.13.

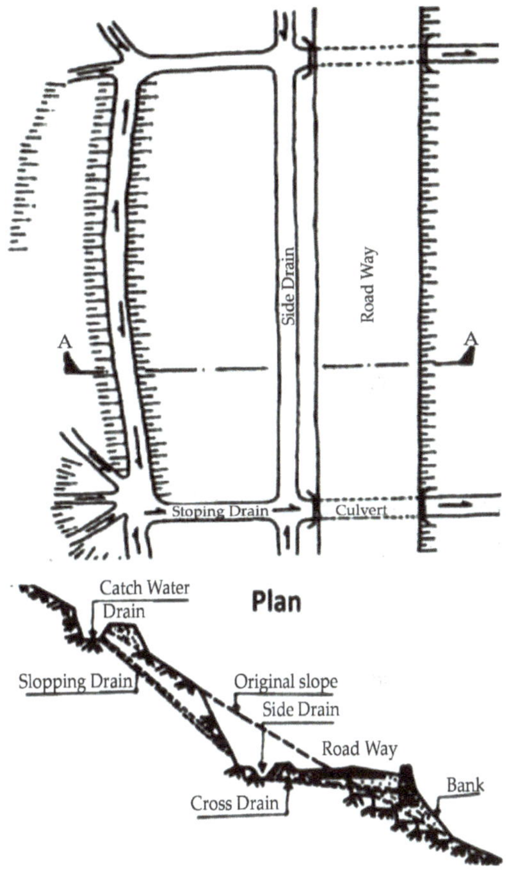

Figure 6.13. *Drainage of water from hill slope.*

Source: https://abhashacharya.com.np/wp-content/uploads/2017/04/Hill-Roads.pdf.

6.3.2.2. Roadside Surface Drainage

The road's hillside is completely lined with side drains. Owing to the formation width restriction, side drains are often designed so that automobiles can use this area to cross in an emergency. The forms could be kerb and channel drains, saucer-shaped, or angular.

6.3.2.3. Cross Drainage

Any hill road must have a cross drainage. The drainage system needs to be buried as deep as feasible beneath the road. To prevent scouring and collect debris, catch pits must be installed at the heads of the small cross drains.

6.3.2.4. Subsurface Drainage

Seepage flow is a significant issue on hill roads. Depending on a number of variables, including the type and depth of the hard stratum, its inclination, the amount of groundwater, etc., groundwater may seep across a hillside above or below the level of the road. There are several ways to control subsurface drainage.

6.3.3. Slope Protection Structures

The steep slopes along hill roads often result in frequent landslides. The primary cause of landslides is the accumulation of shear stresses greater than the soil's shear strength. The least stable portion of the roadbed is the freshly unturfed embankment and cut slopes because the soil on these surfaces is directly exposed to the elements—sun, rain, and wind.

6.3.3.1. Causes of Landslides

- Increase in moisture content of the soil in hill slopes which increases the pore water pressure.
- Alternate swelling and contracting of the soil mass.
- Seepage pressure of percolating groundwater.
- Steeper slopes.
- Human activities like blasting and using heavy vehicles in unstable zones.

6.3.3.2. Preventive Measures

- The highway may be realigned in areas more prone to landslides.
- Construction of retaining walls must be done at places where required
- Adopting easy slopes during the design and construction of the road.
- Treatment of slopes to increase stability conditions.

6.3.3.3. Reinforced Retaining Walls

This particular kind of composite retaining wall reinforcement adds inextensible tensile reinforcement in the form of strips, sheets, grids, or geotextiles to increase fill strength.

It is suitable for hill roads because:

- The fill material is readily available at a cheaper cost.
- The land required for embankment is less.
- Cost-effective, easy to construct, and environmentally friendly.
- It causes less alteration in the natural slope.

6.3.4. River Training Structures

The structural actions made to enhance a river and its banks are referred to as river training. River training is crucial to general flood control, flash flood mitigation and prevention, and other initiatives like guaranteeing a flood's safe passage beneath a bridge. Hill roads near the river could also be dangerous because of various issues it causes.

Problems created by the river:

- Frequent changes in river course.
- Avulsion of one river into another.
- Development of natural cut-off.
- Landslides in catchment -rise in silt load.
- Aggradation of river bed -high flood levels –Flooding
- River instability -change in bed slopes (seismic activity).
- Degradation of river bed downstream of a dam or a barrage.
- Effects of flood embankment on the regime of rivers.
- Effects of extraction of sand and boulders.
- Effects of heavy urbanization along the river banks.

6.3.4.1. River Training Works

The process of building various structures alongside a river to direct and control the flow of the waterway is referred to as "river training works." Helping to regulate and control the river beds is one of the other goals of river training projects.

6.3.4.2. Functions of River Training Works

It includes Guide banks, Marginal bunds, Spur in river training works, etc. The main functions of River Training Works are:

- Providing a non-tortuous approach for the weir.
- Preventing the outflanking of the weir by the river.
- Preventing submergence of additional area due to afflux.
- Deter erosion of the river banks.
- Ensuring a smooth and axial flow of the river water to prevent the river from changing its course.

6.3.4.3. Types of River Training Works

6.3.4.3.1. High water training

- It is used for training the river water for discharge.
- The main goal of high training works is to dispose of the flood water accumulated due to flooding and thus provide protection against a catastrophe.
- It mainly relates to the elevation of the marginal embankments for suitable alignment and flood disposal.
- It is also used for taking appropriate measures for the improvement of the channel improvement.

6.3.4.3.2. Low water training

- It is used for training the river water for depth.
- The main objective of low river training work is to provide a sufficient water depth suitable for inland navigation, especially during dry weather seasons in summer.
- It is achieved through the contraction of the channel width in low water by the use of groynes.

6.3.4.3.3. Mean water training:

- It is also done for training of sediments.
- Mean water training is done for ensuring a smooth disposal of the bed load for maintaining the suspended sediment load as well as the channel size.
- Mean water training is the most important type of River water training.

- In such a river training work, the water should be maintained to alter the cross-section and alignment such that it is appropriate for the young river phase.
- The maturation stage and the old age of the river at which maximum sediment movement occurs are taken into consideration.
- Although the maximum sediment speed is attained in the young stage, this stage lasts for only a short span of time.

6.3.4.4. Components of River Training Works

Some main components of river training works are discussed below.

6.3.4.4.1. Marginal Embankment (dyke or levee)

To contain floodwater between the embankments, the marginal embankment, also known as the dyke, is a trapezoidal earthen embankment built parallel to the river bank.

6.3.4.4.2. Guide Bank

- To prevent erosion of the structure, guide banks are built on both sides of the river. An earthen embankment with curved heads at either end is called a guide bank. These are built both upstream and downstream of the abutments of the hydraulic structure, parallel to the direction of the river's flow.
- They do not harm the approaches as they direct the river's flow through a hydraulic structure.
- It was first designed by Bell and is also called Bell's Bund.
- The Guide bank serves the following purposes.
- It protects the barrage against the effect of scouring and erosion.
- It offers a straight approach to the barrage.
- Helps in preventing the river from changing the course.

Guide banks generally serve two purposes: (see Figure 6.14).

- They aid in preventing the high-water discharge from damaging the bridge's approach embankment. The approach embankments stretch in both directions, perpendicularly connecting the river bank to the guide banks.
- Axial river flow control is made easier with the help of guide banks.

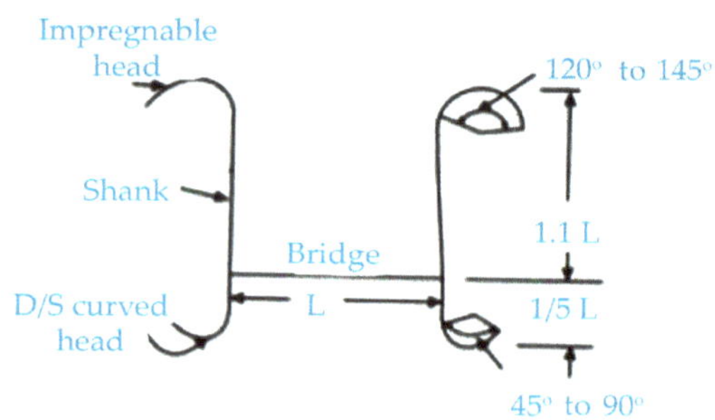

Figure 6.14. *Dimensions of guide bank.*

Source: https://testbook.com/civil-engineering/river-training-works-classification-and-types.

6.3.4.4.3. Spurs

- To prevent the riverbank from eroding, spurs are temporary permeable structures installed at the river's bend.
- Spurs often protrude from riverbanks at a 60- to 75-degree angle to the bank. Their primary purpose is to aid in reducing the flow's velocity.

On its upstream side, where the sediments accumulate, it helps form water pockets.

6.3.4.4.4. Groynes

- The objective of groynes in river training works is tantamount to the spur.
- However, Groynes are impervious permanent structures that are constructed on the curves.
- They are constructed to protect the river bank from erosion.
- Groynes generally make an angle of 60 to 75 with the bank.

6.3.4.4.5. Types of Groynes

1. *Repelling Groynes (see* Figure 6.15).

- Groynes in river training works can be aligned either perpendicular to the bank or in the inclined direction that may be pointing either upstream or downstream.
- When the groyne is pointing upstream, it is called a repelling groyne.
- This groyne has the property of repelling the river flow from the bank.

Figure 6.15. *Repelling groynes.*

Source: https://testbook.com/civil-engineering/river-training-works-classification-and-types.

- This is made possible by the creation of a still pond at the upstream end.
- The river also follows beyond the still pond and the river starts flowing away from the bank.

2. *Attracting Groynes (see* Figure 6.16).

- A Groyne that is pointing in the downstream direction of the river flow it is called an Attracting groyne.
- This Groyne has a tendency of attracting the river flow towards the bank.

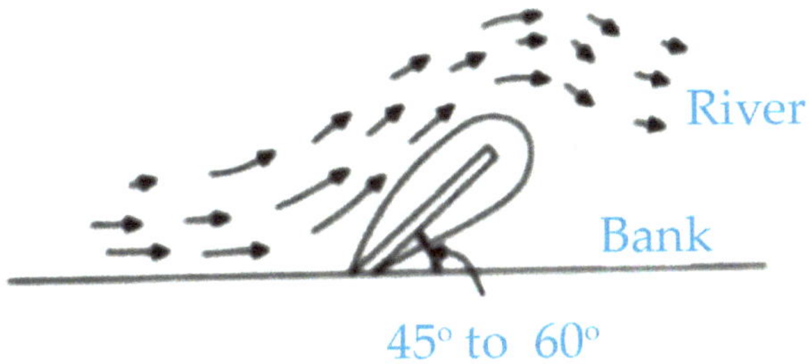

Figure 6.16. *Attracting groynes.*

Source: https://testbook.com/civil-engineering/river-training-works-classification-and-types.

3. *Deflecting Groynes (see Figure 6.17).*

- A Groyne of shorter length that is perpendicular to the bank is called a Deflecting Groyne.
- A Deflecting Groyne deflects the flow locally.

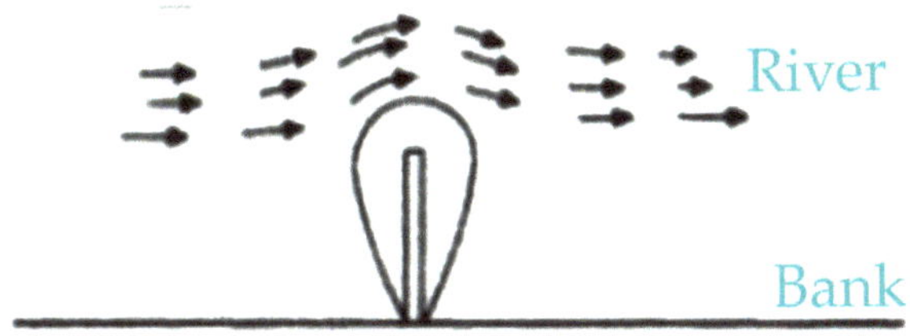

Figure 6.17. *Deflecting groynes.*

Source: https://testbook.com/civil-engineering/river-training-works-classification-and-types.

6.3.4.4.6. Functions of Gryones

Groynes in river training works serve the following purposes:

- They help in the protection of the river bank by restricting the flow of water near the river bank.
- They aid in the creation of a still pond along the bank for silting the area in the immediate vicinity.
- Groynes also train the river to make it flow along a desired course by attracting, deflecting, or repelling the flow.
- They aid in the contraction of the wide river channel in order to improve the depth of navigation depth.

6.3.4.5. Advantages of River Training Works

Some important advantages of river training works are as follows.

- They prevent the changing of the river course.
- They prevent floods.
- River training works to aid in the protection of the river banks.
- They help in sediment disposal and thereby help in maintaining a minimum water depth in the water flow.

6.3.4.6. Disadvantages of River Training Works

The disadvantages are discussed below.

- These may result in the rise of the High flood levels.
- River training works are prone to piping failure.
- These structures can easily be damaged by the river flow

CASE STUDY

ECONOMICAL DESIGN OF HILL ROAD

It is the object of the engineer to establish the shortest, most economical, and safest route between the obligatory points. This should be done keeping the expenditure of transportation, the wear and tear of the vehicle and the annual maintenance cost to the minimum. Selection of a suitable alignment in a hilly region is a complex job when the valley pattern takes a radial form and where the valley converges into a knot of mountains due to the meeting of several mountains. A thorough knowledge of the geological formations of the area is essential to decide the road alignment, as the process of road construction in the mountains disturbs the natural stability conditions. Prevention of soil erosion and stabilization of hill slopes have been major problems in the maintenance of hill roads. The climatic conditions also influence the progress of hill road construction. In summer, very heavy storm occurs in the mountains and about 15 to 20% of the annual rainfall may occur in one storm. The effects of such heavy rainfall on the construction and maintenance of hill roads are serious. At many locations in the hill roads, the landslides and slips occur blocking traffic during the heavy rainfall. A large number of streams cross the road and hence suitable facility for cross drainage is needed.

Proposal

Phase-I: The proposal for stage 1st of this project is: -

- Earthwork in cutting in hilly areas, providing a formation width of 5.50m and carriageway of 3.0m
- 1 x 10m span and 2 x6m span RCC culverts have been provided in addition to 25 No. RCC pipe.
- Crossings.
- Sufficient provision for protection walling has been kept for slope stabilities.
- Provision for pucca drains in built-up areas and Kucha drains in open areas have been kept for surface drainage.

Phase-II: It has been proposed to construct the road by providing GSB course of approved quality material which is to be laid in double layers (300mm) and compacted with proper specification as per IRC. The side slopes shall be properly dressed WBM G-III shall be laid to proper grade camber and compacted with proper specification as per IRC. Necessary screens/binding materials shall be laid /compacted as per IRC SP-20. The Bituminous coat of 50mm thick and OGPC 20mm thick with a seal coat of 6mm thick shall be laid after cleaning and brooming of the surface and the application of prime and tack coats of hot bitumen emulsions shall be laid before laying of macadam/

OGPC. Also, it is proposed to construct breast walls on a hillside and a PCC drain at identified spots.

Phase I

Road Alignment

Using professional surveyors' staff and professional survey equipment, including GPS, Total Station, and automatic/digital levels, the contractor shall conduct a full topographical survey, develop primary survey controls, and establish a control points' network along the project site based on National Datum benchmark to be used in the horizontal and vertical design of the road. The contractor shall develop and set out the Preliminary alignment and verify whether it suits actual field conditions. The alignment shall be modified by the contractor where necessary to give the best possible grade to suit the existing terrain conditions, and minimize the amount of cut and fill.

Topographical Survey

Establishing Survey Control Points Prior to carrying out the topographical and level surveys, permanent control points to be established for both horizontal and vertical control along the full length of all roads. The control points shall be placed at maximum intervals of 500 m and located so that a minimum of two adjacent points are inter-visible. A traverse survey shall be carried out to determine X and Y coordinates. The elevations (Z value) of the control points shall also be determined with reference to the existing National Datum. The permanent control points shall be concreted and the center shall be a stainless steel pin with 12 mm diameter. Each control point shall be marked with its reference number, date of construction, elevation, and coordinates. These points shall be established at ground level in suitable ground conditions in location where they shall not be disturbed by construction works and shall be maintained throughout the contract period.

The following requirements shall be fulfilled during the traverse survey;

- Horizontal angle and distance measurement.
- Horizontal angle shall be observed by 2 rounds of angle on 2 different zero settings,

Leveling

- Levelling shall be started from an existing National Datum benchmark.
- Requirements for the leveling survey shall be as follows
- An automatic level or digital level shall be used;
- Miss-closure of leveling between existing National Datum benchmarks and control points shall not exceed +30 mm D, where D is the distance in kilometers.

A topographical survey shall be carried out for the full length of the road. The extension will cover the existing right of way if any, or a minimum 60m wide strip across the road corridor (extending 30m from the centerline on both sides). All topographical details like existing roads, tracks, drainage structures, buildings, services/utilities (electric, telephone, and water lines), existing road furniture, right-of-way markers, and kilometer posts shall be surveyed. At the bridge location wider area shall be surveyed and the position of all features including the existing bridge, riverbank, and waterline will be taken to accommodate the possible location of the bridge and potentially high approach roads (Figure 1).

Figure 1. *Topographical survey.*

Source: https://www.jetir.org/papers/JETIR1903I27.pdf.

Cross-Section Surveys

Cross-sections shall be taken at 20m intervals and extend a minimum 60m wide strip (extending 30m from the center-line on both sides) along straight sections. The Centerline of the existing road, carriageway limit, shoulders, and embankment toe shall be surveyed carefully. At the curve section of the alignment and where sudden elevation changes are encountered, cross-section will be taken at closer intervals if required to represent the true surface profile:

- Traffic count data
- Climatic conditions
- Existing soil conditions
- Hydrological/geotechnical conditions

All pavements shall be designed in accordance with the AASHTO pavement design standards. The minimum pavement section shall be designed based on a natural subgrade CBR as evaluated from the geotechnical site investigation. The material used for constructing all pavements shall be in accordance with the approved technical specifications and as Directed by the PMGSY. The design of the pavements and shoulders shall be developed consistent with the site-specific geotechnical investigation and submitted for approval to the PMGSY.

Drainage

The Constructor is required to design each drainage structure to be constructed to suit the site-specific requirements. Roadways are to be viewed primarily for use by vehicular and pedestrian traffic and for providing access to property by vehicle and foot. They are not to be viewed as primary for drains and floodways. Public amenity and safety are to be paramount considerations;

- Consideration is to be given to the impact of the proposed drainage system on existing drains and buildings and downstream catchments.
- The drainage system is to be designed to ensure adequate removal of rain, flood, and irrigation water from the roadway and adjacent road structures.
- Diversion of upstream catchments to independent catchments by the construction of a new road will not be permitted. The designer shall ensure that provision of floodways and culverts is made to ensure catchments are not altered and that no downstream catchment is affected detrimentally.
- Environmental considerations are a major consideration. Erosion and sediment control are required.
- No encumbrance of any land designed, or intended to be utilized as a floodway will be permitted.
- During the wet season, the ground becomes saturated and all drainage design should be carried out with the view that there will be 100% runoff from all sites.
- To avoid mosquito breeding, all drainage systems and associated structures should be designed to have no ponding of water.
- Where drainage outlets or outfalls are influenced by tidal action a discharge vs. probability analysis is to be undertaken to ensure that the appropriate annual occurrence probability is being catered for.

All drainage pipes and culverts including headwalls and aprons are to be constructed in accordance with the Standard Drawings and Technical Specification.

Phase –II

Geotechnical Investigation and Report

Site Specific Information Gathering site specific geotechnical information necessary to design and construct the pavement and drainage structures on this project shall be the

Contractor's responsibility. The Contractor shall determine all necessary geotechnical conditions by appropriate field and laboratory investigations and supporting calculations.

CBR Test

Weight of mold = 7.383 kg

Weight of mold + soil sample = 11.842 kg

Volume of mold = 2110 cu-m

Bulk unit weight of soil = 20.699

Dry unit weight of soil = 19.2276 kg/cc

w = 8%

Penetration (mm)	Proving ring Dial reading	Piston load (Kg)
0	0	0
0.5	1.2	11.04
1	2.4	22.08
1.5	3.5	32.41
2	7.5	69.45
2.5	10.4	96.304
4	12.5	115.75
5	13.8	127.788
7.5	19	175.94
10	20.8	192.608
12.5	24.7	228.722

$$\text{CBR at 2.5mm} = \frac{96.304}{1344} \text{x } 100 = 7.165$$

$$\text{CBR at 5mm} = \frac{127.788}{2016} \text{x} 100 = 6.338\%$$

for CBR 7.165% and for light traffic required pavement thickness from graph = 300mm Minimum thickness as per PMGSY Guidelines for hill roads keeping frost action in consideration.

CLASS ACTIVITY

Prepare an estimate for the portion of a road from chainage 14 to 22 from the data given below. Draw typical cross-sections for cutting and banking. The formation width of the proposed road is 12 m, with side slopes 3/2:1 m. cutting and 2:1 in banking.

Chainage	R.L of ground
14	108.6
15	109.25
16	109.4
17	108.85
18	108.55
19	107.25
20	106.8
21	107.15
22	107.2

The road formation is proposed as a uniform falling gradient 1 in 200 passing through G.L at chainage 14 length of one chain = 30 m.

SUMMARY

- The hilly regions generally have extremes of climatic conditions, difficult and hazardous terrains, topography, and vast high-altitude areas. The region is sparsely populated and basic infrastructural facilities available in plain terrain are absent.
- Due to the numerous watercourses present in the hilly regions, there may be the necessity of intense cross drainage works. The alignment should be so selected in such a way that the number of cross-drainage structures required becomes minimal.
- When designing hill roads, the route is located along valleys, hillsides, and if required over mountain passes. Due to complex topography, the length of the route is automatically increased. Due to harsh geological conditions, special structures also must be provided.
- The inclination of folds may vary from horizontal to vertical stratification of rock. These folds often have faults. Limestone or sandstone folds may be interleaved

with layers of clay which when wetted may cause fracturing along their surface. This may result in shear or slip fold.

- In hill roads, a heavy amount of earthwork is required. So to reduce the earthwork for reducing construction costs the gradients selected are close to the maximum. Although steep gradients help in reducing earthwork and length of road, it also causes increased fuel consumption and reduction in operating speed as the vehicles will be on low gears which will use more energy.
- A retaining structure is usually a wall constructed for supporting vertical or nearly vertical earth banks. Retaining walls are constructed on the valley side on the cut hillside to prevent the slide towards the roadway.
- A cross drainage is always required on a hill road. The drainage must be taken under the road as far as possible. At the heads of the small cross drains, catch pits must be provided to collect debris and prevent scouring.
- The term landslides or less frequently, landslides refer to several forms of mass wasting that include a wide range of ground movements, such as downfall deep-seated slope failures, mudflows, and debris flow.

REVIEW QUESTIONS

1. Explain the challenges faced in achieving optimal alignment for hill roads.
2. What factors are considered when selecting a route in hilly terrain?
3. How do land acquisition and environmental considerations impact route location decisions in hilly areas?
4. Describe the construction techniques used for hairpin bends in challenging terrain.
5. Compare and contrast the cross-sectional design of hill roads with that of flat terrain roads.
6. What are retaining structures, and why are they used in hill road construction?
7. Discuss the factors influencing the choice between reinforced soil walls and anchored walls in hilly terrain.
8. Why are drainage structures essential in hill road construction?

REFERENCES

1. Bora, B., et al., (2018). Landslide susceptibility assessment along hill roads using analytical hierarchy process and logistic regression model: A case study from Darjeeling Himalayas, India. *Geocarto International, 33*(5), 471–489.
2. Chakraborty, S., et al., (2015). Landslide susceptibility assessment along hill roads: A Case Study from Eastern Himalayas, India. *Natural Hazards, 79*(2), 1161–1185.
3. Hill, J. R., & Liu, J. Y. (2008). Slope stability analysis of hill roads using geographic information systems: A case study. *Journal of Geotechnical and Geoenvironmental Engineering, 134*(9), 1329–1340.

4. IRC: 52–2001. (2001). *Recommendations About the Alignment Survey and Geometric Design of Hill Road* (2nd ed.). IRC.
5. IRC: 73–1980. (1980). *Geometric Design Standards for Rural (Non-Urban) Highways.* IRC.
6. Khanna, S. K., & Justo, C. E.G., (1971). *Highway Engineering.* Nem Chand & Bros.
7. Marsani, A., & Shrestha, D. K. (2071). *Transportation Engineering Volume–I.* Kathmandu: Divine Print Support, Lagan Tole.
8. Parajuli, P. M. (1999). *Course Manual of Transportation Engineering–I.* Lalitpur, Nepal: IOE, Pulchowk Campus.
9. Rai, R., et al., (2016). Landslide susceptibility assessment along hill roads using gis-based analytical hierarchy process in Rudraprayag district, Uttarakhand, India. *Geocarto International, 31*(4), 437–455.
10. Sharma, P., & Sharma, N. (2014). Special considerations in hill road design. *International Journal of Engineering Research & Technology (IJERT), 3*(1), 1047–1051.

CHAPTER 7

Urban and Rural Roadway Design

LEARNING OBJECTIVES

After studying this chapter, you will be able to:

- Understand design parameters
- Describe geometric alignment
- Explain camber, cross-fall, and superelevation
- Discus standard road configuration

KEY TERMS FROM THIS CHAPTER

Geometric design
Speed management
Approach sight distance (ASD)
Road geometry
Transverse gradient
Roadway infrastructure
Minimum gap sight distance (MGSD)
Vertical alignment
Sight distance
Crossing sight distance (CSD)
Bus rapid transit (BRT)
Busway configuration
Horizontal alignment
Overtaking sight distance (OSD)

INTRODUCTION

For proper operation at the appropriate road speed to occur safely and predictably, the networks of rural and urban streets must have geometric designs. Although there will be differences in design constraints between projects, particularly in the rural-urban boundary areas, the fundamental ideas of good road design remain the same.

Users of the streets should be able to assess the conditions of the road, determine a safe speed and course, and then take appropriate action. They should be able to observe other users, anticipate their likely actions, choose their responses, and take appropriate action when interacting with them. Enough time should be allotted in geometric design for safe decisions and actions. The majority of roading standards' design requirements have historically been centered on ensuring safe operation at different design speeds and road safety. The ability of the road to provide sufficient drainage for surface water is impacted by the geometric design and any changes made thereafter. The impact of flooding from any nearby watercourses must be taken into account, and the carriageway's vertical alignment must be fixed at a suitable level.

The surrounding environment is impacted by a road's alignment, both horizontally and vertically. Both the choice of surfacing material and the road's elevation can have an impact on noise and appearance. The alignment affects the quantity of construction vehicles needed to transport or remove materials from the site, which in turn affects the surrounding communities. The amount of money needed to build the new alignment directly relates to how much material needs to be imported, dug up, or relocated. The largest influence on the amount of material required is geometry, and improper alignments, road widths, or elevation can result in significant cost increases.

The safe operating performance of vehicles and the capacity of the roadway are significantly influenced by the design of the roadway and its alignment, including intersection spacing and control methods. Roadways and their auxiliary movement environments must be planned to minimize their negative effects on the surrounding land while still accomplishing the movement goals outlined in the Roads and Streets Framework. Environments with a high place value may need capacity or speed reductions to ensure the safety of people riding or walking. On high-movement corridors, the focus may be on improving the capacity or travel times of different modes safely and efficiently.

When planning the infrastructure of a roadway, it is crucial to consider the safe and economical ways in which road environment maintenance can be accomplished, thereby lowering the need for traffic management and the expenses and inconveniences that go along with it.

In order to ensure that the road or street is safe and functional for all modes while maintaining the network, the design should incorporate maintenance requirements.

7.1. DESIGN PARAMETERS

7.1.1. Design Speed

The design speed of a road is the maximum speed at which a vehicle can safely travel on that road under good conditions.

7.1.1.1. Determining Factors

The design speed is based on the:

- road and street type
- conditions of the road itself
- conditions of the surrounding land
- maximum speed allowed by law
- volume of traffic
- operating speed of the road, i.e., how fast traffic moves
- In greenfield situations, the road types will be determined by the Roads and Streets Framework in conjunction with the structure plan or precinct plan for the relevant land.

7.1.1.2. Design Speed vs. Speed Limit

In an urban setting, the road's design speed must match or be lower than the street's intended speed limit.

The 85th percentile speed should be used for high-speed urban or rural roads with intended speed limits exceeding 60 km/h, with the design speed being 10 km/h higher for the posted speed checks (Dominique, 2002).

On the other hand, the geometry of rural roads needs to be chosen in a way that ensures consistency over specific types and characters of lengths.

For intersection design and alignment, this design speed is employed. It might be necessary to use a faster operating speed for safety-related design checks. his includes:

- Sight distances
- Clear zones
- Safety barriers
- Separation between users (e.g., Buffer width between the traffic lane and footpath or cycle lane, flush median, and turning bays)

7.1.1.3. Consistency

For drivers to navigate roads safely, they must be geometrically consistent. The design speed must adjust if environmental factors, road alignments, or geographic limitations result in varying operating speeds along the route. Drivers will not be able to react in time if these speed changes are not consistent with typical driver expectations and capabilities.

In order to give drivers of motor vehicles a consistent expectation and, as a result, a desired speed that is not higher than the design speed, the entire road environment—horizontal alignment, intersection spacing, and control, adjacent land use, and street activity, and speed management features—should be designed. Features like a crest curve or a bend with a sharply decreasing radius should not hide changes; instead, they should be easily noticeable.

7.1.1.4. Urban Road Speed Management

It is important to take precautions to ensure that the combination of road geometry and operating conditions can ensure that the Operating Speed does not exceed the intended speed when the intended Operating Speed is to be kept low for safety and the urban design environment. It is possible to lower the Design Speed below 50 km/h if this is consistently shown.

7.1.1.5. Not All Vehicles

Not every vehicle on the road is covered by design speed; for example, cars can move faster than tractor-trailers. In certain regions, e.g., Tractor-trailers may need to operate at a slower speed in areas with steep hills or sharp curves. Make sure that faster cars can safely pass slower ones when building such a road.

7.1.1.6. Road Side Infrastructure

Designers must consider the interplay between operating speeds, visibility, stopping distances, and survivable speed when recommending infrastructure that cars may collide with.

7.1.2. Design Vehicles

7.1.2.1. Purpose

Design vehicles are chosen automobiles whose weight, size, and operational traits are used to create controls for highway design that allow for the accommodation of cars belonging to specific classes.

7.1.2.2. Turning Speed

Design vehicles are chosen automobiles whose weight, size, and operational traits are used to create controls for highway design that allow for the accommodation of cars belonging to specific classes.

Considering the desired deceleration to the turn speed and the speed differential in road design, buses and check vehicles should typically turn at a speed between 5 and 25 km/h.

A large disparity in the speed of oncoming traffic and turning traffic can pose a serious risk to public safety. The operating speed and the design turning speed ought to be in line. For instance, the intended turning speed at a junction on a 60 km/h arterial road might be 25 km/h (Cruzado and Donnell, 2010).

7.1.2.3. Maneuver Speed

Certain maneuvers, like parking, turning in reverse, and using vehicle crossings, call for a swept path speed that is slower than that of intersections. You can use a maneuver speed of up to 3 km/h.

7.1.2.4. Acceleration and Deceleration (Buses)

When approaching or leaving bus stops, extra caution should be used when monitoring speed because the vehicle is probably going to be accelerating or decelerating. This must be taken into account when calculating the tracking turning speed.

7.1.2.5. Swept Path Width

The vehicle's body width plus 0.5 meters of clearance to account for projections and variations in actual vehicle paths will be the design path. The face of a kerb or an adjacent traffic lane (which may also include a kerbside channel) must be cleared.

7.1.2.6. Tracking Profiles

The Transport Design Manual homepage offers a download for the design and check vehicle profiles. Software configurations for intersection and maneuver design should follow the instructions provided by AT design tools.

7.1.2.7. Buses

Every situation where a road is or could be used by buses requires bus tracking to use all common AT bus types as specified by the tracking profiles above.

7.1.2.8. Freight, Dimensional, and Overweight Routes

Where the road is part of a freight, over-dimensional, or overweight route it is a requirement that tracking be undertaken to show the effect this will have on any proposed design. All freight routes, whatever street type, require 19.45 m semi as Design Vehicle and 23m truck & trailer as Check vehicle.

7.1.2.9. Design and Check Vehicles for Each Road Type

The biggest cars that frequently travel on specific roads are known as design vehicles. It is anticipated that they will be able to stay in the designated traffic lane. Larger vehicles that are occasionally anticipated to use a road are known as check vehicles. While they might not always be able to stay in a traffic lane, interactions with other Design vehicles need to happen in a safe and predictable way. The road types and design vehicles listed below must be utilized consistently unless it can be proven that a different design or check vehicle is more appropriate in a particular situation.

7.1.2.10. Principles for Swept Paths

Swept paths ought to be established for the proper design range and vehicle check, utilizing tracking type controls on lane usage and at suitable turning speeds.

- To reduce the chance of wheels damaging the kerb, kerblines should closely follow the swept path and have 0.5 meters of space between them and the wheel track.
- When a check vehicle is permitted to cross into another traffic lane, space must be left for a design vehicle to use that lane's swept path.
- If that clearance is not available, the encounter between a check vehicle and a design vehicle must be at a low speed, with visibility to enable one to stop and allow safe

passage for the other, including for visibility of other users (safe encounter).

7.1.2.11. Other Design Considerations

Use of an alternative design vehicle may be necessary if a Design or Check vehicle is unable to adequately address all planning cases. For example, a local road in a commercial area may need to track larger vehicles suitable for the industry served, residential streets serving as service and delivery routes for supermarkets or other specific approved land uses, or roads leading to boat ramps may need to be designed for cars and large boat trailers.

- It is acceptable to use variable drive speed to track the vehicle if it is not the design vehicle (but not turning wheels at the stop).
- Considerations must be given for emergency vehicle use at all times; however, they are not required to track fully in their lane.
- Where a body-swept path overhangs a kerbline, protection for path users must be provided, usually by a buffer strip.
- At signal-controlled intersections, vehicle-swept paths must have regard to detector location.

7.1.3. Visibility for Safety

The distance a user travels from a point of observation to a feature, an obstruction, or another vehicle on the road is known as sight distance. It is possible to compute sight distance using the following factors: road geometry, vehicle assumptions, driving style, and hazard awareness (Ossenbruggen, Pendharkar, and Ivan, 2001). Sight distance needs to be sufficient for users to react to anticipated or unforeseen changes in the state of the road in a way that prevents injury to themselves, damage to their vehicles, and damage to infrastructure.

While sight distance calculations are necessary to verify geometric design, what matters most are reaction times—the amount of time needed to see, decide, and act, as well as the amount of time needed to flee, slow down, or stop. Design strategies should guarantee that there is enough time, free from interruptions, for safe action.

The line-of-sight envelope that is visible to both the observer and the object as they move along their respective paths is known as the visibility envelope.

7.1.3.1. Safe Road Use

To guarantee that road infrastructure can be used safely by everyone, three specific road user tasks need to be evaluated:

- *Safe Path*: Every road user needs to have clear visibility of the road ahead to identify a safe path to follow and choose an appropriate speed to approach different features.
- *Safe to Go*: They must be able to see approaching users for a sufficient amount of time to determine when it is safe for them to enter a conflict area, such as an intersection or crossing when their path takes them there.
- *Safe Avoidance*: They must be able to react quickly enough to avoid colliding with any unexpected stationary or moving hazards that might be in their path or about to cross it.

7.1.3.2. Different Types of Sight Distance

Austroads lists the different sight distances

that are typically taken into account. They are arranged below by kind of task.

Geometric design also provides guidance on applying sight distances to specific circumstances:

- Sight distances on horizontal curves
- Sight distances on horizontal curves with roadside obstructions
- Headlight sight distance (This should be consulted when assessing safety on unlit roads)
- Horizontal curve perception sight distance

7.1.3.3. Safe Path

Determine which features need a speed decrease from the original design speed. Ascertain that there is enough visibility envelope for observation and slowing down at d=0.25 (preferred).

7.1.3.4. Safe to Go

Verify the ability to halt or yield to a user movement priority by utilizing Approach sight distance (ASD). If observation time permits accurate determination of the intended stop position, the object may differ from the limit line road marking.

- Minimum gap sight distance (MGSD)
- Crossing sight distance (CSD)
- Overtaking sight distance (OSD)
- Intermediate sight distance

7.1.3.5. Safe Avoidance

Check the stopping sight distance of cars and trucks (SSD) for hazards on the road or unexpectedly entering it while maintaining the Safe Path speed.

Instead of just applying Safe Intersection Sight Distance (SISD) to show responses to the various hazard cases, check SSD with increased observation time or increased coefficient d on the approach to intersections, crossings, or complex conflict zones.

7.1.3.6. User Type

In general, the Design Vehicle should be used to evaluate driver safety. In every situation where a heavy vehicle is a design vehicle or a check vehicle where the assessment is likely to differ significantly from the design vehicle, a check should be performed on the heavy vehicle.

The Urban Street and Road Design Guide should be used to evaluate the stopping capabilities and speed of other user types.

7.1.3.7. Sight Distance for Urban Roads

In urban streets, safe road usage is more complicated. There are more opportunities for conflict, more opportunities to change course and speed, and more occasions when it may be necessary to react to a new hazard while already altering course or speed.

7.1.3.8. Observer Position

The observer could be moving or stationary, and the user type will affect the eye height. For eye height, GRD3, GRD4A, and GRD4B parameters should be utilized.

The position should be the likely safe position when the observer is stationary, allowing for a pedestrian or cyclist to be 10 m from the front of a cycle or pram and 2.5 m from the front of a vehicle. When the observer is in motion, the initial point of observation follows the observer's timeline, maintaining a consistent speed or slowing down when approaching an intersection, curve, or other obstacle. The observer may have to make several decisions at intersections with poor visibility. The observer's position should consider expected behavior and whether it should be encouraged or discouraged to reduce the possibility of obstructing a conflict area.

7.1.3.9. Object Position

Be mindful of the size and distance of objects, particularly when approaching at a high speed. These influences identify a risk that probably calls for taking preventative measures.

Fixed features can be vertical or on the ground. While it is usually necessary to observe road markings at a limit line when calculating the speed at which to approach an intersection or crossing, it is possible to take a more thorough approach to define the object to select a speed reduction that will enable a user to stop when needed.

Traffic Control Devices are designed to enhance the identification of features that require slowing down or stopping, rather than replace the need for observing the feature (Fang and Elefteriadou, 2005).

In order to determine what a user must see in order to correctly identify and react promptly, a curve, traffic calming feature, intersection, or crossing may be evaluated as a whole.

The object might be a moving vehicle or another road user. Determining the duration of observation and response requires taking into account the other user's path, behavior predictability, and the amount of time it takes to reach a conflict.

The "object" for Safe to Go could be a void in traffic flow. According to MGSD, the gap must be apparent to the observer for a sufficient amount of time for them to recognize it and decide whether to move forward. If the gap is deemed to be too small, a correction may be made and further action may be taken. Safe Avoidance SSD can be evaluated so that an observer traveling at a constant speed or faster can avoid the priority user path. This needs to be verified in cases where limited observation time or geometry could lead to inaccurate speed estimation.

7.1.3.10. Observation and Reaction Time

The three tasks, the state of the road, and the abilities of various users all affect this. A complex environment necessitates the observation of more features and risk areas, which leads to increased alertness.

If a moving hazard is noticed in the periphery of one's vision, it prompts a quicker response to prevent injury. Certain hazards require more time to observe and make a decision, such as far-off cars on a straight road, curves that pass out of view, intricate intersections, or users who could move, stop, or change direction. When choosing times for calculations, these factors need to be taken into account.

7.1.3.11. Friction and Vehicle

The right design vehicle and the coefficient of friction of the road surface have a significant impact on the sight distance calculations. It is reasonable to assume that standard friction rates of 0.26 for trucks and 0.35 for cars will account for variations in weather and surface conditions over time. Since this performance will not influence the selection of the Safe Path deceleration rate, high-friction surfaces should only be evaluated with an increased coefficient of friction for Safe Avoidance.

7.1.3.11. Hazard Object Check Vehicle

The sideways acceleration that a single-occupant car driver can tolerate under ideal circumstances is higher than the design friction factor. To determine the path speed of a car in the 50%ile approaching a conflict zone, a factor of 0.48g should be applied. This is required to evaluate the possible collision speed for survivability and to allow enough time for observation of such a vehicle, which is considered a moving hazard.

Other features, like traffic calming, also have an impact on path speed. According to standard design guidelines, the initial speed on straight, unobstructed roads can be assumed to be 10 km/h over the posted speed limit.

7.1.3.12. Initial Speed

Generally speaking, the starting speed is considered the road's design speed. On a curved path, this should be adjusted based on sideways acceleration all the way down to the comfort speed.

Another task that is already underway, like slowing down to make a turn, may have an impact on the initial speed. Next, it's important to determine where in the user's time path a second calculation—like Safe Avoidance of an Unexpected Hazard—is required.

In the event that the 85-percentile speed is noticeably higher than the design speed—which could happen on a straight, broad road—a check for safe avoidance may also be required.

The starting speed for Safe to Go can also be determined from stopped conditions, like a border or intersection.

7.1.3.13. Deceleration Rate

The design of a road's geometry shouldn't be dictated by emergency braking. Deviation from the intended slowdown might necessitate mitigation via design. Usually, this entails making sure that no additional departures have an impact on safety locally.

The deceleration or acceleration rate is obtained by multiplying g (9.81 m/s^2) by a coefficient d.

The deceleration rate for the Safe Path design should be d = 0.25 (2.5 m/s^2).

The deceleration rate for safe avoidance depends on the road gradient and sideways friction during turns, and it is based on stronger braking.

Depending on the car and the surface, a check for the maximum friction rate may also be performed to determine the likelihood of stopping to prevent a collision or, if a collision cannot be entirely avoided, reducing speed enough to prevent injury.

7.1.3.14. Acceleration Rate

For Safe Path or Safe to Go, the acceleration of a design car from rest or an initial speed should be taken into consideration. This could be true for an object user, an observer, or both. For every percent grade down, add 0.005, and for every percent grade up, subtract 0.005 (0.0025 when the speed exceeds 50 km/h). This is how coefficient d should be adjusted.

7.1.3.15. Final Speed

The ultimate speed can be halted for a give-way or signal control, the comfortable speed for a low-radius bend, or the maneuvering speed for turning at an intersection.

At the end of acceleration, the final speed for Safe to Go might be the Design speed.

7.1.3.16. Conflict Zone

Determine the region enclosed by user path dimensions that overlap. Unless one user gives way to the other in a safe manner, two users may collide within this zone. As each user approaches the Conflict Zone, sight distance is calculated along their time path. Every user who needs to make way for another is given a Safe to Go assessment.

7.1.3.17. Conflict Zone Clearance

When deemed safe to proceed, an observer and an object may cross a conflict zone at different times with no harm coming to either party.

The user might, however, be stopped or blocked and unable to exit the conflict zone as planned. For this, a safe avoidance assessment is required.

7.1.3.18. Collision

If a collision cannot be avoided, design should aim to make sure that users are not seriously injured when it does happen. When users don't take safe preventative measures, this could happen. Because of this, it is ideal for conflict zones to be found where user kinetic energy and speed are lowest.

7.1.3.19. Road Side Clearance Envelope

There must be no obstructions in the visibility envelope that a specific sight distance creates. The height of the vegetation must be kept to no more than 600 mm above the ground.

To prevent them from being hit by cars, this is in addition to the clearance envelope for vegetation and roadside structures.

7.1.4. Design for Maintenance

Where utilities, and manholes in particular, have been given a departure to be in the roadway then chambers and lids must not:

- Be placed in the wheel track
- At the deceleration or acceleration zones of an intersection
- In the middle of an intersection

It is necessary to take into account and show that regular upkeep, like resurfacing, can be completed without requiring a lot of traffic control. In order to avoid complete road closures, it will be necessary to review the larger network and determine whether the designed road widths, lengths, and intersections are appropriate. Additionally, any detours that are required must be short and take routes that can handle the diverted traffic.

7.1.5. Network Utilities

Only network utility pipework and ducting that is considered to be of a transmission style—that is, no local connections to nearby property lots or catch pits—will be permitted in the roadway. All local or distribution utilities must be installed in the area between berms and walkways, away from the road. Transportation will only take into account a deviation from these guidelines if the applicant can distinctly show that locating the utility network inside the roadway is an absolute last resort and that all other options, including deviations from other pertinent codes, have been explored and determined to be unfeasible.

The minimum depth for utilities to be within the roadway, as determined by an authorized Transport Departure from Standard, must be one meter or beyond the edge of the road formation, whichever is higher. The impact of compaction-induced vibration on the formation and the possible harm to the utility network must be considered when evaluating depth.

7.2. GEOMETRIC ALIGNMENT

7.2.1. Horizontal Alignment

The constraints that will be applied to the developed alignment can be identified once the design speed has been established. These could include:

- Level constraints caused by the site fronting the road;
- Headroom clearance at structures;
- Overhead and underground services;
- Drainage requirements;
- Neighborhood planning;
- Land take; and
- Any public commitments

Any pre-agreed departures are taken into consideration when creating an initial alignment. If more relaxations and departures are needed to get the best alignment, they can be established. There are more limitations and a need for more thorough justification when searching for departures close to intersections.

- There may be a requirement for curve widening.
- Negative chainage systems should not be used; start the main alignments before the start of the work.
- Corner kerblines at intersections should have positive chainages which relate to the minor or side road.
- The ideal alignment across any structure is straight. Avoid small radii and transitions.

7.2.1.1. Legibility

The horizontal alignment of the road should match that of other roads of the same type. Make sure users have enough time to notice and comprehend changes when the Type or a constraint (such as a tighter radius bend) necessitates them, and then adopt a Safe Path speed that is suitable for the alignment. Steer clear of alignments that obstruct important features or hazards around curves that encourage fast approaching that makes it difficult to see and respond to the hazard.

7.2.1.2. Urban Road Alignment

7.2.1.2.1. Design Speeds 50km/h or Less

Straights and circular curves may serve as the foundation for the horizontal alignment of any urban road with a design speed of 50 km/h or less. Vehicle tracking will be used to establish or verify lane lines and kerblines.

7.2.1.2.2. Design Speeds Greater than 50km/h

The designer may include transition curves appropriate for the higher design speed on roads that might always have a design speed greater than 50 km/h.

7.2.1.2.3. Local and Collector Roads

Apart from the previously mentioned design approach, traffic calming measures might include horizontal alignment.

7.2.1.3. Rural Road Alignment

7.2.1.3.1. Safe Alignments

Rural roads should have consistent alignment throughout, or there should be a distinct transition point between changes in geometry and character.

Hazards should be clearly marked, and if necessary, safety barriers or signs should be added. Vertical alignment shouldn't hide bends and intersections.

7.2.1.3.2. High-Speed Rural Roads

The Austroads Guide to Road Design should be followed when designing a geometry where the posted speed is 60 kph or above.

7.2.2. Vertical Alignment

A vertical alignment should guarantee that no features, like bends, intersections, or crossings, are hidden behind a crest curve and that an approaching user can see and understand them clearly enough to modify their approach path and speed accordingly.

The design of the vertical alignment must be appropriate for the site location.

Sag / Crest curves: Designed suitably to attain the minimum gradient change rate required for design speed. If necessary, reduce from a starting Sag K value of 9 and a Crest K value of 10. Crest curves are also used to achieve visibility for Safe Stopping Distance.

7.2.3. Longitudinal Gradients

7.2.3.1. Road Drainage Design

Acceptable road drainage criteria serve as the foundation for the minimum acceptable longitudinal gradient.

7.2.3.2. Preferred Gradient

To provide path users with the greatest possible accessibility, no more than five percent of the road network should be included in the network. If topography makes this impossible, a maximum grade of 8% over the course of the route might be allowed. Roadside pedestrian routes requiring treatment will need to accommodate gradients steeper than 8%. Only when suitable alternate accessible path routes are available should one choose to use these steeper gradients.

7.2.3.3. Maximum Gradient

Although gradients shouldn't be any steeper than 8%, they can be in places where topographical restrictions apply. When deviating from the standard, the maximum gradient shouldn't be higher than 12.5% if the road is designated as a public one.

7.3. CAMBER, CROSS-FALL, AND SUPER-ELEVATION

Typically, a road surface slopes in two ways: longitudinally, along the centerline of the road, and transversely, perpendicular to the centerline. Various terms are used to describe the transverse slope, such as camber, crossfall, superelevation, can't, and cross slope.

7.3.1. Overview

7.3.1.1. Carrying Water

Rainfall is transported from the road surface to one or more of the edges of the road via transverse fall in addition to longitudinal fall. In the absence of this, vehicles may aquaplane as a result of the obstruction of the tire tread by standing water.

7.3.1.2. Cross-Fall and Camber

When water is removed from the surface of a road, it is referred to as cross-fall or camber.

- **Cross-fall:** One side of the road receives the conveyance of surface water.
- **Camber:** Surface water is transferred from the road's center to its edges. Camber is a downward curve that moves from a road's center toward its edges rather than a straight drop. A "balanced carriageway" is a term frequently used to describe a road with camber.

7.3.1.3. Transverse Gradients

The angles of the fall in a camber or cross-fall are referred to as transverse gradients. Seldom is the degree of upward curvature specified, and most geometric modeling software does not allow for it.

7.3.1.4. Superelevation

Super-elevation is the process of using transverse fall to "bank" a portion of the lateral forces of a car's movement into downward pressure, which improves safety and comfort for passengers as they round bends. Transition curves should always come before and after super-elevation.

7.3.2. Design Considerations

7.3.2.1. Cambers and Cross-Falls

7.3.2.1.1. Minimum Transverse Gradient

All sealed roads should typically have minimum transverse gradients of 3% toward the outer edge of the road due to Auckland's high levels of rainfall. On the other hand, the camber or cross-fall may vary from 2 to 4% in cases where current features prohibit this. Transverse gradients with a minimum of 4% toward the road's outer edge should typically be used on unsealed roads (Figure 7.1).

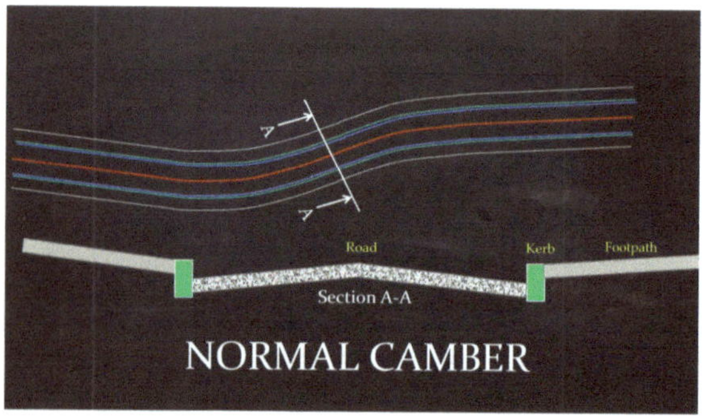

Figure 7.1. *Normal road camber.*

Source: https://www.gcelab.com/blog/type-of-road-camber-superelevation-advantages/.

7.3.2.1.2. Maximum Transverse Gradient

For sealed roads, the maximum transverse gradient is 5%, and for unsealed roads, it is 6%. A departure from Transport must be approved for greater transverse gradients.

7.3.2.1.3. Transverse Gradient in Relation Longitudinal Gradient

In order to encourage drainage towards the road channel, a transverse gradient steeper than 3%, or even up to 5%, is preferred on steep roads.

In order to promote sheet flow to the road edge, crossfall cannot be less than 3% in areas where the longitudinal gradient is less than 1%.

7.3.2.1.4. Intersection

The minor road should be built to grade into the major road's channel line at intersections, with the major road's camber taking precedence. The minor road's crown shouldn't extend into an arterial or collector road's traffic lane if this cannot be accomplished. On roads with design speeds higher than 50 km/h, adverse cross-fall, or falling toward the outside of a bend, will not be allowed.

7.3.2.2. Super-Elevation

Super-elevation is not allowed on collector or local roads, and it's unlikely that arterial roads in cities will need it.

In areas where there is a high volume of pedestrian traffic, the maximum super-elevation should be restricted to 5%. Super-elevation is necessary on curves where the longitudinal gradient is steeper than 8% in order to reach the road's design speed.

7.4. ROADWAY CONFIGURATIONS

There are more design options available for where to place the segregated busway within a particular roadway than may first appear. Busway alignment, also referred to as configuration, minimizes potential conflicts with turning cars, stopping taxis, and unloading delivery trucks, which is essential to fast and efficient operations. As a result, The Bus Rapid Transit (BRT) Standard gives the highest points to the following configurations: two-way busways in the middle of the road, two-way busways that run next to an edge condition, such as a waterfront, and bus-only corridors, such as a transit mall, as these minimize conflicts that occur at the curb the most. Less points are given to a two-way busway that runs beside a one-way street.

Figure 7.2. *Curitiba utilizes a variety of roadway configurations. Each street's design depends on the local circumstances.*

Source: https://brtguide.itdp.org/branch/master/guide/roadway-and-station-configurations/roadway-configurations.

Because it is unlikely that pedestrians will anticipate traffic coming from the opposite direction, safety concerns led to the decision to drop the point. Even fewer points are given to one-way busways that are located in the median of one-way streets, and even fewer are awarded to those that run along the curb of one-way streets. Virtual lanes receive the lowest point values.

Throughout its length, a corridor may take on various configurations. Similar to numerous other BRT-related design choices, there isn't a single, ideal roadway configuration. A lot is dependent upon the local conditions. Before entering the downtown area, Johannesburg, South Africa, has a two-way median-aligned busway that divides into one-way, median-aligned busways that travel on one-way streets. In Curitiba, Brazil, BRT is limited to using center lanes, both side lanes and streets. Curitiba basically modifies the layout of the road to suit the specific circumstances on the designated section of road (Figures 7.2 and 7.3).

Figure 7.3. *TransJakarta utilizes buses with doorways on both sides of the vehicle in order to service both median and curbside stations. ITDP.*

Source: https://brtguide.itdp.org/branch/master/guide/roadway-and-station-configurations/roadway-configurations.

7.4.1. Median Busways

Placing the busway in the center two or median is the most popular choice. A median busway alignment receives full points under the BRT Standard. This is because of alleys, parking lots, and other obstructions, turning cars on a roadway nearer the curb have fewer conflicts with the central verge. Furthermore, even though delivery cars and taxis need to have access to the curb, the middle portion of the road is typically clear of these obstacles. A central station that serves both busway directions is also made possible by the median location. A single station that serves both directions of travel receives full points under the BRT Standard's "Center Stations" metric, making it simpler to switch between routes or directions. Customers can choose from a variety of routing options from a single-station platform when using a median station. The construction of a single station is less expensive for infrastructure than building separate stations for each direction.

7.4.2. Curbside Busways

Conventional bus lanes are usually found at the curbside, but BRT rarely places the busway on the sides of the road. Because curb lanes seldom work as intended, the BRT Standard gives this configuration no points for "Busway Alignment." Curb lanes significantly reduce the capacity of the system by interfering with turning traffic, and stopping taxis, delivery vehicles, and non-motorized traffic. It is challenging to reach capacities of over 5,000 patrons per hour in each direction if turning vehicles regularly obstruct busway operations. Curbside busways increase the possibility that a single taxi picking up a passenger, a police officer momentarily parked, an accident, or a turning vehicle stuck behind a large volume of pedestrians crossing the street could stop traffic on the entire busway.

7.4.3. Side-Aligned, Two-Way Busway Configuration

Although curb-aligned busways typically don't work well because turning conflicts with mixed traffic occur, two-way busways placed alongside the road can function well for specific sections of the road. There might not be any turning conflicts over extended distances if a roadway is surrounded by open space, green space, or water (such as an ocean, bay, lake, or river frontage). In this scenario, side alignment might even be preferable to median alignment. Side-aligned busways that are close to such an edge condition receive the highest points under the "Busway Alignment" metric according to the BRT Standard. The lack of access to development along a specific corridor edge, such as a park or an airport boundary wall, is crucial in determining this kind of alignment.

Other side-aligned options can still be taken into consideration in cases where there isn't an edge condition of that kind. Adjacent to a two-lane, two-way general traffic roadway in Lima is a two-lane, two-way busway. Although they can present challenges, intersections alongside-aligned busways can be resolved with the help of roundabouts and traffic signals.

7.4.4. Fixed Guided Busways

Given that a BRT vehicle is normally 2.6 meters wide, a lane that is only marginally wider than this might be adequate. A driver needs a road width of about 3.5 meters to safely stay in their lane during normal operation, and 3 meters at the station because they will be slowing down to pull up next to the boarding platform. However, a lane width of three meters is feasible if a vehicle is physically restrained by a guidance mechanism.

Adelaide, Australia; Bradford, UK; Essen, Germany; Leeds, UK; and Nagoya, Japan all have BRT systems that use physical guidance systems. The car stays in its designated lane thanks to a guidance wheel mounted on the side. The Netherlands has also utilized a small ditch in the roadbed for brief stretches with fair success. It is also feasible to use magnetic or optical guidance systems.

When there is significant value in narrowing the lane width by about 0.9 meters, a fixed-guide way system may be something to think about. Other benefits of guidance systems include increased operating speeds and safer vehicle operation. The main drawback is the additional infrastructure expense for the guidance track and side wheel. An additional maintenance concern is that guide wheels are prone to breaking off when the bus docks improperly at a curbside stop outside of the guided busway sections (Figure 7.4 and Table 7.1).

Figure 7.4. *Nagoya, Japan's elevated BRT line utilizes a mechanical guidance system to reduce the required roadway width.*

Source: https://brtguide.itdp.org/branch/master/guide/roadway-and-station-configurations/roadway-configurations.

Table 7.1. *Advantages and Disadvantages of Guided Busway Systems*

Advantages	Disadvantages
Higher speeds (reduced travel times) are achievable within safety standards	Increases busway construction costs considerably
Permits construction of narrower busway lanes	Increases vehicle costs
Contributes to a more permanent image of the busway	Reduces flexibility with regard to the type of vehicles that may utilize the busway
Allows construction of lanes without paving the center strip, which can be grassed to give a "green" and softer appearance.	Speed advantages of guided busways are only realized when the distances between stations are quite significant
Self-enforcing as general traffic is unable to utilize the busway	Difficult for the recovery vehicle to remove the stranded bus from the guideway
The permanence of the busway, as the roadway cannot easily be opened to mixed-traffic operations	Ongoing maintenance of bus guidance wheels on bus front axle

Source: https://brtguide.itdp.org/branch/master/guide/roadway-and-station-configurations/roadway-configurations.

Furthermore, some system developers have chosen not to pave the center of the lane because busways do not necessitate vehicle lane changes. The presence of grass or dirt underneath the bus can aid in mitigating engine noise; reports of noise reductions of up to 40% have been made using this method.

Figure 7.5. *The guideway allows for hands-free driving on the Nagoya-Yutorito Line.*

Source: https://brtguide.itdp.org/branch/master/guide/roadway-and-station-configurations/roadway-configurations.

Even in cases where roller guides are not being used, other busway developers are thinking about not paving the middle of the lane. Given that guided buses will experience less variation in their lateral movement, the paved strips for non-guided buses will probably be wider than those for guided buses. The viability of this strategy and the financial benefits of not paving the center-lane area will vary depending on local building standards and expenses. Sometimes local contractors might not know how to use this construction method well. Nevertheless, any possible cost savings should be taken into account, as the busway's paving may be the system infrastructure item with the highest cost. The driver of a vehicle does not actually need to steer when traveling along a guided busway. The car can be driven "hands-free" in theory because the guideways forbid any turning motions. In certain systems, like the one in Nagoya, Japan, safety issues arise when the BRT vehicle departs the guideway. A collision could happen if the driver does not re-engage the physical steering for whatever reason. Therefore, in Nagoya's case, a forced stop is executed at the guideway's exit to prompt the driver to resume using physical steering (Figure 7.5).

7.4.5. Grade-Separated Busways

In narrow right-of-way configurations, grade separation is an option, as is running the BRT corridor underground or on an elevated roadway. It is also an option at extremely busy intersections and roundabouts. Another option to think about is grade separation, which allows you to avoid difficult terrain or water. Since constructing grade separations is so costly, it is typically done in key areas where the separation significantly enhances operations. On the other hand, grade-separated busways can also span an entire corridor, such as São Paulo's Expresso Tiradentes, which is an elevated road. Since a grade-separated busway is a fully exclusive right-of-way that is entirely isolated from mixed traffic, it receives maximum points under The BRT Standard's "Busway Alignment" metric.

The engineering suitability of the base materials and physical terrain for tunnels or elevated structures must always be taken into account. From an engineering and financial perspective, underpasses and tunnels may not be feasible due to high water tables or hard bedrock. Similarly, the cost of securely siting pillars for elevated structures can be greatly increased by soft soils. Thus, whenever grade separation is being considered as a possibility along specific BRT corridor segments, an engineering and cost-feasibility analysis should be carried out.

Elevated busways are not only very expensive (up to five times as much as at-grade infrastructure), but they can also divide an urban area and have an aesthetic impact on the neighborhood.

7.4.6. Transit Malls and Transit-Only Configurations

"Bus-only" or "transit-mall" corridors are useful choices for ensuring that public transportation has top priority. These corridor sections can be found throughout a corridor, like the Orange Line in Los Angeles, USA, but they are usually used in central locations where space constraints make it difficult for private automobiles and public transportation to share an area. The "Busway Alignment" metric of The BRT Standard gives maximum points to transit mall configurations because they constitute a fully exclusive right-of-way that is totally segregated from mixed traffic. When there are only two lanes of available road space along a crucial corridor, transit malls are often a practical solution. Thus, sections that have a road width of just seven meters might be suitable for a transit mall. During public transportation operating hours, private cars, motorcycles, and trucks are prohibited from both the corridor segment and its surrounding areas. That being said, a one-way transit mall can function in as little as three meters of space, as the Quito Trolebús' Plaza del Teatro section demonstrates.

Transit malls are especially suitable when the public transportation system boosts business and smoothly fits into the current land-use patterns. In these situations, the transit mall creates a peaceful, traffic-free street environment. A limited number of patrons are allowed entry to stores and street amenities in transit malls. As a result, transit malls are usually found in areas with strong retail sales. An atmosphere that is conducive to street activity and pedestrians is promoted by the absence of mixed traffic.

Buses typically travel slower in these areas due to the open interaction between pedestrians and the public transportation service in a typical commercial transit mall. If not, mishaps might happen or the system would make the public area less beneficial. But the Plaza del Teatro section of the Quito Trolebús circumvents this issue by dividing the pedestrian zone from the busway physically. Although this division lowers the possibility of collisions, it also makes the streetscape less sociably pleasing for pedestrians. TransMilenio in Bogotá limits the top speed in its transit mall to 13 km/h, although the system as a whole operates at a faster pace.

When there is a lot of foot traffic along a transit mall, the street's overall quality may suffer from the presence of public transportation vehicles. The volume of pedestrian traffic on the Oxford Street corridor in London has outgrown the available footpath space, making conditions challenging. In this instance, it might be preferable to give pedestrians the entire space currently occupied by taxis and public transportation vehicles. Therefore, a street may be better used as a "pedestrian mall" than a "transit mall" at certain pedestrian volumes.

Figure 7.6. *An exclusive busway runs under the Mater Hill Hospital in Brisbane.*

Source: https://brtguide.itdp.org/branch/master/guide/roadway-and-station-configurations/roadway-configurations.

Certain cities, like Quito and Bogotá, have bus-only corridors installed. Pittsburgh, Pennsylvania, USA; Ottawa, Canada; and Brisbane, Australia have all created bus-only corridors across specific sections of their respective roads (Figures 7.6 and 7.7).

However, transit-only corridors are not only found in busy commercial and retail areas. Certain busways, for instance, are effectively limited-access roads that are only intended for bus use. In primarily residential areas, the West Busway in Pittsburgh, Pennsylvania, USA, travels along a bus-only corridor. The busways in Pittsburgh and Brisbane are situated alongside extensive green spaces. As a result, no residential driveway opens straight onto the hallway. If not, these plans probably wouldn't be as successful.

Figure 7.7. *Transit mall in Utrecht, Netherlands. Deliveries are made very early in the morning, or via side streets.*

Source: https://brtguide.itdp.org/branch/master/guide/roadway-and-station-configurations/roadway-configurations.

The most significant obstacle to the successful operation of transit malls and other transit-only corridors may be the accessibility for locals and delivery vehicles. The desire of certain merchants for 24/7 delivery access presents a technical and political challenge to the establishment of a transit mall. Some retailers might also be concerned about losing direct customer access via private vehicles and on-street parking. Thus far, empirical evidence suggests that pedestrian malls and transit malls typically enhance both retail sales and property values. As a result, retailers occasionally object to initial vehicle restrictions.

Establishing delivery access for stores during non-transit hours is a typical solution. As a result, merchants can move bulky items in the late night and early morning. Smaller items can usually be delivered whenever you want by using delivery services and carts that operate from the pedestrian area. However, in some places, this might present a safety risk for freight deliveries at night, so it needs to be properly addressed.

Conflicts arise most often when people want private vehicles to have access to their properties and parking spaces in a residential area. When neighboring parking garages are established and access is granted outside of public transportation's regular operating hours, these conflicts can occasionally be resolved. A transit mall's successful completion is probably going to need careful political wrangling in the cases of residential access and store deliveries.

7.4.7. One-Way Pairs Configuration

Rather than building a transit mall, cities usually think about dividing public transportation into two distinct routes, usually running parallel to each other. As a result, the public transportation system runs in two one-way pairs. This arrangement is occasionally referred to as a couplet. Usually, in this situation, at least one mixed-traffic lane can be kept.

The effect on parking, truck deliveries, and mixed traffic is the main benefit of dividing the route. Private automobiles can still enter corridor properties directly in some capacity. Furthermore, because this configuration closely resembles the current bus routes, it may be more well-liked by the operators already in place. In its central sections, Guayaquil, Ecuador, has effectively implemented a split-route system, with one-way pairs running down the middle of the street. Similar configurations of median-aligned, one-way pairs can also be found in the downtown area of Johannesburg, South Africa. The BRT system's two directions are combined again in a more traditional two-directional configuration outside the denser city center.

However, because of the transfer penalty that customers must pay if they need to transfer to a different line or go in a different direction, as well as the possibility of customer confusion regarding which station to board for which direction, this configuration only receives half the points under the BRT Standard's "Busway Alignment" metric.

7.4.8. Bi-directional One-Lane Configuration

A brief section of a narrow busway could run on a single lane in certain exceptional circumstances with lower demand and advanced technology. As a result, service for both directions would be offered sporadically by a single lane. In order to prevent two vehicles from attempting to use the one-lane segment simultaneously, a unique traffic control system is typically used.

Most famously, the single-lane operation is utilized in Eugene, Oregon, USA. Due to its restriction to brief road segments and low bus frequencies, this option is effective. At strategic locations for passing, the busway splits into two lanes and an advanced signaling system holds oncoming buses. Eugene's Lane Transit District has been able to avoid most conflicts through simple scheduling because of these low frequencies.

With this kind of design, the longer the one-lane operation, the greater the chance that the system as a whole will be disrupted. Additionally, in systems where vehicle frequencies are high and customer demand is high, this option is unlikely to be feasible.

The single-lane operation, however, may be utilized in certain situations to get around obstructions spanning brief road sections. Planners may decide against an otherwise perfect corridor because of impassable obstacles like a narrow historic street, a single-lane tunnel, or a bridge. In such circumstances, single-lane operation may be an option to take into account.

7.4.9. Virtual Busways

A virtual busway is a single bus lane that is in the middle of a road that is used for both directions of traffic, much like the bi-directional, one-lane configuration. The distinction lies in the fact that the buses alternately use the lane based on the direction and necessity of queue jumping in the hallway. The bus will enter the virtual lane as it gets closer to the intersection. In order to maintain a relatively free-flowing lane, the traffic light will feature a separate phase for public transportation vehicles that will permit BRT vehicles to exit the virtual lane and enter the general traffic lane ahead of other vehicles. When the bus approaches the intersection, it usually stays in the general traffic lane until the virtual lane is once again allocated to its direction of travel.

A virtual busway has been successfully implemented in Rouen, France. The route features a three-lane cross-section that accommodates a virtual lane in the median, a general traffic lane in each direction, and on-street parking. Virtual lanes are allocated toward signalized intersections on approach, providing public transportation vehicles with unhindered access to both the signalized intersection and a bypass or queue-jump lane. By locating an island stop close to the intersection, the stops are found before the signalized intersection stop line.

During peak hours of the corridor's operation, the amount of congestion (and queue backup) in the general traffic lanes will determine how much of each direction of travel is dedicated to within the corridor.

It might not be able to secure these virtual lanes with delineators to guarantee self-enforcement because of their placement in the median and the narrow width of the road corridors. Because of this, the effectiveness of these lanes will rely heavily on manual enforcement, which could raise operating expenses. Additionally, systems with relatively low demand are typically best suited for virtual busways.

7.4.10. Contra-Flow Busway

Apart from the various road layouts, system designers have the choice of "with flow" or "counter-flow" bus movements. "With flow" refers to the cars moving in the same direction as the mixed traffic in the lanes that are next to them. When vehicles operate in the opposite direction of mixed traffic, it is referred to as "counter-flow." When the doorways on the current buses require the bus to drive on one side, "counter-flow" is occasionally utilized. Around the world, a variety of traditional bus systems make use of counter-flow bus lanes. Counter-flow designs are frequently used to deter private automobiles from using the bus lane. However, if private vehicles still choose to enter the area, the counter-flow lane might just cause congestion on the busway.

Increased pedestrian accidents are a potentially serious issue with "counter-flow" setups. Pedestrians may unintentionally cross into a hazardous area because they are not used to looking in the direction of the counter-flow lane.

In BRT systems, counter-flow systems are typically not used, especially because of worries about pedestrian safety. For a brief period, Quito used counter-flow movements for the Ecovía corridor because the only cars that were available had doorways on the wrong side. But as soon as the manufacturer's new cars arrived, Quito changed the corridor back to "with flow" movements (Figure 7.8).

Figure 7.8. *Bangkok employs contra-flow design. ITDP.*

Source: https://brtguide.itdp.org/branch/master/guide/roadway-and-station-configurations/roadway-configurations.

7.4.11. Mixed-Traffic Operation

A BRT system can run in mixed traffic on some corridor segments. There may be no need for dedicated infrastructure at that point if the corridor is not congested and future congestion can be managed. Additionally, a lot of cities are creating "direct service" systems, in which services are intended to travel in dedicated lanes on the trunk infrastructure as well as off the trunk corridor, frequently in mixed traffic. Pittsburgh, Pennsylvania, USA; Guangzhou, Lanzhou, and Changzhou, China; Ottawa, Canada; Cleveland, Ohio, USA; and an increasing number of other cities worldwide are home to direct-service systems.

However, many systems operate in mixed traffic in exactly the locations—downtowns, where heavy traffic may be—where specialized infrastructure is required. There just isn't the political will to limit access to mixed traffic. This decision will negatively affect travel times, system control, and the overall perception of the system if the link is congested. Consequently, mixed-traffic operations receive a score of 0 points in The BRT Standard's "Busway Alignment" metric.

The buses run in mixed-traffic lanes close to the Usme terminal of the Bogotá TransMilenio system. Two factors led to this design decision: (1) limited right of way and road space (two lanes in each direction); and (2) relatively low levels of mixed traffic. The BRT system coexists with mixed traffic in the Usme terminal area, where there is little to no congestion, meaning that it has little effect on public transportation operations. The mixed-traffic operation in this instance barely affects system performance. Mixed traffic, however, may lead to more bus conflicts with other road users; in particular, past experience has shown that bus-motorcycle accidents are more common.

In contrast, a mixed-traffic area in the city center of Changzhou, China, is traversed by the BRT corridor. Multiple four-phase intersections have a significant detrimental effect on BRT speeds (Figure 7.9).

In some cases, mixed-traffic operation may be required for a BRT vehicle to navigate around a flyover or other obstacle. When the BRT vehicle needs to move to the center median, it may have to temporarily share the road with cars coming down from the flyover. Although this situation is not ideal in terms of travel time and system control, congestion typically occurs before reaching the bottleneck or flyover, rather than at the obstacle itself. By offering separate lanes for public transport vehicles leading up to the flyover, they can bypass traffic congestion with minimal impact on overall travel time.

Figure 7.9. *Changzhou, China, operates its BRT system along mixed-traffic lanes at a crucial segment of the corridor, and thus travel times and system control are negatively affected. ITDP.*

Source: https://brtguide.itdp.org/branch/master/guide/roadway-and-station-configurations/roadway-configurations.

Therefore, it is likely possible to tolerate brief and specific periods of mixed-traffic operation without jeopardizing the system's overall functionality. Longer stretches of mixed-traffic operation, however, may make the BRT system identical to a regular bus system. Such a design affects the system's psychological perception in addition to performance and operational control. The primary physical characteristic that distinguishes a BRT vehicle as a superior mode of transportation is the dedicated, priority lane allocated to it. Customers are able to form a mental map of the system in their minds thanks to the segregated lanes. Eliminating this segregation from a sizable portion of the system significantly reduces BRT's metro-like characteristics and decreases its appeal to riders who are using it voluntarily.

Queue-jump lanes are an option for mixed-traffic operations; they help provide some sort of priority during peak hours to prevent cars from becoming stuck in traffic (Figure 7.10).

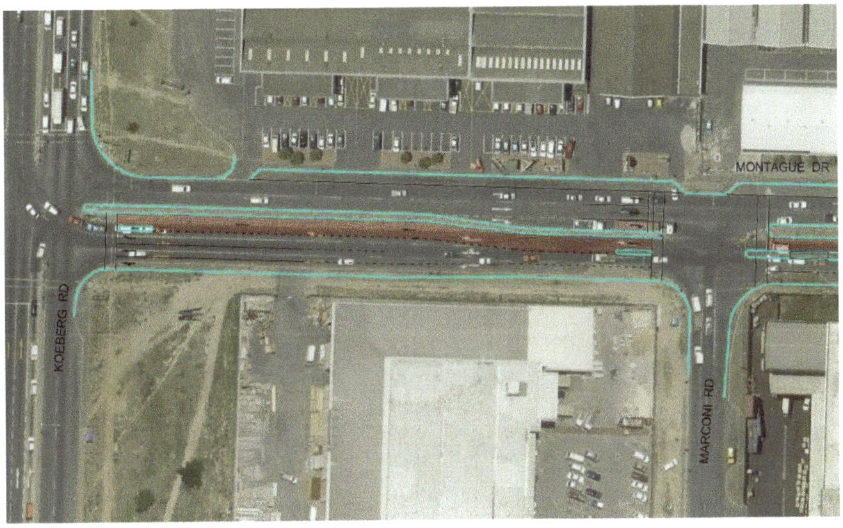

Figure 7.10. *A section of median queue-jump lane in Cape Town, South Africa.*

Source: https://brtguide.itdp.org/branch/master/guide/roadway-and-station-configurations/roadway-configurations

One can find queue-jump lanes in the center of the hallway or on the curbside. The location of these lanes is influenced by various factors, such as turn lanes, median breaks, driveway spacing and frequency, etc. The challenge of general traffic vehicles intruding into queue-jump lanes and the enforcement challenges arising from the need to cross the designated lane in order to access driveways, turn lanes, and median breaks are problems related to queue-jump lanes.

There are situations where general traffic is allowed to use the queue-jump lanes during off-peak hours, with access to the lanes being restricted only during peak hours. This should not be encouraged since it lessens the road space's exclusivity and increases the likelihood of violations during peak hours.

CASE STUDY

THE INTEGRATED DESIGN OF URBAN ROAD INTERSECTIONS: A CASE STUDY

A thorough evaluation of an urban road intersection's design is always necessary because it must both serve its intended functions and fit the unique characteristics of the location in which it will be implemented. Analyzing the infrastructure's quality and functionality is essential, particularly in urban areas. This includes considering its integration into the surrounding environment, potential user exploitation, viability, and technical-economic utility.

The design decisions must be carefully evaluated in light of the overall goals of the planned works, taking into account their coherence, geometric-functionality, dimensionality, landscaping, and architectural compatibility.

Examining the infrastructure's quality and usefulness is also essential, taking into account its viability and technical-economic convenience. These issues are occasionally created by the designers after the primary design decisions have been made; other times, certain issues are omitted to favor other issues like geometric, functional, economical, and so forth.

Methodological approach

The selection of the best option should result from an ever-more-complex optimization process; to follow this ethically correct course, they typically require technical and computational tools to integrate and naturally take into account different aspects of interest. The design tools that are currently available enable the collection of these aspects in an extensive and integrated analysis, even in the early phases of selecting the optimal solution, resulting in a high degree of definition and a constrained commitment to the necessary research. We specifically refer to software for designing road infrastructure, which is now widely used but is still largely unutilized in feasibility studies and early design phases.

Instead, a fresh, all-encompassing methodological strategy is required. Here, we describe a case study of an urban road intersection in Orte (VT, Italy), for which several software programs have been used to determine the optimal geometric-functional solution. For the case study that was presented, the optimization process was directed toward enhancing functionality and safety.

The main "logical steps" to determine the most appropriate solution, were:

- analysis of the present situation: location, road network, and territory;
- identifying needs and analysis of critical aspects;

- choice of the best possible solution;
- functional testing;
- geometrical modeling;
- environmental audit;
- technical and economic analysis

The current state of the intersection

Despite being a small to medium-sized town, Orte serves as a major hub for infrastructure networks (primarily for rail, but also for roads) primarily due to its convenient location and easy access to major destinations. The ability to swiftly enter Rome in particular has a significant draw for intermodal traffic; in fact, many users actually board the train at Orte station. Due to this fact, a significant number of vehicles are driven toward or toward the railroad in accordance with the time that trains depart and arrive.

The primary issue with the local road system is that the infrastructure is not up to par during peak hours. As a result, the regularly congested traffic conditions lead to critical environmental and technical issues in this area, which is significant from a landscaping and historical perspective.

In actuality, Orte is a very old town, having been a member of the Etruscan Federation. It reached its zenith of prosperity in the late Middle Ages, a period marked by numerous archaeological discoveries. Furthermore, the town is situated in a valuable and pleasant area, with the urban center in particular looking very suggestive.

Identifying needs and analysis of critical

Given that all flows must pass through the intersection in order to reach the train station or, if they are approaching it from the direction of their destinations, the intersection under consideration is situated at the intersection of short- and medium-distance traffic.

At the moment, the intersection lacks the necessary working equipment to effectively direct traffic flows and guarantee safe, secure travel. Because of the node's interference with pedestrian pathways and extremely limited visibility of conflict points, even cyclists and pedestrians experience extreme discomfort. In order to improve the features of an existing intersection, it is necessary to have accurate information about traffic flows. To this end, field surveys were conducted using continuously recording video cameras. Subsequently, the data underwent analysis, taking into account three vehicle categories: cars, trucks, and motorcycles, and determining the matrix O/D for each.

As anticipated, heavy traffic flows were observed in the morning and in the early afternoon and evening close to the railway station. Particularly in the latter case, the intersection amply demonstrates the complete absence of traffic demand; users find the lengthening travel times and reduction in Level of Service (LOS) intolerable. In addition, long lines and traffic congestion cause serious pollution issues as well as safety hazards from illegal overtaking.

Choosing the optimal solution

The goal of road design should be to guarantee the regular and safe movement of vehicles and pedestrians. The most efficient layout will depend on a number of factors, including budget, traffic patterns, the morphology of the area, the presence of buildings or other constraints, and the size and geometry of the intersection.

The best option in the case under consideration is to replace the current linear at the grade junction with a 36 m main diameter roundabout in order to increase safety and functionality while also lowering noise levels and pollution levels in the environment. However, there may be some inconvenience for pedestrian traffic in urban areas; therefore, it is important to take the drawbacks of the longer path into account.

Functional Testing

The roundabout is one of the most used design solutions to approach and solve problems related to the functionality of urban and suburban intersections. Three fundamental aspects are involved:

- The capacity (indicative of the efficiency in terms of vehicle flow stream)
- The level of service (linked to the delays accumulated during all maneuvers);
- The user's safety.

It is evident that a roundabout's capacity and service quality are dependent on two key elements: the geometric properties of the node and the traffic flow that operates within it.

In terms of safety, compared to a traditional grade-crossing intersection, a roundabout reduces the conflict points between various traffic flows. Other intriguing features of roundabouts include their wide field of vision, which facilitates a clear understanding of intersection operations and the control view of vehicles, and their lowered speed and small angle of incidence, which lessen the severity of consequences in the event of a collision between vehicles.

The design characteristics of the roundabout – mainly related to the inscribed circle diameter, the geometry of entry and exit legs, and the size of islands – are identified, within this framework, from the operating conditions and safety expectations.

The functional verification of the intersections is carried out through the traffic micro-simulation models, which simulate the behavior of vehicles within a predefined road network and are used to predict the likely impact of changes in traffic patterns resulting from traffic flow variability or changes to the physical environment. Thus, it is possible to evaluate the effects of design choices and for controlling traffic on local-scale road networks.

The software used for simulations, in this case study, is the Sidra Intersection® 4.0. This program allows, among other things, the evaluation of the average delay per vehicle, i.e., the average waiting time for a vehicle wishing to merge in the traffic flow of an intersection and the overall level of service (LOS) of the node, that is the quality indicator of vehicular flow.

Geometric Modeling of the intersection

A geometric 3D model of the intersection has been created, in order to verify the suitability of the design choice, depending on the consistency and compatibility of the work (point of view geometrical, functional, spatial, and architectural), both with the existing environment and land and with conditions of future use of infrastructure.

The operational steps considered are:

- construction of the model of the terrain surface (DTM) from the available maps (celerimetric survey, tables of land-use plan, and technical maps);
- geometrizing horizontal alignment of the roads of the intersection;
- geometrizing vertical alignment identifying the grade tangents and vertical curves;
- assembly of typical sections (lanes, shoulders, drainages, in cut or fill, etc.) one for each alignment;
- extruding of road solids and geometrical adjustment of terrain surface;
- rendering and visualization of intersection by means of suitable Software
- The geometric model of intersection has allowed us to carry out the design verifications, in terms of sight distances and environmental impact of the work.

Economic Analysis

The economic benefits and costs of replacing the actual three- three-stop-controlled intersection with a roundabout were compared. Economic benefits were based on the difference in road user costs among the two alternatives; for the roundabout, the construction costs were calculated. Road user costs were considered as composed of time delays and fuel consumption (due to idle fuel, stop and go, speed-change running, etc.) and they were calculated by a micro-simulation program. The economic benefits and costs can be estimated for the entire time frame of reference that extends from the beginning of construction work of the roundabout until the end of its life cycle.

Regarding the estimation of road user's benefits and costs, it was developed two separate economic analysis: the first on fuel consumption, and the second comprising also the average delay experienced by users, monetized properly.

Conclusions

The design of road infrastructures is a complex activity because of significant interactions among various involved components: environment, users, costs, etc. The procedures to perform an integrated design process, especially for road intersections, require three-dimensional geometric modeling as well as other computational and functional analysis; for these purposes can be useful specific software and other technical tools. A case study referred to an urban intersection in a particularly valuable territory was presented. In this case, various specialistic analyzes were needed, because the current status of the road junction is critical, especially for pollution and environmental damage.

The selected design solution was that to substitute the existing linear three-leg intersection with a 36 m main diameter roundabout. The results of functional analyzes demonstrate that the LOS improves in all traffic conditions. In addition, it is possible to estimate a strong reduction of noise and pollution in the infrastructure, with significant advantages on the environmental quality.

Regard to construction costs, the economic analysis demonstrates that after a few years (respectively about seven or four years if users delay were considered or not) the investment for roundabouts will be refund.

CLASS ACTIVITY

Create a class activity to explore the differences between urban and rural roadways can be an engaging way to teach students about transportation infrastructure, road safety, and the environmental and social impacts of road design.

SUMMARY

- The geometric design of the urban street and rural road networks is important to ensure that correct operation at the right road speed occurs safely and predictably.
- The design of the roadway and its alignment including intersection spacing and methods of control play a significant role in the safe operational performance of vehicles as well as the capacity of the roadway.
- The design speed of a road is the maximum speed at which a vehicle can safely travel on that road under good conditions.
- In the urban environment, the design speed of the road shall be the same or less than the intended speed limit of the street.
- Design vehicles are selected motor vehicles with the weight, dimensions, and operating characteristics used to establish highway design controls for accommodating vehicles of designated classes.
- Sight distance is the distance along the path that the user takes from a point of observation to a feature, obstacle, or another road user. Network utility pipework and ducting will only be allowed in the roadway where it is deemed to be of a transmission style, i.e., no local connections to adjacent property lots or catch pits.
- An initial alignment is prepared, taking into account any pre-agreed departures, and then further relaxations and departures can be established if required to achieve the best alignment.
- In addition to longitudinal fall, transverse fall is used to carry rainfall from the road surface to the edge or edges of the road. Without this, standing water may cause aquaplaning of vehicles due to the blocking of the tire tread.
- The location of the segregated busway within a specific roadway is a design decision that offers more options than might be immediately apparent.
- Grade separation, where the BRT corridor either runs on an elevated roadway or underground, is an option within narrow right-of-way configurations, as well as an option at very busy intersections and roundabouts.
- As an alternative to the transit mall, cities frequently consider splitting each direction of public transport service between two different (typically parallel) roads.
- Similar to the bi-directional, one-lane configuration, a virtual busway is a single bus lane in the middle of a roadway used by both directions of travel.
- The degree to which the direction of dedication is allocated to each direction of travel along the corridor will be dependent on the levels of congestion (and queue backup) in the general traffic lanes during the peak periods of operation of the corridor.

REVIEW QUESTIONS

1. Differentiate between design speed and speed limit.
2. Discuss about speed management in urban areas.
3. How does geometric design affect the alignment of highways?
4. Explain geometric design standards for urban roads and streets.
5. What are the guidelines for road drainage?
6. Write short notes on median busways, curbside busways, and fixed guided busways.
7. Discuss the advantages and disadvantages of guided busway systems.

REFERENCES

1. Carlsson, A., & Tapani, A. (2006). Rural highway design analysis through traffic micro-simulation. In *Proceedings of the 5th International Symposium on Highway Capacity and Quality of Service*, Vol. 2, Tokyo (pp. 249–258).
2. Cruzado, I., & Donnell, E. T. (2010). Factors affecting driver speed choice along two-lane rural highway transition zones. *Journal of Transportation Engineering, 136*(8), 755–764.
3. Donnell, E., Porter, R., & Shankar, V. (2010). A framework for estimating the safety effects of roadway lighting at intersections. *Safety Science, 48*(10), 1436–1444.
4. Fang, C., & Elefteriadou, L. (2005). Some guidelines for selecting microsimulation models for interchange traffic operation analysis. *Journal of Transportation Engineering, 131*(7), 535–543.
5. Obeidat, M. S., & Rys, M. J. (2016). Effective Light Source for Illuminating Overhead Guide Signs and Improving Roadway Safety. *Journal of the Transportation Research Forum, 55*(1), 103–119.
6. Ossenbruggen, P. J., Pendharkar, J., & Ivan, J. (2001). Roadway safety in rural and small urbanized areas. *Accident Analysis & Prevention, 33*(4), 485–498.
7. Steger, C. (1998). An unbiased detector of curvilinear structures. *IEEE Transactions on Pattern Analysis and Machine Intelligence, 20*(2), 113–125.
8. Steger, C., Glock, C., Eckstein, W., Mayer, H., & Radig, B. (1995). Model-based Road extraction from images. In *Automatic Extraction of Man-Made Objects from Aerial and Space Images* (pp. 275–284). Birkhauser Verlag Basel.
9. Van Houten, R., & Van Houten, F. (1987). The effects of a specific prompting sign on speed reduction. *Accident Analysis & Prevention, 19*(2), 115–117.
10. Vosselman, G., & de Knecht, J. (1995). Road tracing by profile matching and Kalman filtering. In *Automatic Extraction of Man-Made Objects from Aerial and Space Images* (pp. 265–274). Birkhauser Verlag Basel.

11. Walle, D. V. (2002). Choosing rural road investments to help reduce poverty. *World Development, 30*(4), 575–589.
12. Wiedemann, C., & Hinz, S. (1999). Automatic extraction and evaluation of road networks from satellite imagery. *International Archives of Photogrammetry and Remote Sensing, 32*(3–2W5), 95–100.

CHAPTER 8

Maintenance and Repair of Roads

LEARNING OBJECTIVES

After studying this chapter, you will be able to:

- Understand the importance of regular road maintenance and repair
- Identify the different types of road maintenance
- Know the causes of pavement failures
- Perform special repair of flexible & rigid pavements

KEY TERMS FROM THIS CHAPTER

- Proactive maintenance
- Potholes
- Economic growth
- Snowplows
- Culverts
- Ultraviolet rays
- Alligator cracks
- Thermal cracking
- Debris
- Livelihoods
- Pavement
- Ditches
- Drainage
- Asphalt cracking
- Rutting

INTRODUCTION

Road maintenance is a potentially dangerous job for both workers and other road users because it involves a variety of organizational and technical skills and is done on roads that are used by traffic. Therefore, in order to achieve effective and safe operations, it is imperative that each category of personnel involved in road maintenance activities receive the proper formal and on-the-job training. Typically, the foreman plays a significant part in both the initial and continuing training of staff members. Since roads represent a significant national investment, upkeep is necessary to keep them in good working order, guarantee safe travel at the right speed, and minimize costs to road users. Inadequate or delayed maintenance will result in higher final repair costs, more expenses and inconvenience for road users, and a decrease in safety. Thus, road maintenance is a crucial task that needs to be completed promptly.

8.1. THE IMPORTANCE OF REGULAR ROAD MAINTENANCE AND REPAIR

The road system, which provides access to residences, companies, and necessary services, is an important component of contemporary society. Regular road maintenance and repair keeps roads safe and functional for all users, including cyclists, pedestrians, and drivers. We will go over the advantages of routine road upkeep and repair in this section.

8.1.1. Safeguards Road Users

All users of the road can be kept safe with the help of routine road maintenance and repairs. This includes mending broken road signs, patching potholes, and enhancing visibility at crosswalks. The overall safety of road users is increased and the likelihood of accidents is decreased with a well-maintained road network.

8.1.2. Extends Road Life

A road's lifespan can be increased through proactive maintenance and repair, which lessens the need for expensive and disruptive complete road rebuilds. Frequent inspections and maintenance can find and fix minor issues before they get worse and cost more to fix.

8.1.3. Improves Road Conditions

The general state of roads can be greatly enhanced by routine maintenance and repair. This can involve patching potholes, clearing out trash and litter, and repaving uneven spots. Better road conditions lessen vehicle wear and tear and provide a smoother, more comfortable ride for both drivers and passengers.

8.1.4. Enhances Road Network Efficiency

The efficiency of the entire road network is increased by well-maintained roads. This involves easing traffic jams and enhancing the flow of traffic, which can save drivers' time and lessen the need for new road construction. Frequent upkeep and repair can also lessen the possibility of road closures, which can significantly affect nearby communities and businesses.

8.1.5. Improves Community Livelihoods

Local communities can benefit from having a well-maintained road network. Better road conditions can support local businesses by drawing in more clients and spurring economic expansion. Regular road upkeep and repair can also facilitate easier access to vital services like social services, healthcare, and education.

8.1.6. Healthcare Access

Better access to healthcare is one of the main advantages of routine road upkeep and repair. Inadequate road conditions can make it more difficult for emergency vehicles to reach people in need and for individuals to get to medical facilities. Maintaining and repairing roads regularly can facilitate access to healthcare services and increase the dependability of emergency response times.

8.1.7. Education Access

Keeping roads clean and in good repair can also increase accessibility to education. Rural schools and universities frequently depend on the road system to get employees and students to and from the campus. Inadequate road conditions can cause delays and make it harder for employees and students to get to their learning environments. Ensuring that staff and students can access educational facilities safely and efficiently can be achieved through routine road maintenance and repair.

8.1.8. Social Services Access

Apart from healthcare and education, consistent road upkeep and restoration can enhance the availability of social services. Social services, like social housing, libraries, and community centers, are frequently found in isolated locations and depend on the road system to make them accessible to people in need. Maintaining and repairing roads regularly can guarantee that these services are available and dependable, enhancing the general standard of living in nearby communities.

To put it briefly, maintaining and repairing roads regularly is crucial to enhancing community access to necessary services. The lives of people who live in rural or remote areas can be significantly impacted by improved access to social services, healthcare, and education. All road authorities should prioritize routine road maintenance and repair in order to guarantee that communities have access to vital services when they need them.

8.2. THE DIFFERENT TYPES OF ROAD MAINTENANCE

Our world is connected by roads, but sadly, they can also be the cause of many issues. This explains why there needs to be so much road maintenance. Numerous factors are involved, which is why a great deal of work is frequently done. Find out what constitutes each type of road maintenance and what it involves below. Three categories—routine, periodic, and emergency—are used to classify different degrees of road maintenance.

8.2.1. Pothole Patching

Snowplows and all the salt used on the roads to melt ice and snow during the winter are the main causes of potholes, which are terrible for your cars. The process of patching a pothole involves clearing the hole of all rocks and debris before adding patching materials. In cities with cold climates, potholes are frequently patched. Filling the holes is a temporary solution that keeps traffic moving safely and smoothly until the road is eventually repaved.

8.2.2. Clearing Pavement

Making sure that there is nothing on the road that could endanger drivers is another aspect of road maintenance. Anything from fallen trees to debris that could be hazardous if left on the road can be considered this. A lot of road clearing may also be caused by bad weather. Snowplows are used in the winter to remove extra snow from the ground. Spreading salt on the roads to melt ice is another wintertime preventive measure that was previously mentioned. It may also be necessary to remove various car parts that are on the road following an accident or in an emergency. Slab sawing is the best technique when it comes to removing concrete from roads and bridges. Anything that could endanger motorists on the road has been removed from the pavement.

8.2.3. Clearing Ditches and Culverts

For safe passage, culvert pipes and ditches must also be cleared. Maintenance workers typically use a water hose to empty ditches on the side of the road of any unwanted materials. Tree branches and other larger, heavier items must be manually or mechanically removed from ditches and culverts. When ditches and culvert pipes are removed correctly, there is no interference with their normal operation.

8.2.4. Proper Drainage

Flooding is another weather factor that can cause problems for the road. Roads must have adequate drainage in order to be ready for future flooding. If driving on the road is too dangerous, a temporary closure of the road may also be required. At this point, temporary signage will be erected to direct traffic away from the inundated road.

8.2.5. Shoulder Grading

Regular maintenance is also required for the shoulders of roadways. It is required to use a grader in order to complete this task. This shoulder technique aids in leveling and reshaping the shoulder. Typically, this work is done on rural roads without sidewalks in rural areas. Even though they aren't always the road itself, shoulders are still a crucial component that must be maintained.

8.2.6. Traffic Sign and Road Marking Repair

Road markings and traffic signs are necessary for traffic on roads, although they occasionally sustain damage. In order to preserve driver safety on the road, damaged signs and road markings—whether from a storm or an accident—need to be repaired. In addition to repairs, new signs may need to be installed for traffic signs. In addition, if the lane boundaries and roadway lines have faded and become less visible over time, road marking repair may necessitate repainting them. Temporary traffic cones and barricades will be erected on the road to direct drivers if construction is underway and vehicles need to be rerouted. To guarantee the safety of every driver on the road, all of these precautions must be taken.

8.2.7. Roadway Lighting

It can be risky to drive at night, especially on the highway where many vehicles are traveling quickly. Roadway lighting greatly improves visibility, and its maintenance is crucial. It's imperative that lights are in good working order and that any issues are fixed right away.

8.2.8. Bridges

Another extremely common and significant aspect of road maintenance is bridge work. A common area of work is on the bridge's expansion joints. This section of the bridge permits traffic to flow above the structure of the bridge. To ensure that traffic on the bridge is safe, maintenance is also performed on the sealants. Additional bridge maintenance tasks include painting, caulking, cleaning, lubricating bearings, clearing debris, caulking cracks, and other preservation-related tasks.

8.2.9. Vegetation Control

Grass, trees, and other vegetation can be found on the sides of most roads and highways as well as in the medians that separate them from boulevards. Because of this, keeping all the vegetation under control so that it doesn't start to harm the road or the driver is another part of road maintenance. A low-hanging tree branch that might be obstructing a sign, vegetation obstructing or limiting side road visibility, and trees that are too close to the road are a few examples of vegetation that impede road or driver visibility.

8.2.10. Resurfacing

Because of the heavy traffic and severe weather they experience year-round—from the sun to the snow—roads deteriorate quickly. Roads will get a new surface when it reaches a certain point. Asphalt overlay is possibly the most popular technique. This creates a smoother, new layer on the road that makes driving safer.

8.2.11. Expansion

As more people drive on roads every day, the amount of use for them increases. Some roads—particularly highways—are widened to handle the increased volume of traffic and commuters. Although there is actual traffic congestion, it can be avoided with a little upkeep. A road's expansion entails the addition of new lanes to boost traffic flow and travel efficiency. There is more to road maintenance than what you may think or already know. The various forms of road maintenance are required to guarantee the efficiency and safety of daily travel.

8.3. PAVEMENT FAILURES

Numerous factors, such as water intrusion, stress from large vehicles, expansion and contraction from seasonal temperature changes, and sun exposure, can lead to pavement failure. Maintaining appropriate care, such as crack and asphalt sealing, is essential to stop cracks from forming or growing larger.

8.3.1. Types of Failures in Flexible Pavements Due to Exposure

8.3.1.1. Weather Conditions

Change in weather imposes distress on flexible pavements.

8.3.1.2. Use of Chemical and Salts in Colder Climates

Potholes and other problems are caused by frost heave and the stripping of asphalt by snow and ice.

8.3.1.3. Ultraviolet Rays

The pavement becomes brittle due to oxidation caused by the ultraviolet rays. The temperature of the pavement can reach 140°C on a hot, bright day. This is the liquid asphalt's softening point. As a result, the pavement will spread and shift. The pavement will shrink as a result of the drop in temperature. The primary cause of the initial cracking is this expansion and contraction.

8.3.1.4. Water (Natural Rain and Irrigation)

Water can seep into the base and subgrade through the cracks, causing structural damage.

8.3.1.5. Vehicle Loads and Petroleum

The integrity of the pavement is weakened by the vehicles' fuel spills. This raises the binder's softening point. There is a chance that a parked car will leak brake or gasoline, liquefying the asphalt. This causes the binder to become separated from the rock, possibly resulting in softer sections. Therefore, it is important to treat oil spots on parking lots promptly.

8.3.1.6. Aging of Flexible Pavements

Aging is a pavement's life cycle degradation. As a result, oxidation and crack formation are greatly accelerated. The degree of the aging effect can be lessened by slight deterioration detected early on. The rate of deterioration increases with temperature and UV exposure.

8.3.2. Types of Failures in Flexible Pavements Due to Distress

The distress faced by the pavement can be of two types:

- Environmental distress, and
- Structural distress

8.3.2.1. Environmental Distress in Flexible Pavements

Environmental factors are the external influences that have an impact on pavement performance. These include issues related to aging, chemicals, snow, and water. Distress of this kind is seen from above down. Such issues can be resolved with a superficial application. These consist of skin-parching, chip seals, crack sealing, and seal coating. A hot mixed overlay may occasionally be applied to the surface as part of the treatment.

8.3.2.2. Structural Distress in Flexible Pavements

Physical failures on the pavement and sub-base are classified as structural failures. These structural breakdowns are the result of overloading, wet subgrade, frosting effect, or inadequate design standards. Distress of this kind arises from below. The only way to treat these is to remove them from the affected area and replace them, or R & R. Alternatively, repaving entails completely clearing, grinding, and pulverizing the surface before repaving.

8.3.3. Types of Failures in Flexible Pavements Due to Structural Distresses

Some of the structural distresses which can cause failures in flexible pavements are discussed below.

8.3.3.1. Cracking

Because asphalt cracks easily and can be caused by a wide range of factors, it accounts for a large portion of pavement failure issues. An asphalt surface will naturally crack with time as a result of soil settling and exposure to the elements, but incorrect mix design during pavement construction can also lead to some less common but equally common cracking problems.

8.3.3.1.1. Alligator Cracking

Map cracking is another name for alligator cracks. This is a result of the asphalt concrete's fatigue failure. Such distress is evident in the form of a sequence of interconnected cracks. The asphalt surface (base) is where the tensile stress is at its highest. This is the location of the fractures, i.e., the region under the highest tensile stress. Over time, a longitudinal crack parallel will spread and eventually reach the surface. Stress concentration and repeated loading will facilitate the connection of the individual cracks. These will have an alligator skin-like appearance or resemble chicken wire. We call this the "alligator cracking." Another name for it is the "crocodile cracking." These cracks are only seen in places where there is constant traffic loading. A major structural distress is alligator cracking. Rutting occurs later on in response to this distress. The pavement has developed alligator cracks, as seen in Figure 8.1.

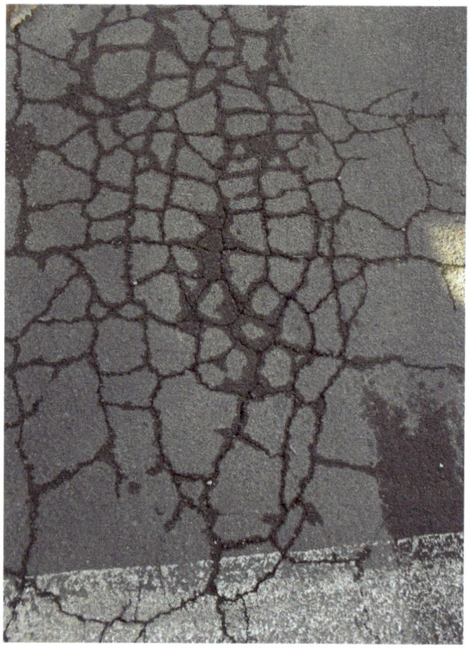

Figure 8.1. *Alligator cracking.*

Source: By Bidgee – Own work, CC BY-SA 3.0, https://commons.wikimedia.org/w/index.php?curid=7367880.

8.3.3.1.2. Edge Cracking

Heavy traffic, inadequate base material, and water damage are the main causes of edge cracking, which develops along a road's edge.

8.3.3.1.3. Block Cracking

Seasonal temperature variations cause the asphalt to expand and contract, which leads to block cracking. If the mix design of an asphalt surface is too rigid and does not allow for variations in seasonal density, block cracking will occur.

8.3.3.1.4. Joint Cracking

Cracks tend to develop in asphalt overlay projects where a flexible concrete base is laid down. As the concrete sub-base expands and contracts over time, cracks will form along the joints of the concrete (Figure 8.2).

Figure 8.2. *Cracking along the road axis possibly due to poor construction and formation of a cold joint.*

Source: By ŠJ⊠, Wikimedia Commons, CC BY-SA 3.0, https://commons.wikimedia.org/w/index.php?curid=16672803.

8.3.3.1.5. Transverse Cracking

Extreme temperature changes, improper paver operation, and settling or shifting base material are the main causes of transverse cracks that run across asphalt roadways.

8.3.3.1.6. Linear Cracking

The most common cause of linear cracking, which runs parallel to the road, is pavement fatigue, although it can also result from weak spots in a parking lot or from joints in the pavement.

8.3.3.2. Potholes

Potholes are created on road surfaces where a portion of the material has broken away, disrupting the surface. Another name for these is kettles. These are referred to as "chuckholes" in the West. The primary cause of pothole formation is pavement fatigue.

Figure 8.3. *Example of a pothole reappearing on a newly patched roadway, also showing the transition between crocodile cracking and the pothole, with water dried up, in a road on the Isle of Wight.*

Source: By Editor5807 – Own work, CC BY 3.0, https://commons.wikimedia.org/w/index.php?curid=11776596.

Alligator cracking is the result of fatigue cracking interlocking with one another. Under constant loading and stress, these pieces that have separated from the pavement's cracks will become loose and fall out. The pavement will develop a pothole as a result. The freezing and thawing action caused by the water trapped in the pothole in cold weather causes more stress and the spread of cracks. As soon as a pothole appears, the distress increases, and more pavement must be removed. This rate of distress expansion will increase if there is water entrapment. The pothole may enlarge to a width of several feet. They don't go into great detail. Large potholes have caused damage to the tires of the vehicle (Figure 8.3).

8.3.3.3. Depression

There are localized patches of pavement that are lower in elevation than the surrounding pavement level. These lowering are pavement-based indentations. They are mostly only noticeable after it has rained and they are filled with water. In parking lot construction as well as overlays, depressions in flexible pavements are a very common distress. These depressions may have developed during construction or they may have been brought on by the foundation soil settling from ongoing loads. When creating a depression in flexible pavement for an airfield, various degrees of severity are taken into account.

8.3.3.4. Rutting

Rutting is the term for channeled depressions in an asphalt surface that develop over time as a result of erroneous base construction and weight limits being exceeded. Tire ruts are caused by heavy vehicles compacting the asphalt on a road over time. If an asphalt surface develops rutting, the base needs to be fixed or redesigned to accommodate more traffic.

8.3.3.5. Raveling

Degradation of an asphalt top coat and

ongoing water intrusion are the causes of raveling. Water and sunlight will continue to erode the asphalt surface after the top layer of bitumen has worn away, breaking the bond that holds asphalt bitumen and aggregate rock together. An asphalt surface that has raveled will no longer be impermeable and will allow water to seep through. Water intrusion into an asphalt surface will lead to more cracks and eventual pavement failure.

8.3.3.6. Swelling of Flexible Pavements

These are long and gradual waves of distress that can reach up to ten feet in length. This swelling distress is marked by a bulge in the pavement surface. Surface cracking occurs after swelling. The primary cause of swelling in the flexible pavement is frost action in the subgrade, leading to soil swelling.

8.3.4. Types of Failures in Flexible Pavements Due to Environmental Distresses

8.3.4.1. Bleeding in Flexible Pavements

Bleeding is the term used to describe the process by which an asphalt binder film forms on the pavement's surface. A reflective surface that resembles shiny glass will result from bleeding. There will be bubbles in the layer that resemble blisters. The resulting asphalt binder will have a sticky texture. Bleeding occurs as a result of asphalt binder expanding into aggregate voids during hot weather and then filling them again later. They stay on top of the pavement as such because the bleeding process cannot be reversed in cold weather. The following conditions may be the cause of the bleeding:

- Excessive asphalt binder in the mix
- Excessive application of the binder during surface treatment
- Lower air void content – no adequate voids for the bitumen to penetrate

8.3.4.2. Block Cracking in Flexible Pavements

Another name for this is thermal cracking. Blocks are breaking as a result of the cracking. The pavement splits into nearly rectangular pieces due to the interconnectedness of these cracks. Each rectangle's dimensions can range from one foot by one foot to ten feet by ten feet. This covers a sizable portion of the pavement. However, these are seen in places with little traffic. Because it is a result of exposure to the environment, this phenomenon is known as thermal cracking. The effects of aging and temperature are the potential causes.

8.3.4.3. Bumps and Sags

Bumpy pavement is defined as a small, localized area of pavement that has experienced upward displacement. These are brought on by the pavement's instability factor. The formation of bumps is influenced by multiple factors. They may even be brought on by the concrete slabs bulging or buckling. These failures are seen in areas where asphalt pavement is placed over concrete pavement. Frost heaves, which produce bumps as a result of expansion, are another factor in bump formation. It is oxidation that causes the crack edges to become pronounced. Pavement bumps can also be caused by plant roots growing beneath the pavement. The pavement surface's displacement or settling is the primary cause of the sags. Sags are localized, sharp, and tiny. Sags in the pavement can cause large or long dips in it.

8.3.4.4. Edge Cracking in Flexible Pavements

Pavements with unconfined asphalt are known to experience edge cracking. The pavement's edges will begin to yield during the compaction process, particularly if there are no curbs or edge barriers to act as a barrier. With age, the edges will yield, oxidize, and become brittle. The 'C'-shaped edge cracking that forms along parking lots, roadways, and street edges is visible.

8.3.4.5. Joint Reflection Cracking

These are cracks visible in the flexible layer covering a hard pavement. These cracks are caused by the hard pavement joints that are beneath.

8.3.4.6. Raveling

The hot mixed asphalt will gradually disintegrate as the aggregate particles are dislodged, moving from the surface downward. We refer to this failure as raveling. The loss of bonding between the aggregate particles and the asphalt binder causes this dislodgement. There are instances when dust particles coat the aggregates, preventing bonding. As a result, the aggregate will bind with the dust instead of the binder.

8.3.4.7. Cold Joints in Flexible Pavements

These are the longitudinal joints that the asphalt pavement has formed. When hot mix asphalt is poured next to an already-existing pavement, this failure happens. This type of failure is most frequently seen in parking lots, inverted crowns, and low-traffic areas. There will be a difference between the two layers as a result of the variations in plasticity and temperature. As a result, there will be a longitudinal joint created between the laid asphalt mats. When considering other pavements, the longitudinal joint has a lower density. These longitudinal joints, also referred to as the "cold joint," will eventually allow water intrusion. It shortens the pavement's lifespan by making it rougher.

8.4. SPECIAL REPAIR OF FLEXIBLE & RIGID PAVEMENTS

Increased vehicle traffic and shifting environmental factors will shorten pavement lifespans and ultimately cause them to fail, regardless of the caliber of pavement material and design. To improve the service life of pavements, it is important to understand the types and causes of failure of both rigid and non-rigid pavements. This knowledge will help in the development of appropriate repair and rehabilitation programs. The kind of distress determines how stiff pavements should be repaired or restored. The most prevalent characteristic of rigid pavement is cracking, and fatigue cracking is thought to be the main reason behind this failure.

The main factor determining how many road repetitions are necessary to cause fatigue cracking is the stress ratio between the concrete's modulus of rupture and flexural tensile stress. Some of the reasons why rigid pavements fail include pumping, faulting, spalling, shrinking, polished aggregate, pop-outs, degradation of the joint load transfer system, linear cracking, durability cracking, corner break, alkali-aggregate reaction, and blow-ups.

Hard concrete surface texturing, diamond grinding, dowel bar retrofitting, joint repair, partial and full depth repairs, crack sealing, and so forth. These are the methods most frequently employed to restore or rehab stiff road pavements.

The infrastructure for roads was greatly expanded by the quick rise in rigid road construction, but it was not maintained and eventually fell into disrepair. Because the damage is frequently so extensive, routine maintenance is no longer sufficient. Rehabilitation or even reconstruction work is required to fully restore the roads, and this work has a lifecycle cost that is three to seven times higher than that of preventive maintenance strategies.

If rigid pavements are properly designed, built, and maintained, they can last for a very long time. When timely repairs and rehabilitation are performed, these pavements can last for the duration of their intended service life and even longer. Pavements of all kinds deteriorate over time, but the pace of deterioration in rigid pavements is significantly slower than in flexible pavements.

Maintaining rigid pavement involves more than just implementing specific techniques; it is a mindset that drives financial planning and asset management within an agency. Pavement management should be customized to address the specific distress of each road agency's system efficiently and cost-effectively. This includes utilizing a range of treatments and repairs to prolong the lifespan of rigid road pavements.

The majority of pavement designs have two or more performance times. At some point during its design life, a pavement is rehabilitated to an acceptable level from its initial state of serviceability. Depending on the state of the rigid pavement that is currently in place, this rehabilitation process may need to be repeated multiple times.

For rigid road pavement, an operational methodology or technique for managing, directing, and controlling the prudent use of maintenance resources is called a Maintenance Management System (MMS). Doing the right thing in the right place at the right time is the foundation of an effective MMS. For the benefit of end users, modern rural roads and highways must be run effectively and efficiently. The performance of the road drainage system and overloading are essential for maintaining the rigid pavement.

At the Ministry of Road Transport and Highways' headquarters, a client-server computer system platform for the Road Management System must be set up (MoRTH). For the suggested planning model, software for GIS, HDM, image processing, terrain modeling, software, and AutoCAD might be needed. It might take a creative team of consultants with experience in each of these fields to create the system from scratch, customize it, and then deliver it to the client.

8.4.1. Distresses of Rigid Pavement

The distresses that occur in rigid pavements include potholes, scaling, blow-ups, mud pumping, spalling, loss of sealant, cracking (longitudinal, transverse, reflection, and sympathetic), corrugation, joint deficiencies, and faulting. During the data collection procedures, however, only cracking, potholes, joint deficiencies, and faulting were taken into consideration.

8.4.2. Pavement Condition Index (PCI)

- PCI is a numerical indicator of overall pavement condition
- It takes into account the roughness and distress of pavement

It is used for:

- As a trigger to plan maintenance and rehabilitation activities
- Identification of immediate pavement maintenance and rehabilitation needs
- Ranking of roads for maintenance and rehabilitation
- Developing road maintenance for short- and long-term plans, and budget
- Policy for minimum service level for various roads' Pavement Maintenance System (PMS).

8.4.3. Treatment Methods for Flexible Pavements

Many factors, such as water intrusion, stress from large vehicles, expansion and contraction from seasonal temperature changes, and sun exposure, can lead to flexible pavement failure. To stop cracks from forming and growing, it is critical to maintain proper maintenance practices like crack and asphalt sealing. When a vehicle turns or brakes, forces are applied, which causes cracks to form. Weakness in the base or subgrade layers or instability in the asphalt mix are the main causes of distortions in asphalt pavements.

8.4.3.1. Alligator/Fatigue Cracking

Avoiding the common causes can go a long way toward preventing fatigue cracking. For instance, fatigue cracking can frequently be avoided by lessening overloading on asphalt pavement or enhancing drainage. The main components of prevention are creating a subbase and pavement that can withstand the anticipated traffic loads and having adequate drainage to keep water out of the subbase.

Increasing the asphalt layer's depth is a smart way to avoid overloading, which is a major contributor to fatigue cracking. Some researchers claim that pavements that are thicker or stronger than a particular threshold could theoretically support an endless number of loads without exhibiting structural flaws like fatigue cracking. Long-term performance pavements, or perpetual pavements, are the names given to these pavements (LTPP).

It is important to identify the primary cause of the distress when fixing pavement damaged by fatigue cracking. However, it is frequently difficult to pinpoint the exact cause, which makes prevention equally challenging. The structural composition of the pavement and the presence or absence of subsurface moisture should be ascertained by excavating a pit or coring the pavement and subbase as part of any investigation. The degree and severity of the cracking also affect the type of repair that is required. Preventative crack filling is a good procedure that can help stop potholes from forming in the future for minor cracks. Early on, using crack sealant to seal cracks prevents additional subgrade degradation from moisture infiltration. Small sections can be fixed by excavating the damaged area and replacing it with a fresh base and asphalt surface. A structural asphalt overlay or full reconstruction is required to ensure structural integrity once the damage has advanced or the affected area is large and extensive. A tack coat applied over paving fabric, milling the damaged asphalt, or applying crack sealant to first seal cracks are some examples of proper repair techniques. The finished repair is then covered with a hot mix asphalt overlay.

8.4.3.2. Bleeding

The asphalt binder film on the pavement's surface may be removed or reduced by the following repair techniques, but they might not address the underlying issue that resulted in the bleeding.

- Using coarse sand to remove extra asphalt binder can frequently stop minor bleeding.
- Significant bleeding can be fixed by using a heater planer or a motor grader to remove extra asphalt. Resurfacing might be required if the final surface is overly rough.

8.4.3.3. Block Cracking

Strategies depend upon the severity and extent of the block cracking:

- Low severity cracks (< 1/2 inch wide). Crack seal to prevent (1) entry of moisture into the subgrade through the cracks and (2) further raveling of the crack edges. HMA can provide years of satisfactory service after developing small cracks if they are kept sealed.
- High severity cracks (> 1/2-inch-wide and cracks with raveled edges). Remove and replace the

cracked pavement layer with an overlay.

8.4.3.4. Corrugation and Shoving

To find the source of failure, a heavily shoveled or corrugated pavement should be examined. One of two types of repair strategies typically applies:

- Shoving or small, localized corrugated areas. Take out the damaged pavement and apply a patch.
- Large areas that are shoveled or corrugated are a sign of general HMA failure. Take out the overlay and damaged pavement.

8.4.3.5. Depression

Depressions, defined as small localized areas, should be examined to identify what caused the failure (such as subgrade settlement or frost heave). To repair a pavement depression, the affected area of pavement should be removed, the poor subgrade dug out and replaced, and then patched over to restore the surface.

8.4.3.6. Joint Reflection Cracking

The degree and severity of the cracking determine the strategy to be used:

- Low severity cracks (< 1/2-inch-wide and infrequent cracks). Crack seal to prevent (1) entry of moisture into the subgrade through the cracks, and (2) further raveling of the crack edges. In general, rigid pavement joints will eventually reflect through an HMA overlay without proper surface preparation.
- High severity cracks (> 1/2-inch-wide cracks that are numerous). The layer of cracked pavement should be removed and covered with an overlay.

8.4.3.7. Longitudinal Cracking

A crack or discontinuity in a pavement that runs generally parallel to the pavement centerline. Longitudinal cracks may occur as a result of poorly constructed paving lane joints, thermal shrinkage, inadequate support, reflection from underlying layers, or as a precursor to fatigue cracking. Longitudinal cracking that occurs in the wheel path is generally indicative of alligator cracking.

8.4.3.8. Patching/Patching Deterioration

In and of itself, patches are a repair action. They can only be eliminated from the surface of a pavement by a structural overlay or non-structural overlay.

8.4.3.9. Polished Aggregates

Apply a skid-resistant slurry seal or BST or overlay.

8.4.3.10. Potholes

There are two types of pothole patching techniques: semi-permanent and temporary. When the weather is not conducive to a longer-term fix, temporary patching is the only option. It typically involves applying a cold mix asphalt patching compound quickly to temporarily restore pavement smoothness. In semi-permanent patching, the failed area's perimeter is carefully rebuilt to blend in with the surrounding pavement. A hot-mix asphalt fill is typically used in addition to the replacement of the proper base materials.

Several repair methods, such as throw-and-roll, semi-permanent, spray injection, and edge seal, are included in the Federal Highway Administration's (FHWA) summary of best practices. The FHWA recommends spray injection, throw-and-roll, semi-permanent, or edge seal procedures as the best patching methods outside of the winter. The throw-and-roll method may be your only choice in the winter. Similar techniques for patching potholes are provided by the Council for Scientific and Industrial Research in South Africa.

8.4.3.10.1. Materials

The two main types of aggregate and binder used in asphaltic patch materials are hot mix and cold mix. Some agencies use hot mixes, which are made at nearby asphalt plants. The FHWA handbook lists three different kinds of cold mixes that are made at a nearby asphalt plant: (1) using the aggregate and binder that are on hand; or (2) in accordance with the guidelines established by the organization that will be using the mix. A proprietary cold mix that is produced in accordance with a specified standard is the third kind.

8.4.3.10.2. Throw-and-roll Repair

The most basic method is the throw-and-roll technique, which works best as a temporary fix in situations where material placement is difficult to control, like during the winter. It is made up of:

- Placing the hot or cold patch material into a pothole
- Compacting the patch with a vehicle, such as a truck
- Achieving a crown on the compacted patch of between 3 and 6 mm

Because of its ease of use and speed, this method is frequently employed, particularly when the material is subjected to unfavorable water or temperature conditions. For longer-lasting effects, it can also be used when the pothole is clean and dry. Eaton et al., pointed out that quick fixes have a high failure rate and can cost up to five times as much as well-done repairs. They only support this kind of repair when inclement weather makes other methods impractical.

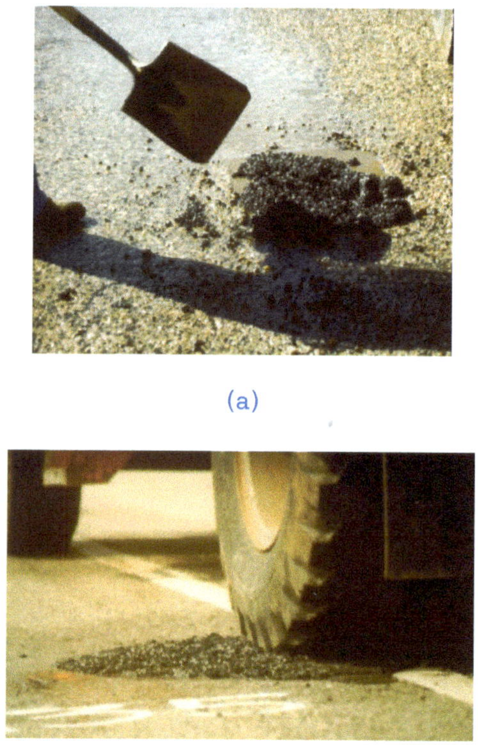

(a)

(b)

Figure 8.4. *(a) Throw-and-roll pothole repair procedure—material placement. (b) Throw-and-roll pothole repair procedure—compaction of patch.*

Source: By T.P. Wilson and A.R. Romine – http://www.fhwa.dot.gov/publications/research/infrastructure/pavements/ltpp/99168/99168.pdf, Public Domain, https://commons.wikimedia.org/w/index.php?curid=31236052.

Researchers at the University of Minnesota Duluth have experimented with combining asphalt with iron ore that contains magnetite. The combined asphalt is then heated using ferromagnetic resonance, which involves heating the asphalt with microwaves at a particular frequency. A compound containing one to two percent magnetite was used in the mixture. The team found that a patch of material could be heated to 100°C (212°F) in about ten minutes (Figure 8.4).

8.4.3.10.3. Semi-permanent Repair

If complete roadway replacement is not feasible, the FHWA handbook lists the semi-permanent repair technique as one of the best for patching potholes. It is made up of:

- Removing water and debris from the pothole
- To ensure that the vertical sides of the repair are in sound pavement, make clean cuts along the edges of the potential patch area. (Eaton, et al., advise applying a bituminous tack coat to the exposed cavity before applying the patch material.)
- Placing the hot or cold patch mix material
- Compacting the patch with a device that is smaller than the patch area, e.g. vibratory rollers or a vibratory plate

Although this repair method yields long-lasting results, it is more labor- and equipment-intensive than the throw-and-roll or spray-injection methods (Figure 8.5).

Figure 8.5. *Semi-permanent pothole repair procedure—small vibratory compactor for larger repairs.*

Source: By Kenneth Allen, CC BY-SA 2.0, https://commons.wikimedia.org/w/index.php?curid=14436051.

8.4.3.10.4. Spray-injection Repair

Spray-injection is recommended by the FHWA handbook as a cost-effective substitute for semi-permanent repair. But specific equipment is needed for it.

The process involves clearing the pothole of water and debris, applying a tack coat of binder on its sides and bottom, and then blowing asphalt and aggregate into it. Once the cover aggregate has been added, the patched area is covered with a layer of aggregate.

Compaction is not necessary after this step (Figure 8.6).

Figure 8.6. *Spray-injection device for pothole repair—all-in-one repair unit in Montreal, which allows the operator to work from the comfort and safety of the cab.*

Source: By Jean Gagnon – Own work, CC BY-SA 3.0, https://commons.wikimedia.org/w/index.php?curid=15277429.

8.4.3.10.5. Edge Seal Repair

The edge seal method is recommended as an alternative to the aforementioned techniques in the FHWA manual. It entails:

- Completing the "throw-and-roll" procedures.
- Once the repaired area has dried, overlap the pavement and the patch by applying a ribbon of asphaltic tack material over the patch edge.
- Sand is applied to the tack material to stop tires from tracking.

In this process, it might take two visits to apply the tack coat while waiting for any water to dry. The tack material serves to seal the patch to the surrounding pavement and keeps water from seeping through the patch's edge.

8.4.3.10.6. Efficacy of Repair Methods

The "throw-and-roll technique proved as effective as the semi-permanent procedure when the two procedures were compared directly, using similar materials," according to a study funded by the FHWA. It also discovered that, when employing high-quality materials, the throw-and-roll method is typically more economical. Furthermore, it was discovered that, depending on the equipment operator's skill level, spray injection repairs were just as successful as control patches.

8.4.3.11. Rutting

To identify the underlying reason for failure, a severely rutted pavement should be examined (e.g. inadequate compaction, rutting in the subgrade, inadequate mix design, or tire wear with studding). In general, minor ruts that are less than 1/3 of an inch deep can be ignored. Deeper grooves in the pavement should be leveled and covered.

8.4.3.12. Slippage Cracking

Crescent- or half-moon-shaped cracks with two ends pointing in the direction of traffic are known as slippage cracks. The impacted area is removed and replaced.

8.4.3.13. Stripping

The loss of bond between aggregates and asphalt binder, known as stripping, usually starts at the bottom of the HMA layer and moves upward. Investigating a stripped pavement is necessary to identify the underlying reason for failure (i.e., how did the wetness enter?). Generally speaking, any subsurface drainage problems must be fixed before replacing the stripped pavement.

8.4.3.14. Transverse (Thermal) Cracking

Cracks perpendicular to the centerline or laydown direction of the pavement are known as transverse or thermal cracking. often a form of thermal cracking. The degree and severity of the cracking determine the appropriate strategy.

- Cracks of low severity (< 1/2-inch-wide and infrequent cracks). Crack seal to prevent (1) entry of moisture into the subgrade through the cracks, and (2) further raveling of the crack edges. HMA can provide years of satisfactory service after developing small cracks if they are kept sealed.
- High severity cracks (> 1/2-inch broad with multiple fractures). Remove the cracked layer of pavement and cover it with an overlay.

8.4.3.15. Water Bleeding and Pumping

To find the source, water bleeding or pumping should be examined. Subgrade drainage needs to be improved if a high water table or inadequate drainage is the issue. A fog seal or slurry seal may be used to reduce water infiltration if the issue is a porous mix (in the case of water bleeding).

8.4.3.16. Weathering/Raveling

A raveled pavement should be investigated to determine the root cause of failure. Repair strategies generally fall into one of two categories:

- Localized, tiny patches of ravel. Take out the cracked pavement and repair it.
- Large, ravaged areas are a sign of a general HMA breakdown. Take out the overlay and damaged pavement.

8.4.4. Treatment Methods for Rigid Pavements

The primary reasons why rigid pavements fail due to faulting are material eroding or pumping beneath the pavement, creating voids beneath the pavement slab that lead to settlement. Slab edges curl due to variations in moisture content and temperature. Pavement failure is caused by inadequate soil support. Hydraulic binders, which stabilize the soil, can help stop moisture-induced soil deterioration. Erosion caused by subterranean water pipe leakage weakens the soil beneath the pavement. The pavement may develop a hole or a small area of void as a result.

8.4.4.1. Blowup (Buckling)

A localized upward slab movement and shattering at a joint or crack is called blowup or buckling. It usually happens in the spring or summer and is caused by the slab expanding too much in hot weather.

Potential Causes: During the winter (e.g. PCC slabs shrink in the winter, resulting in broader joint openings. In the event that incompressible material—such as rocks or soil—fills these openings, the PCC slab will expand later on during hot weather. g. spring, and summer) could result in significant compressive strains. In order to release the tensions, the slabs may buckle and break if they are sufficiently high. The following can quicken blowup.

- Joint spalling (reduces slab contact area and provides incompressible material to fill the joint/crack).
- D cracking (weakens the slab near the joint/crack area).
- Freeze-thaw damage (weakens the slab near the joint/crack area).

8.4.4.2. Corner Break

A corner break is a crack that runs through the entire PCC slab and intersects the joints near the corner, typically defined as within approximately 2 meters (6 feet). These cracks are caused by high stresses at the corner, which result from repeated loads, lack of support, inadequate transfer of load across joints, curling stresses, and warping stresses.

8.4.4.3. Durability Cracking (D Cracking or Alkali Silica Reaction ASR)

A cluster of closely spaced, crescent-shaped cracks near a joint, corner, or crack is called durability cracking. It results from the big aggregate in the PCC slab expanding due to freeze-thaw cycles. Durability cracking is not specific to pavement PCC; rather, it is a general PCC distress.

"D" cracking is indicative of a general aggregate freeze-thaw problem. Although a full-depth patch or partial-depth patch can repair the affected area, it does not address the root problem and will not, of course, prevent "D" cracking elsewhere.

8.4.4.4. Faulting

The definition of faulting is the variation in elevation across a crack or joint that is typically connected to undoweled JPCP. Typically, the approach slab is elevated above the leave slab because of pumping, which is the most frequent cause of faulting. When the average faulting in the pavement section reaches approximately 2.5 mm (0.1 inch), faulting becomes apparent. Consideration should be given to diamond grinding or other rehabilitation techniques when the average faulting reaches 4 mm (0.15 in).

It is not necessary to repair faulting heights of less than 3 mm (0.125 inch). A dowel bar retrofit is a possibility if there is a fault in an undoweled JPCP between 3 mm (0.125 inches) and 12.5 mm (0.5 inches). When there is more than 12 mm (0.5 in) of faulting, complete reconstruction is usually necessary.

8.4.4.5. Joint Load Transfer System Deterioration

Joint dowels cause a transverse crack or corner break, which is known as joint load transfer system deterioration. The impacted joint load transfer system must be removed and replaced, and the affected area must then be fully patched.

8.4.4.6. Linear (Panel) Cracking

By using crack sealing, slabs with a single, narrow linear crack can be fixed. Generally, a full-depth patch is necessary when there are multiple linear cracks.

8.4.4.7. Patching

In and of itself, patches are a repair action. They can only be eliminated by replacing the slab or applying an overlay.

8.4.4.8. Polished Aggregates

Areas of PCC pavement with polished aggregates have aggregate particles in the portion of the aggregate that is visible on the surface traffic applications made repeatedly. The rough, angular particles that protrude from a pavement usually get polished as it ages. If the aggregate is prone to abrasion or experiences excessive wear from studded tires, this may happen more quickly.

Repair: Diamond grinding or overlay.

8.4.4.9. Popouts

Popouts are tiny PCC fragments that separate from the surface and leave tiny divots or pockmarks behind. Popouts have diameters between 25 and 100 mm (1 and 4 inches) and depths between 25 and 50 mm (1 and 2 inches). Popouts are typically caused by inadequate aggregate durability. Several things, including the following, can contribute to poor durability.

- Poor aggregate freeze-thaw resistance.
- Expansive aggregates.
- Alkali-aggregate reactions.

Repair: It might not be necessary to repair isolated, low-severity popouts. A partial-depth patch can typically be used to repair larger popouts or a collection of popouts.

8.4.4.10. Pumping/Water Bleeding

Material moving beneath the slab or material being ejected from underneath the slab as a result of water pressure is known as pumping or water bleeding. When a PCC slab deflects under load, water that has accumulated beneath it will pressurize. This pressurized water can do one of the following:

- Under the slab, move around.
- Go under one slab and then under another that is nearby. Faulting results from the movement of this kind.
- Step out onto the pavement's surface from beneath the slab. As a result, the base, subbase, and/or subgrade material beneath the slab gradually disappear, reducing the structural support.

Water buildup under the slab. A high-water table, inadequate drainage, panel cracks, or shoddy joint seals that let water seep through to the underlying material are a few possible causes of this.

Repair: In order to eliminate any deteriorated slab areas, the pumping area should first be repaired with a full-depth patch. Secondly, dowel bars should be considered in order to improve the load transfer across any large transverse joints that the repair has created. Third, since a sizable portion of the slabs' underlying base, subbase, or subgrade may have been removed by the pumping, thought should be given to stabilizing any slabs close to the pumping area. Lastly, the cause of inadequate drainage or the water source should be dealt with.

8.4.4.11. Punchout

Punchout is the splitting of a localized slab section into multiple pieces. Usually only a problem with CRCP. May be a sign of a specific building flaw, like poor consolidation. It can be brought on in CRCP by steel corrosion, insufficient steel, or shrinkage cracks that are either too wide or too close.

8.4.4.12. Reactive Aggregates Distresses

Reactive aggregate distresses are a pattern or map of reactive aggregate-induced crazing or cracking on the surface of PCC slabs. Reactive aggregates are those that, when exposed to specific chemical compounds, either expand or develop expansive byproducts. Distress of this kind is a sign of subpar overall characteristics. It is most frequently the outcome of an alkali-aggregate reaction.

Repair: Replace the slab for large areas of scaling, or apply a partial-depth patch for smaller areas of scaling.

8.4.4.13. Shrinkage Cracking

Hairline cracks that do not occur at joints and are created during PCC setting and curing are referred to as shrinkage cracking. They typically don't go all the way through the slab's depth. If shrinkage cracks appear in an uncontrollably manner, they are regarded as distress either too close together in CRCP or outside of contraction joints in JPCP). Since all PCC shrinks as it solidifies and cures, rigid pavement should anticipate shrinkage cracks and take precautions to prevent them. Uncontrolled shrinkage cracking, however, may be a sign of sawed contraction joints too late.

- If contraction joints in JPCP are sawed too late, the PCC may have already cracked in an unfavorable place. Bad design of the reinforcing steel.
- Proper design of the reinforcing steel in CRCP should cause shrinkage cracks every 1-2 m (4-10 ft) improper method of curing.
- Slab surfaces shrink and crack too quickly if they are allowed to dry too quickly.
- High PCC early strength. High early-strength PCC may be used to try to open a newly built or rehabilitated section to traffic as soon as possible. This kind of PCC shrinks more quickly and completely than ordinary PCC made from unmodified Type 1 Portland cement, and it can have a high heat of hydration.

8.4.4.14. Spalling

The term "spalling" refers to the breaking, chipping, or cracking of joint or crack edges usually happens within 0.6 m (2 feet) of the joint or crack edge. Potential reasons include:

- Excessive stresses at the joint or crack brought on by the expansion that follows the infiltration of incompressible materials (may also result in blowups).
- PCC disintegration due to "D" cracking or freeze-thaw action.
- PCC at a joint that is weak because of insufficient consolidation during building. This can occasionally happen at a construction joint if two things happen: (1) the final bit of slab volume is filled in with low-quality PCC, or (2) dowels are inserted incorrectly.
- Dowel corrosion or misalignment.
- High volume of traffic.

Repair: Partial-depth patches are typically effective in repairing spalling that is less than 75 mm (3 inches) from the crack face. A full-depth patch should be used to repair any spalling that extends farther than approximately 75 mm (3 inches) from the crack face, as this could indicate possible spalling at the joint bottom.

8.4.5. Effect of Different Maintenance Strategies

The majority of pavement designs have two or more operating times. A pavement is built with an initial level of serviceability and, at some point in its design life, is rehabilitated to an acceptable level. Depending on how badly the current pavement deteriorates over time, this rehabilitation process may be repeated multiple times.

8.4.6. Maintenance Management Systems (MMS)

MMS is an operational methodology or technique for overseeing, managing, and regulating the prudent use of maintenance resources. Doing the right thing in the right place at the right time is the foundation of an effective MMS. Modern roads and highways must satisfy customer-defined requirements in order to function effectively and efficiently for the benefit of all users. The effectiveness of road drainage is essential to maintaining pavement performance.

8.4.7. Pavement Management Systems (PMS)

PMS constitutes the following aspects:

- Pavement Condition Data
- Maintenance Standards
- Economic Analysis
- Programming
- Control of Works and Feedback.
- Budgetary Analysis

8.4.8. Assessment of Maintenance Needs

Maintenance needs are assessed every year as part of planning and considering the following factors:

- Traffic
- Survey
- Visual rating
- Roughness measurements
- Axle Load Survey

Some rigid pavement repairs and maintenance techniques:

- Concrete pavement restoration (CPR)
- Crack sealing
- Cross crack stitching (Longitudinal and Transverse crack)
- Diamond grinding for rectification of pavement faulting
- Dowel bar retrofit
- Isolated, partial, or full-depth concrete repair
- Joint resealing
- Repair of edge spall and corner
- Breaks of slabs
- Slab lifting or slab jacking or under-sealing
- Ultra-thin white topping
- Evaluate Structural Adequacy
- Causative factors that cause distress
- Select the course of maintenance
- Patching
- Sealing
- Re-surfacing
- Re-texturing
- Diamond grinding

8.4.9. Emerging Repair Materials

High-strength and high-performance concrete using fly ash and silica fume, synthetic resins, magnesium phosphate cement, magnesium oxy-chloride cement, silicon sealant, poly-urethane sealant, sulfur, and sand mortars are the new repair materials for the upkeep of rigid pavements.

8.4.10. Strategic Repair and Rehabilitation Techniques

The saying "build pavements and forget them" has been bandied about a lot. For highway rigid pavements, reconstruction is a type of rehabilitation because the current pavements are becoming more deteriorated and require more frequent maintenance. Recent years have seen advancements in the creation and application of materials for repair and rehabilitation that are set up quickly and hardened quickly.

CASE STUDY

REPAIR OF A CRACKED ROAD SURFACE

Successful test installation of TenCate Polyfelt® PGM-B in Nijverdal.

Challenge

Systems of asphalt interlayers must perform the roles of reinforcement, barrier, and stress relief. Asphalt reinforcement has been achieved using a variety of raw materials. An increasing number of people agree that e.g., Because of their relatively low modulus, PET yarns have specific limitations when it comes to asphalt reinforcement. Glass fibers have been used extensively in the past, but TenCate Polyfelt® PGM-B, a novel material made of high modulus basalt fibers, has now demonstrated its suitability.

Solution

Using very stiff components to absorb stress is an effective way to prevent cracks. The key factors influencing success are adequate stiffness and good bonding. With its exceptional tensile stiffness and flawless bitumen bond, basalt is a novel mineral material that offers the best reinforcing qualities. Therefore, if the multi-layered asphalt system is successful in transferring the necessary forces from the asphaltic matrix to the interlayer and vice versa, it can be reinforced by suitable interlayers.

Benefit

Compared to traditional methods, TenCate Polyfelt® PGM-B offers a higher quality surface, longer maintenance intervals, and a more economical method of road maintenance.

- High tensile strength / higher stiffness
- Better affinity to bitumen
- 3 functions in one product

CLASS ACTIVITY

ACTIVITY: GENERAL ROAD INSPECTION AND REMOVAL OF OBSTACLES.

Scope and Purpose

- Throughout the entire road, check the carriageway, roadside areas, and drainage to make sure there are no significant damages or obstructions. It is important to notify the relevant road authority right away if significant damages have happened.
- Items that are obstructing the drainage system or the carriageway, such as fallen trees, rocks, boulders, or bushes, should be removed and properly disposed of at a location that is designated by the engineer, far enough away from the road. Erosion and silting-related soil removal are dealt with in a different activity.
- If obstructions are not regularly removed, they can clog the drainage system and harm the surrounding areas and the road itself.
- Road users may also be put in danger by obstructions.

Procedure

The inspection is done while riding a bicycle or on foot, noting any obstacles. On-foot inspections of side drains and culverts are required. Any significant damages should be reported to the appropriate authorities. When conducting an inspection, it is preferable to remove small, easily transportable objects right away.

Request that any materials that the locals have used for construction or agriculture that have gotten into the road or are clogging drains be removed. Inform the appropriate authority in writing if the aforementioned materials are not removed by the time of the next inspection. Report any landslides to the other members of the maintenance team right away, and then work with the team to remove them. If you can, make the necessary corrections; if not, notify the other members of the maintenance team if water is spilling onto the road or corroding road structures. The material that has been deposited shouldn't interfere with the drainage system, traffic flow for cars or pedestrians, farms, residences, canals, or streams. The removed material should, if needed, be taken to an appropriate disposal location. Trash needs to be gathered in bags and put in the proper place.

Note:

- Inspection should be done frequently and specifically after heavy storms.
- This activity is performed as an individual task system.

Quality Assurance

Ensure the inspection of the entire road is done on a regular basis and after each heavy storm.

Environmental

When disposing of obstacles on the spot, disposal should be adhered to in order to avoid littering.

Health and Safety

Wearing reflective clothing is imperative for the inspector's safety from oncoming vehicles. Make sure the bicycle is parked off the carriageway, ideally at the outer end of the shoulder, if a foot inspection is needed.

Labor

Semi-skilled Labor.

Material

Not required

Tools and Equipment

- Measuring tape
- Shovel
- Pickaxe
- Machete/cutlass
- Reflective clothing
- Raincoat
- Hard hat
- Bicycle
- Wheelbarrow
- If possible, a mobile phone

Performance Indicator

There is no debris on the road's pavement, shoulders, or drains. Any urgent problems are noted in the notebook, reported to the appropriate authority (the maintenance team or the engineer), and fixed.

SUMMARY

- The road network is a crucial aspect of modern society, providing access to homes, businesses, and essential services. Maintaining and repairing roads on a regular basis ensures they are safe and functional for all road users, including drivers, cyclists, and pedestrians.
- Roads connect our world, but unfortunately, they can be the source of a lot of problems. This is the reason why so much maintenance on roads is necessary. Many elements are at play, which is why a lot of work is routinely performed.
- There are various levels of road maintenance, which are broken up into three categories: routine, periodic, and emergency.
- Pavement failure is caused by a number of variables including, water intrusion, stress from heavy vehicles, expansion and contraction from seasonal temperature changes, and sun exposure. It is important to keep up with proper maintenance like crack and asphalt sealing to prevent cracks from spreading or forming.
- The outside influence that affects the pavement performance is categorized under environmental factors. These include snow, chemicals, water, and problems with aging. These types of distress are observed from the top down.
- Asphalt cracking by far makes up for the majority of pavement failure issues because of its tendency to spread and the wide variety of elements that can cause cracking.
- The phenomenon of formation of a film of asphalt binder over the surface of the pavement surface is called bleeding. The occurrence of bleeding will give a shiny glass-like reflecting surface. The layer will have bubbles which are seen as blisters. The asphalt binder formed will be sticky in nature. The filling of asphalt binder into the aggregate voids during hot weather conditions and their expansion in later situations will result in bleeding.
- Regardless of the quality of pavement material and design, an increase in vehicular traffic and changing environmental conditions will reduce the service life of pavements and ultimately result in their failure. The causes and types of failure of pavements in general and rigid pavements, in particular, should be understood in order to initiate proper repair and rehabilitation programs to enhance their service life.
- Potholes, Scaling, Blow-ups, mud pumping, spalling, Loss of Sealant, Cracking (longitudinal, transverse, reflection, sympathetic), corrugation, joint deficiencies, and faulting are the distresses occurring in rigid pavements, however, the considered distresses were Cracking, Potholes, joint deficiencies and faulting in the data collection procedures).
- Flexible pavement failure is caused by a number of variables including water intrusion, stress from heavy vehicles, expansion and contraction from seasonal temperature changes, and sun exposure. It is important to keep track of proper maintenance like crack and asphalt sealing to prevent cracks from forming and

spreading. The cracks form due to the forces applied by the turning or braking motion of vehicles. Distortions in an asphalt pavement are caused by instability of an asphalt mix or weakness of the base or subgrade layers.

- The main causes of failure in rigid pavements due to faulting are the pumping or the erosion of material under the pavement, resulting in voids under the pavement slab causing settlement.
- Most Pavement Designs involve two or more performance periods. A pavement is constructed at an initial serviceability level and is rehabilitated to an acceptable level at some time during its design life. This process of rehabilitation may be repeated several times depending on the condition of the existing pavement deteriorating from time to time.
- The emerging repair materials for the maintenance of rigid pavements are high-strength and high-performance concrete using fly ash and silica fume, synthetic resins, magnesium phosphate cement, magnesium oxy-chloride cement, silicon sealant, poly-urethane sealant, sulfur, and sand mortars.

REVIEW QUESTIONS

1. What causes slippage cracking?
2. What is alligator cracking? How do you fix an alligator crack?
3. What causes asphalt slippage?
4. What causes reflective cracking?
5. What are the treatments for concrete pavement?
6. What are the different methods of structural evaluation of flexible pavements?
7. What are the special repairs in flexible pavement?
8. What is used to repair roads?

REFERENCES

1. Adams, P. W. (1991). *Maintaining Woodland Roads*. Corvallis, OR: Oregon State University Extension Service.
2. Banister, D. (2012). Transport and economic development: Reviewing the evidence. *Transport Reviews, 32*(1).
3. Dormidontova, T. V., Pavlova, L. V., & Dimeev, A. M. (2020). Restoration of highways after repair of communications. *National Association of Scientists, 1*(29(56)), 41–44. http://dx.doi.org/10.31618/nas.2413-5291.2020.1.56.226.
4. Eaton, R. A., Joubert, R. H., & Wright, E. A. (1989). *Pothole Primer—A Public Administrator's Guide to Understanding and Managing the Pothole Problem* (Special Report No. 81–21). US Army Corps of Engineers—Cold Regions Research & Engineering Laboratory. http://hdl.handle.net/11681/12121 (accessed on 16 July 2024).

5. Gallego, E., Moya, M., Ayuga-Téllez, E., García, A. I., & Ayuga, F. (2016). The influence of management and construction methods in the repair costs of Spain's low-volume road network. *Journal of Agricultural Engineering, 47*(2), 118. http://dx.doi.org/10.4081/jae.2016.537.
6. Gogolev, V., & Barysheva, M. (2021). International and national practice for estimating cash expenditures of road replacement, repair and maintenance. *SHS Web of Conferences, 112*, 00041. http://dx.doi.org/10.1051/shsconf/202111200041.
7. Gould, E., Parkman, C., & Buckland, T. (2013). *The Economics of Road Maintenance*. RAC Foundation, Adept.
8. Heggie, I. G. (1998). *Commercial Management and Financing of Roads*. Washington, D.C: World Bank.
9. Jonbi, J., Tjahjani, A. R. I., Tinumbia, N., Pattinaja, A. M., & Haryono, B. S. (2018). Repair of rigid pavement using micro concrete material. *MATEC Web of Conferences, 195*, 01014. http://dx.doi.org/10.1051/matecconf/201819501014.
10. Kayondo-Ndandiko, L. M. (2012). *Geographical Information Technologies–Decision Support for Road Maintenance in Uganda* (Doctoral thesis). Blekinge Tekniska Högskola, Sektionen för planering och mediedesign. http://urn.kb.se/resolve?urn=urn:nbn:se:bth-00539 (accessed on 16 July 2024).
11. Paliichuk, T. V. (2020). The principle of "positive amount" in the road sector: Decentralization of roads. *Business Inform, 12*(515), 369–377. http://dx.doi.org/10.32983/2222-4459-2020-12-369-377.
12. Peters, O. O. (2009). *The Challenges of Road Development and Maintenance in Nigeria: Paper Presented at the Federal University of Technology, Akure*. Akure, Nigeria: School of Engineering and Engineering Technology.
13. Skorseth, K. (2000). *Gravel Roads: Maintenance and Design Manual*. Washington, D.C.: U.S. Dept. of Transportation, Federal Highway Administration.
14. Skorseth, K. (2005). *Gravel Roads: Maintenance and Design Manual*. Brookings, SD: South Dakota Local Transportation Assistance Program.
15. United States. Federal Highway Administration. (2015). *Gravel Roads: Construction & Maintenance Guide*. Washington, D.C.: U.S. Department of Transportation, Federal Highway Administration.
16. Vlasov, V., Bogumil, V., & Baytulaev, A. (2021). Digital infrastructure and telematics systems for monitoring the maintenance of roads. *INFRA-M Academic Publishing LLC*. http://dx.doi.org/10.12737/1014643.
17. Wasike, W. S. K. (2001). *Road Infrastructure Policies in Kenya: Historical Trends and Current Challenges*. Nairobi, Kenya: Kenya Institute for Public Policy Research and Analysis.
18. Wilson, T. P., & Romine, A. R. (1999). *Materials and Procedures for Repair of Potholes in Asphalt-Surfaced Pavements—Manual of Practice* (FHWA Report No. FHWA-RD-99-168). Washington, DC: Federal Highway Administration.

CHAPTER 9

Highway Engineer and the Development Process

LEARNING OBJECTIVES

After studying this chapter, you will be able to:

- Define the role of the highway engineer
- Discuss highway planning and development processes
- Find out a strategy to improve the feasibility

KEY TERMS FROM THIS CHAPTER

Environmental assessments
Materials engineering
Environmental scientists
Highway engineer
Transport professionals
Water resources engineers
Construction industry
Professional Engineer (PE)
Continuing education
Geographic Information Systems
Geometric design
Urban planners
Intelligent transportation systems
Effective communication
Highway designers
Transport planners
Engineers-in-Training (EIT)
Geographic information systems
omputer-aided design (CAD)
Autonomous vehicles (AVs)

INTRODUCTION

Highway engineers are responsible for a wide range of tasks, from planning and design to construction and maintenance, and they play a crucial part in the development of transportation infrastructure. Highway engineers participate in environmental assessments and feasibility studies in the early phases to ascertain the impact and feasibility of proposed projects. Analyzing traffic patterns, environmental aspects, and socioeconomic ramifications is necessary to make sure the project satisfies community needs and complies with legal requirements. The design phase, during which engineers create comprehensive plans and specifications, starts after the viability has been determined. The layout and dimensions of the road are the main focus of geometric design, which makes sure the road is effective and safe for all users. Because it entails the planning of bridges, tunnels, and other vital infrastructure, structural design is also very important. By using materials engineering, construction materials are guaranteed to meet the necessary sustainability and durability standards. To integrate the highway with the current infrastructure and minimize any negative effects, engineers must work in conjunction with other experts during this phase, including urban planners, environmental scientists, and government representatives.

Highway engineers supervise the design plans' implementation during the construction phase. In addition to making sure that work complies with safety regulations and planned specifications, their duties also include overseeing construction sites. This entails overseeing finances, liaising with contractors, and making sure deadlines are fulfilled. In order to keep the project moving forward, engineers must also handle any unforeseen problems that crop up during construction, such as weather or geological difficulties, and make decisions in real time. Highway engineers are involved in the upkeep and management of the highway system after construction is complete. To make sure the infrastructure stays secure and operational, this entails routine inspections, maintenance, and upgrades. To plan for future improvements, engineers must consider factors such as wear and tear, traffic growth, and technological advancements. This continuous procedure is necessary to increase the highway's lifespan and make adjustments for evolving transportation requirements. In addition, highway engineers need to keep up with new developments in transportation engineering technology and procedures. The importance of innovations like intelligent transportation systems (ITS), which combine automated and electronic systems for improved traffic control and safety, is rising. Engineers need to be aware of these technologies and how to incorporate them into both new and old infrastructure.

9.1. ROLE OF HIGHWAY ENGINEER

Highway engineers play a variety of roles that call for a strong grasp of project management, engineering principles, and regulatory compliance. They must strike a balance between the needs of the public, environmental protection, and technical requirements, making sure that every project makes a valuable contribution to the larger transportation system. Highway engineers collaborate and communicate effectively with a wide range of stakeholders during the development process, which is another important factor. By doing this, they contribute to the development of infrastructure that fosters economic expansion, increases mobility, and raises community standards of living.

Plans and specifications are created by the highway engineer for projects involving the construction or renovation of roads. The precise path the road will take, the number of lanes required, how this road interacts with other roads, and how traffic flow will be maintained must all be determined by the engineer before a road structure can be designed. Numerous factors are taken into account when designing a road system, such as impact attenuators, guide rails, drainage, parking, traffic calming, pedestrian and bike paths, highway signs, and road markings. In addition to writing technical reports, conducting construction analyzes, and taking part in public meetings, the highway engineer studies traffic.

Other responsibilities typically include:

- Performing roadway geometric design for horizontal and vertical alignment from preliminary to final design.
- Preparing technical designs in accordance with design standards using roadway design software.
- Preparing plan and specification documents including base sheet, details, quantities, special provisions, and estimates.
- Coordinating with reviewers to ensure accuracy and quality of work.
- Managing and maintaining project documentation.
- Design systems or structures using sustainable materials, such as porous pavement or bio-retention structures.
- Preparing design for drainage and runoff.
- Preparing designs that do not damage natural plant and wildlife environments.

9.1.1. Become a Highway Engineer

A combination of formal education, real-world experience, and professional certification is needed to become a highway engineer. Usually, the adventure starts in high school with a solid foundation in science and math, where courses like calculus, physics, and chemistry are crucial. After that, prospective highway engineers ought to enroll in an accredited university and work toward a bachelor's degree in civil engineering or a related discipline. The basic engineering concepts covered in this undergraduate program include fluid mechanics, materials science, and structural analysis. Specialized courses in transportation engineering and highway design are also included. It is encouraged of students to take part in cooperative education or internships during their undergraduate studies. Through the development of technical skills, application of theoretical knowledge to real-world projects, and understanding of the daily responsibilities of highway engineers, these practical experiences offer invaluable hands-on training. Additionally, networking with industry professionals can be advantageous for future employment prospects and is provided by internships.

Prospective highway engineers need to get licensed in order to work as professionals after earning a bachelor's degree. To achieve this in the US, one must pass the Fundamentals of Engineering (FE) exam, which measures one's grasp of fundamental engineering concepts. Candidates are classified as Engineers-in-Training (EIT) or Engineering Interns (EI) upon passing the FE exam. Next, under the guidance of a licensed professional engineer (PE), they must obtain four years of relevant work experience. They can hone their skills in project management, design, and construction oversight during this time, as well as gain a deeper understanding of the field. Upon obtaining the requisite work experience, the subsequent course of action involves clearing the Principles and Practice of Engineering (PE) examination, which concentrates on particular facets of civil engineering, including transportation. A high degree of competence is demonstrated by passing this exam and earning the title of Professional Engineer (PE), which is frequently necessary for senior positions and leadership roles in engineering firms and government organizations.

For highway engineers to stay up to date on new developments in technology, materials, and procedures, continuing education is essential. To keep their licenses, PEs are required by many states to complete a specific amount of hours of continuing education. Professional associations that offer workshops, seminars, and conferences to promote learning and professional development include the American Society of Civil Engineers (ASCE) and the Institute of Transportation Engineers (ITE). A variety of soft skills must be developed by successful highway engineers in addition to their formal education and license. Effective and safe transportation system design requires strong analytical and problem-solving skills. Since highway engineers frequently work with architects, urban planners, contractors, and government officials, effective communication and teamwork skills are also crucial. To oversee projects, manage budgets, and make sure deadlines are met, one needs strong leadership and project management abilities.

9.1.2. Highway Engineer's Duties and Responsibilities

Include a wide range of activities aimed at guaranteeing the design, planning, building, and upkeep of roads that are sustainable, safe, and efficient. Highway engineers first conduct thorough planning and feasibility studies (Muller RH., 1996). This entails analyzing traffic patterns—both past and present—as well as the environmental and socioeconomic effects of suggested projects. To evaluate data and project future transportation requirements, they employ geographic information systems (GIS) and traffic simulation software. These studies assist in identifying the best sites and layouts for brand-new highways or upgrades to already-existing ones. Highway engineers draft comprehensive plans and specifications that describe the alignment, grading, and drainage of the roadway during the design phase. They guarantee that the design conforms to safety regulations and can handle a range of vehicle kinds and traffic volumes. Another crucial component is structural design, which includes the layout of overpasses, tunnels, bridges, and other supporting structures. Engineers are required to choose building materials and techniques that satisfy environmental, safety, and durability requirements. Additionally, they incorporate intelligent transportation systems (ITS) to improve safety and traffic control.

Highway engineers supervise the construction process to make sure the project follows the authorized plans and specifications. They manage the project budget and schedule by coordinating with builders, contractors, and other relevant parties. Engineers inspect the site to keep an eye on quality and progress and to quickly resolve any problems that may arise. To keep the public and employees safe, they must make sure that safety rules are followed. This stage frequently entails resolving unforeseen difficulties, like geological problems or unfavorable weather, which calls for engineers to modify plans and make prompt, well-informed decisions. Highway engineers are in charge of maintaining and running roads after construction is complete. This includes doing routine inspections to spot and fix wear and tear issues like cracks, potholes, and structural damage. To minimize traffic disruption, engineers oversee repair work and create maintenance schedules. In addition, they assess the highway's performance and suggest improvements or additions to handle growing traffic or emerging technologies.

Highway engineers need to stay up to date on the latest technological developments and industry best practices. Continuing education is crucial because it enables them to apply creative solutions and boost productivity. To broaden their knowledge and connect with colleagues, they take part in workshops, conferences, and professional organizations. Engineers who are dedicated to their professional development are better equipped to tackle new and developing issues, like incorporating autonomous vehicles into the transportation system or using sustainable building techniques. For highway engineers, effective communication is an essential ability. They have to communicate technical information succinctly and clearly to a range of stakeholders, such as the public, contractors, and government representatives. In reports, meetings, and public hearings, engineers frequently present their conclusions and suggestions. To make sure that transportation projects are in line with more general community objectives and environmental regulations, they must also work in conjunction with urban planners, environmental scientists, and other experts.

9.1.3. Technology Changing the Role

Highway engineers' work is being fundamentally changed by technology, which is revolutionizing the way they plan, design, build, and maintain transportation infrastructure. The development of building information modeling (BIM) and computer-aided design (CAD) software, which enables engineers to produce incredibly accurate and detailed digital representations of highways, is one important development. These tools improve collaboration between multidisciplinary teams, allow more accurate simulations, and improve visualization. Today's engineers can identify possible design problems early on, cutting down on errors and saving money and time. Additionally, geographic information systems (GIS) have grown to be essential tools for highway engineers, offering them a broad range of spatial data analysis capabilities. In order to support more strategic decision-making, GIS enables the integration of multiple datasets, including land use, traffic patterns, and environmental impacts. Engineers can more efficiently plan for future growth, determine the best routes, and evaluate the effects on the environment by analyzing and visualizing large, complex data sets using GIS (Smith J et al., 2014).

With the use of robotics and automated machinery, technology is increasing accuracy and efficiency during the building phase. Tasks like paving and grading can be completed by autonomous construction equipment with extreme precision, decreasing human error and boosting output. Drones are used for site surveys and inspections; they provide engineers with real-time aerial data to monitor progress, evaluate site conditions, and make sure safety regulations and design specifications are being followed. Highway engineers' role is being reshaped by another technological advancement called Intelligent Transportation Systems (ITS). It improves traffic management and safety by integrating communication and electronic technologies into the transportation infrastructure. For example, real-time adjustments to traffic conditions by smart traffic signals improve flow and lessen congestion. Road sensors are able to track traffic patterns and speeds, identify incidents, and supply information for preventive maintenance. Now that these technologies are available, engineers need to take them into account when designing new roads to make sure that they work with smart infrastructure. Highway engineering is also being impacted by the rise of autonomous vehicles (AVs). Roads must be designed by engineers to support autonomous vehicles (AVs). This may entail adding features like AV-friendly lanes, sophisticated signage, and vehicle-interactive communication systems. A better understanding of traffic patterns and infrastructure performance is made possible by the data produced by AVs, which facilitates more intelligent planning and maintenance choices.

Another area that technology is significantly influencing is sustainability. To lessen the environmental impact of transportation projects, highway engineers are increasingly utilizing sustainable building materials and techniques. Heat island effects can be lessened and stormwater runoff can be managed with the use of technologies like permeable pavements, recycled materials, and green infrastructure solutions. These sustainable practices are modeled and optimized by engineers using software, guaranteeing that environmental objectives are met without performance being sacrificed. Additionally, sensors and Internet of Things devices are gathering data on road conditions and infrastructure health to enable predictive maintenance, which is made possible by advanced data analytics and machine learning. Engineers can prevent failures and optimize resource allocation by using this data to forecast when and where maintenance is required. By being proactive, infrastructure longevity is increased and safety is enhanced. Highway engineers now work in a more dynamic, technologically advanced field rather than performing traditional design and construction duties. Large-scale dataset analysis, cutting-edge technology integration, and the use of sophisticated software are now skills that engineers need to possess. Engineers must be always learning and adapting because technology is advancing so quickly that they must keep up with the newest techniques and tools. Through the adoption of these technologies, transportation engineers can create and manage transportation networks that are safer, more intelligent, and adaptable to the changing demands of society.

9.1.4. Bring the Highway Engineering Career

A career in highway engineering combines technical proficiency with significant contributions to the infrastructure of society, making it a dynamic and fulfilling path. A solid educational foundation is the first step in the journey, usually entailing a bachelor's degree in civil engineering or a closely related discipline. Essential knowledge in subjects like mathematics, physics, materials science, and engineering principles is provided by this academic background. Your expertise can be further enhanced by specializing in transportation or highway engineering through advanced degrees or elective courses. In this field, gaining real-world experience is essential. During college, internships and cooperative education programs provide practical experience and exposure to real-world projects. Through these opportunities, you can put your theoretical knowledge to use, hone your technical skills, and gain an understanding of the daily duties of a highway engineer. You can acquire industry standards and best practices by working under the guidance of seasoned professionals, which will lay a strong foundation for your future career. For highway engineers, getting their license is a big accomplishment that usually requires passing the Fundamentals of Engineering (FE) exam and several years of relevant work experience. You can sit for the Principles and Practice of Engineering (PE) exam to become a licensed Professional Engineer (PE) once you've had enough experience. This certification denotes a high degree of competence and dedication to the field and is frequently required for senior positions.

Highway engineers work on a range of projects in the workplace, from designing and planning roads to managing their construction and upkeep. They ensure accuracy and efficiency in infrastructure design by using cutting-edge software such as CAD and BIM. In their designs, engineers have to take future traffic demands, environmental effects, and safety regulations into account. They oversee projects, communicate with contractors, and make sure that rules and specifications are followed when construction is underway. In this quickly changing field, it is imperative to pursue ongoing professional development. You can stay up to date on the newest techniques and technologies by participating in lifelong learning through continuing education programs, certifications, and professional associations like the American Society of Civil Engineers (ASCE). In addition to improving your skills, attending conferences, seminars, and workshops gives you the chance to network with peers and business executives. You can further advance your career by specializing in cutting-edge fields like autonomous vehicle infrastructure, sustainable design, or intelligent transportation systems (ITS). These specializations use cutting-edge technologies to solve contemporary problems, which increases the market value of your experience. For instance, knowing ITS can help you create more intelligent traffic control systems, and having a solid understanding of sustainable design can result in infrastructure solutions that are friendly to the environment. For highway engineers, teamwork and communication are essential. Working well with a wide range of stakeholders—including public servants, environmental scientists, urban planners, and government officials—is essential. Project alignment with broader community goals and regulatory requirements is ensured through clear communication. As your career develops, having strong leadership and project management abilities will also be crucial since they will allow you to oversee teams, control spending, and make sure deadlines are fulfilled.

9.1.5. Workplace/Work Environment for Highway Engineer

A highway engineer works in a variety of environments, including offices and field sites, each with its own set of duties and difficulties. Highway engineers work on planning and design projects in the office. To produce intricate designs and models of roads, they employ cutting-edge software such as geographic information systems (GIS) and computer-aided design (CAD). In this environment, engineers must possess a high degree of technical proficiency, accuracy, and attention to detail in order to analyze data, run simulations, and make sure that designs adhere to safety regulations. In order to create comprehensive transportation solutions that seamlessly integrate with the current infrastructure and community needs, engineers frequently collaborate with urban planners, environmental scientists, government officials, and other stakeholders. Collaboration is a key component of the office environment. Another essential facet of a highway engineer's work is fieldwork, which provides a dynamic and frequently difficult environment. Engineers visit construction sites, supervise construction activities, conduct surveys, and carry out inspections to make sure that projects are carried out in compliance with safety regulations and design specifications. Practical problem-solving abilities and the capacity to act quickly and intelligently in the face of unforeseen circumstances, such as difficult geology, inclement weather, or construction hold-ups, are prerequisites for fieldwork. To safeguard employees and the general public, safety procedures must be strictly followed. Safety is the top priority in the field.

There is a lot of interaction between the work environment, contractors, and construction crews. In order to keep projects on schedule and within budget, highway engineers must effectively coordinate activities and resolve conflicts. Since engineers are frequently in charge of overseeing schedules, budgets, and resource allocation, this part of the work requires strong leadership and project management abilities. Additionally, highway engineers are essential to the upkeep and functionality of the current road network. This entails evaluating road conditions, planning maintenance or improvements, and carrying out routine inspections. In order to prevent major problems before they arise and increase the lifespan of infrastructure, engineers use data from sensors and monitoring systems to prioritize maintenance tasks and put predictive maintenance strategies into practice. Highway engineers' work environments are becoming more and more shaped by technological advancements. Engineers need to be adept at using contemporary technology and data analytics in order to integrate intelligent transportation systems (ITS), such as smart traffic signals and real-time traffic monitoring. It takes ongoing professional development, such as going to conferences, workshops, and training sessions, to stay up to date with these advancements. Furthermore, presenting project plans and updates to stakeholders and the community is a common part of highway engineers' public engagement roles. This calls for the capacity to address public concerns, incorporate feedback into project designs, and communicate complex technical information in a clear and understandable manner.

9.1.6. Career Path Progression for Highway Engineer

A career in highway engineering offers a well-defined path from entry-level work to leadership positions at higher levels, with chances for specialization, growth in responsibility, and professional development. Those with a bachelor's degree in civil engineering or a similar discipline usually start as entry-level engineers. They support senior engineers during this early phase by helping with data collection, initial design work, and project documentation. Bypassing the Fundamentals of Engineering (FE) exam, many engineers aim to become certified as Engineering Interns (EI) or Engineers-in-Training (EIT) in order to progress. With this certification, they can get real-world experience working under the supervision of Professional Engineers (PEs) who are licensed, which helps them get ready for licensing. In order to become licensed Professional Engineers (PEs), engineers must complete the Principles and Practice of Engineering (PE) exam after gaining a certain amount of work experience, usually four years. Upon obtaining their license, engineers generally go on to become project engineers. Serving in this role, they supervise particular projects or sections of more extensive projects, managing duties like project design, contractor coordination, and standard and regulation compliance.

Engineers may become senior engineers as they accumulate more experience and show leadership potential. Senior engineers take on more complex projects and manage larger teams. They use their practical experience and in-depth knowledge of engineering principles to solve difficult issues and make important choices. They frequently serve as mentors for junior engineers, offering advice and assistance as they advance in their careers. As engineers advance, they may eventually take on project management responsibilities, supervising entire projects from inception to conclusion. Strong project management abilities, including scheduling, budgeting, and client communication, are necessary for this position. To guarantee project success and requirement compliance, project managers communicate with clients, stakeholders, and regulatory agencies on a regular basis. Another path to career advancement in highway engineering is specialization. Engineers can specialize in areas like environmental engineering, traffic engineering, pavement design, or intelligent transportation systems (ITS). Roles as subject matter experts, consultants, or leaders in specialized fields of the industry can result from specialization.

Experienced engineers may advance to executive leadership positions at the top of the career ladder, such as chief engineering officer, vice president, or director of engineering. In these roles, engineers set the direction of their companies and contribute to industry-wide standards and initiatives. They are also in charge of organizational leadership, business development, and strategic planning. Highway engineers gain from networking opportunities, professional development, and ongoing education as their careers advance. They participate in conferences, workshops, and professional organizations to stay up to date on industry advancements, regulations, and best practices. Highway engineers are kept at the forefront of their profession by this constant growth and adaptation, which promotes innovation and excellence in transportation infrastructure.

9.2. HIGHWAY PLANNING AND DEVELOPMENT PROCESSES

Creating a transportation plan for an urban area is the process of transportation planning. It is a continuous process that aims to address the local population's transportation needs and, with the help of consultations with all pertinent parties, find and implement a suitable plan to fulfill these needs. There are several stages to the process. Politicians must determine the overall locations of the transportation corridors and networks that should be developed first, the amount of money that should be allotted to the various programs, and the mode or modes of transportation that should be utilized within them when creating transportation policies at the administrative and political levels. Professional planners and engineers work below this level to define in detail the networks and corridors that make up each of the given systems that have been chosen for development at a higher political level. This is the level where a "transportation study," as it is commonly called, is conducted. It describes the physical characteristics and modal mix of the system needed to meet the region's transportation needs, whether they be road, rail, bicycle, or pedestrian-based. It also forecasts future population and economic growth and the amount of potential movement within the area.

9.2.1. Travel Data

Gathering historical traffic data for the targeted geographic area is the first step in the planning process. Growth rates from previous years serve as a reliable predictor of volumes to be expected over a selected future period, such as 15, 20, or 30 years. The process then starts with determining what kind of transportation plan or suite of plans is most appropriate, along with the size and location of the plan or group of plans in question, if these figures show the need for new or upgraded transportation facilities. People's need to travel from one place to another to carry out the daily tasks that comprise their lives creates a demand for highway schemes.

The level of this demand for travel depends on a number of factors:

- The location of people's work, shopping, and leisure facilities relative to their homes
- The type of transport available to those making the journey
- The demographic and socio-economic characteristics of the population in question

9.2.2. Highway Planning Process

When the highway planning process takes place within a large urban area and other transport options such as rail and cycling may be under consideration alongside car-based ones, the procedure can become quite complex and the workload involved in data collection can become immense. In such circumstances, before a comprehensive study can be undertaken, one of a number of broad strategy options must be chosen:

- The land-use transportation approach
- The demand management approach
- The car-centered approach
- The public transport-centered approach

9.2.2.1. Land Use Transportation Approach

This approach views the control of land use planning as the key to reducing transportation demand. A growing number of commuters are using their own cars to get to work because they live in small satellite towns or the suburbs of large conurbations but work in or close to the city center. Due to this, there is no traffic on the roads, necessitating the introduction of significant improvements to public transportation in addition to more road space. Effective land use control mechanisms include land use strategies like placing jobs near to large residential areas and actively preventing urban sprawl, which tends to make commuters more dependent on their private vehicles.

9.2.2.2. The Demand Management Approach

Rather than building new road links, the demand management approach involves making future plans by improving demand management on the current road network. Demand management strategies include tolling heavily traveled highway segments, perhaps only during rush hours, and encouraging carpooling, in which commuters willingly attain high car occupancy rates in order to save money or in order to meet employer-imposed targets set by the planning authority. Allowing private vehicles with multiple occupants to travel in bus lanes during peak hours or offering them discounted parking at their destination are two ways to encourage the use of carpooling.

9.2.2.3. The Car-Centered Approach

Many major US cities, most notably Los Angeles, have supported the car-centered approach. It aims to build larger, better roads—whether they be intra-urban or inter-urban connections—to accommodate anticipated rises in traffic demand. Prioritizing the construction of road connections within and between the major urban centers is typically part of this strategy. The best use of the available road space is achieved by taking steps like providing drivers with in-car information about areas of congestion along their planned route and installing cutting-edge traffic control technology at every intersection.

9.2.2.4. The Public Transport-Centered Approach

The strategy will highlight the value of bus and rail-based improvements as the recommended means of meeting the growing demand for transportation in the public transport-centered approach. Advocates of this strategy highlight its benefits for the environment and society, citing decreased air pollution and noise levels, improved fossil fuel efficiency, and increased accessibility to transportation for those unable to afford a car. The effectiveness of this tactic, though, will depend on transportation planners' capacity to persuade more drivers of private automobiles to switch to public transportation during rush hour. This will reduce traffic on the highways as more people drive during rush hour. Only when a hygienic, cozy, frequent, and reasonably priced public transportation service is offered will such an outcome be possible.

9.2.2.5. Transportation Studies

Whatever the nature of the proposed highway system under consideration, be it a new motorway to link two cities or a network of highway improvements within an urban center, and whatever planning strategy the decision-makers are adopting (assuming that the strategy involves, to some extent, the construction of new/upgraded roadways), a study must be carried out to determine the necessity or appropriateness of the proposal. This process will tend to be divided into two subsections:

- A transportation survey to establish trip-making patterns
- The production and use of mathematical models both to predict future transport requirements and to evaluate alternative highway proposals.

9.2.2.6. Transportation Survey

The responsible transport planners first determine the exact boundary that will be used for the study. The land-use activities in the study area serve as the foundation for the majority of transport surveys, which also entail inventorying the current trip patterns and taking socioeconomic factors that influence them into account. The process of creating a profile of the origin and destination (OD), mode of transportation, and purpose of each journey undertaken within the study area is used to ascertain travel patterns. Household surveys are used to collect the OD data for travels that start within the study area. These can be completed with or without assistance from the interviewer. In the former instance, also known as a personal interview survey, the interviewer documents the respondent's responses. In the latter case, also known as a self-completion survey, the interviewee fills out a questionnaire on their own without the help of an interviewer. The typical procedure involves sending the questionnaire to the interviewee, who mails it back or arranges for it to be picked up once all the questions have been answered.

Interviewing trip planners as they pass through the "cordon" at the study area's boundary allows researchers to gather OD information for trips that start outside the study area, travel through its exterior, and end inside the study area. These are known as "intercept surveys," in which respondents are questioned about the beginning and end of their trip while they are traveling. The suitability of the current infrastructure, the land use activities in the study area, and the socioeconomic status of the local population should all be gathered through a transportation survey. It is typically necessary to know the amount of traffic on the current road network, as well as the journey speeds, the proportion of heavy-duty vehicles utilizing it, and approximate vehicle occupancy rates. Office and factory floor areas and employment figures will indicate the current levels of industrial/commercial activity for each designated zone within the study area; census data and recommendations on housing densities will indicate the population size. Each zone will need to conduct a personal household-based survey of some kind to find out how household income affects how frequently people travel and what kind of transportation they choose.

9.2.3. Developmental Processes

Comprise a number of organized and related steps meant to ensure that transportation infrastructure is planned, designed, built, and maintained in a way that makes it easier for people to move around safely and effectively. These procedures are essential to the construction of highways and are critical to maintaining the network of transportation's sustainability, safety, and functionality. Comprehending and proficiently handling these procedures is imperative in ensuring the completion of highway projects that fulfill community requirements, foster economic expansion, and augment the general standard of living. Transportation planning, the initial phase of the highway development process, entails determining long-term transportation goals and objectives, evaluating the current state of transportation needs, and developing strategies to meet those needs. Comprehensive data analysis, stakeholder participation, and consideration of variables like population growth, land use patterns, economic trends, and environmental concerns are all necessary for this stage. After that, feasibility studies and needs assessments are conducted to determine whether the suggested highway projects are feasible and to find any potential flaws in the current infrastructure. In order to evaluate the feasibility and possible advantages of suggested projects, this stage frequently involves traffic forecasting, economic analysis, and evaluation of social and environmental implications.

In order to ensure compliance with environmental regulations and minimize negative environmental effects, environmental assessment and permitting are essential parts of the highway development processes. Evaluating possible environmental impacts of highway projects, such as noise pollution, habitat disruption, and pollution of the air and water, is known as environmental assessment. Getting a permit entail asking regulatory bodies for permission to carry out a project and making sure that suitable countermeasures are in place for any negative effects on the environment. In order to create efficient mitigation plans and guarantee legal compliance, this stage necessitates cooperation with environmental experts, regulatory bodies, and stakeholders. In highway development processes, route selection and alignment are essential phases that entail assessing potential routes considering various factors, including topography, land use, environmental effects, and community preferences. Route selection is informed by geotechnical investigations and geological assessments, and the best option is determined by risk assessment and cost-benefit analysis. The design and engineering phase starts as soon as a route is selected, with an emphasis on pavement design, geometric design, and the inclusion of safety features and accessibility considerations. This phase entails meticulous planning and cooperation between environmental specialists, engineers, architects, and other experts to create comprehensive design plans that satisfy stakeholder expectations and regulatory requirements.

In order to obtain the land required for the construction of highways, right-of-way acquisition, and land acquisition are crucial. This entails negotiating contracts with landowners, abiding by rules and laws, and reducing the negative effects on impacted properties and neighborhoods. The highway development process heavily relies on funding and financing since project construction and maintenance depend on budgeting, financial planning, and locating funding sources. Utilizing creative financing methods and public-private partnerships can help maximize resources and project completion. In order to guarantee the successful completion of highway projects, procurement, contracting, scheduling, and budgeting are coordinated during the crucial phases of construction and project management. Enforcing construction standards involves the implementation of quality control and assurance measures, and managing project complexities effectively requires coordination with regulatory bodies, utility providers, and other stakeholders. In order to maintain the infrastructure's integrity and functionality over time, operation and maintenance procedures are put in place after construction. The longevity, efficiency, and safety of the highway depend on routine maintenance, emergency response, and performance monitoring.

Public participation and stakeholder communication are essential to promoting transparency, resolving issues, and fostering community trust throughout the highway development process. Public meetings, hearings, and outreach initiatives encourage communication and teamwork, guaranteeing that the needs and interests of all parties involved are taken into account during the highway development process. The highway development process is ultimately a cooperative endeavor that involves coordination between public and private sector partners, government agencies, and communities in order to provide transportation infrastructure that supports social justice, environmental sustainability, and economic prosperity while enhancing mobility and safety.

9.2.4. Production and Use of Mathematical Models

Now that all the data is gathered, models are created to convert the data on current travel patterns and land-use profiles into a profile of the study area's future transportation needs. The creation of a transportation model involves four steps: traffic assignment, modal split, trip distribution, and trip generation. Based on the type and intensity of land-use activity within each zone, the first stage estimates the number of trips generated by each zone. The second establishes a pattern of trip making between each zone by distributing these trips among all possible destinations. The actual route that each trip maker takes within the network is then ascertained, following the mode of transportation that each trip maker used to finish their journey. The upcoming chapter provides a detailed description of each of these four phases. Together, they make up the transportation demand analysis process, which is essential to highway engineering. In an effort to forecast demand for cars and other forms of transportation, it makes an effort to characterize and explain both past and present travel behavior.

9.2.4.1. The Decision-Making Process in Highway and Transport Planning

Planning for highways and transportation can be defined as the process of making decisions regarding the long-term viability of a particular transportation system. The choices pertain to projecting future demand, examining the connections and interplay among various transportation modes, assessing how the proposed system will affect current and future land uses, assessing the system's effects on the economy, environment, society, and politics, and evaluating the institutional framework needed to carry out the plan. Most people believe that transport planning is a logical process, i.e., a logical and systematic process for selecting from competing project proposals during the planning phase. It entails combining the processes of decision-making and information gathering. Table 9.1. provides a summary of the five steps in the rational planning process.

Table 9.1. Steps in the Rational Decision-Making Process for a Transportation Project

Step	Purpose
Definition of goals and objectives	To define and agree the overall purpose of the proposed transportation project
Formulation of criteria/measures of effectiveness	To establish standards of judging by which the transportation options can be assessed in relative and absolute terms
Generation of transportation alternatives	To generate as broad a range of feasible transportation options as possible
Evaluation of transportation alternatives	To evaluate the relative merit of each transportation option
Selection of preferred transportation alternative/group of alternatives	To make a final decision on the adoption of the most favorable transportation option the chosen solution for implementation

Source: https://books.google.co.in/books?id=7pgtCwAAQBAJ&printsec=copyright#v=onepage&q&f=false.

Generally speaking, administrators and professionals in the transportation industry adhere to the principles of rational planning and use this process in the manner described below. However, the rational process is a subset of the larger political decision-making system, and it directly interacts with it both during the goal-setting phase and when the preferred option is chosen. To increase the amount of acceptance of the proposal under consideration, political and community representatives who have a direct stake in the transportation plan under examination must provide input in both scenarios. The process of evaluating and selecting the most suitable transport option is a crucial step in the transport planning process, provided that the rational model is a fundamental component and that all available options and criteria have been determined. In general, there are two types of evaluation processes. The first set of techniques calls for the evaluations to be made exclusively in monetary terms. They evaluate the proposal under consideration solely in terms of its economic implications. The second category comprises a collection of broader-based methodologies that enable the evaluation of an extensive array of decision criteria, including environmental, social, and political in addition to economic ones. Evaluations can take many different forms, including non-monetary and monetary ones. Economic evaluations refer to the first set of techniques, while multi-criteria evaluations refer to the second.

Many steps must be taken in order to evaluate transportation proposals. In the end, these are meant to make the determination regarding their authorization clear. Choosing the best location for a proposed highway or ranking the various transportation options included in a state, regional, or federal strategy are crucial aspects of the planning process. During the approval process of a highway scheme by the government, two additional assessments are particularly significant to the overall approval process for a particular project proposal. These are conducted in addition to traffic studies that evaluate the future traffic flows that the proposed highway will have to accommodate.

- A monetary-based economic evaluation, generally termed a cost-benefit analysis (CBA)
- A multi-criteria-based environmental evaluation, generally termed an environmental impact assessment (EIA)

Public participation in the decision-making process is a requirement that is layered on top of the evaluation process. The benefits of this potentially time-consuming process include actively involving all pertinent interest groups in the decision-making process and providing planners with insight into public concerns about the proposal. When the process is done right, it should provide decision-makers with some comfort knowing that everyone who will be impacted by the development has been appropriately consulted before the construction phase begins.

9.2.4.2. Economic Assessment

Within the US, both economic and environmental evaluations form a central part of the regional transportation planning process called for by federal law when state-level transportation plans required under the Intermodal Transportation Efficiency Act 1991 are being determined or in decisions by US federal organizations regarding the funding of discretionary programs.

9.2.4.3. Environmental Assessment

A highway project's environmental and social impacts must be considered in addition to its economic assessment. This evaluation area is part of the proposal's environmental impact assessment (EIA). The National Environmental Policy Act of 1969 established the Environmental Impact Assessment (EIA) as a federal law in the United States. This act mandated that all federally funded projects that were anticipated to have a significant negative impact on the quality of the human environment undergo an environmental assessment. Since then, state governments have also enforced this law.

9.2.5. Public Consultation

Public hearings are held for major trunk road schemes to allow interested parties to participate in the process of determining the optimal location for the highway as well as its basic need. For federally funded highways in the US, at least one public hearing will be required if the proposal is seen to:

- Have significant environmental, social, and economic effects,
- Require substantial way leaves/ rights-of-way, or
- Have a significantly adverse effect on property adjoining the proposed highway.

The state highway agency representative argues for the necessity of the proposed roadway during the hearing format. They also discuss the roadway's potential effects on the environment, society, and economy, as well as the steps they have taken to try to minimize those effects. Additionally, the agency must accept public submissions and engage with the public at different points during the project planning process. Public consultation is a requirement of the planning process in the United Kingdom. The consultation process is focused on choosing the preferred route from the alternatives that are being examined once the necessity of the scheme has been established. Public consultation will still be conducted to evaluate the proposal in comparison to the "do-minimum" option when only one feasible path can be determined. A consultation document detailing the plan in plain language, along with a general summary of its cost and potential environmental and social ramifications, is sent to everyone who has a legitimate interest in the proposal as part of the public participation process. Most consultation documents include a pre-paid questionnaire asking the public to rate the relative merits of the various alignments that are being considered.

9.2.6. Important Technical Terms Related to Roads

Some important terms having immediate concern with highway engineering are discussed below:

- **Road**: A public thoroughfare over which vehicles, cyclists, pedestrians, etc. may lawfully move from one place to another is called a road or highway. Usually, the term highway is used for an important road of national or state importance in a country.
- **Traffic**: The vehicles, cycles, carts, pedestrians, etc. traveling together on a road constitute the traffic.
- **Footway**: The portion of roadway of an urban road reserved only for pedestrians is called footpath, sidewalk, or footway.
- The minimum width of the sidewalk should be 1.5 meters.
- **Cycle track:** The portion of roadway of an urban road reserved only for bicycles is called cycle track.
- The minimum width of the cycle track should be 2 meters.
- **Motorway:** The portion of roadway of an urban road reserved for use only by high-speed and power-driven vehicles is called motor way, express way, or super highway.

The land width for an expressway is recommended as 50 to 60 meters. The function of the expressway is to cater to the movement of heavy volumes of motor traffic at high speeds.

9.2.7. Importance of Highway Transportation Development

The method of detail distribution between residences, businesses, factories, etc. is through highway transportation. Only roads are capable of transporting cargo to and from airports, ports, and train stations. Given the importance of roads in all regions of a nation, it is appropriate to compare them to the arterials in an individual's body. Similar to how arteries keep blood flowing, which maintains human health, roads increase a country's wealth by facilitating the movement of people and goods. Thus, it is clear that roads play a major role in a country's development and well-being. In actuality, the nation's economy depends heavily on its roads.

The importance or necessity of highway transportation can be easily judged from the following purposes or advantages of roads:

- They facilitate the conveyance of people, goods, raw materials, manufactured articles, etc. speedily and easily in the different parts of a country.
- They act as the only source of

communication in regions of high altitudes i.e., in mountainous regions.

- They help in growth of trade and other economic activities in and outside the villagers and towns by establishing contact between towns and villages.
- They help in providing efficient distribution of agricultural products and natural resources all over the country.
- They help in the price stabilization of commodities due to the mobility of products all over the country.
- They help in the cultural and social advancement of people and making the villagers active and alert members of the community.
- They help in promoting the cultural and social ties among people living in different part of a country and thus strengthen the rational unity.
- They help in providing improved medical facilities quickly to human beings, especially to those who live in rural areas.
- They provide more employment opportunities.
- They enhance land value and thus bring better revenue.
- They serve as feeders for airways, waterways, and railways.
- They help reduce distress among the people, caused due to famine, by supplying them food and clothing quickly.
- They help in maintaining better law and order in a country.
- They play a very important role in the defense of a country during war days.

Lastly, it can be said that roads are the symbol of a country's progress and thus development made by any country can be judged by the quality and network of its road system.

9.2.7.1. Classification of Roads

Roads are classified into different categories:

- According to location
- According to importance
- According to traffic
- According to tonnage

9.2.7.1.1. Classification of Roads According to Location

According to location and financial responsibility, non-urban roads in India are classified into the following five categories: -

- National Highways (NH)
- State Highways (SH)
- Major District Roads (MDRs)
- Other District Roads (ODRs)
- Village Roads (VR's)

This classification of roads was done as per recommendations made in the Nagpur plan finalized the Indian Roads Congress in 1943. This classification is, therefore, popularly known as the IRC classification of roads.

The different categories of roads according to this classification are discussed below: –

- **National highways:** The primary thoroughfares that connect the nation's major regions, international highways, state capitals, etc. referred to as National Highways (NHs). These roads serve as the nation's primary thoroughfare and are crucial for military purposes. The carriageway on national highways should be at least two lanes wide. They ought to have contemporary surfacing. The central government is in charge of building and maintaining these roads.

- **State highways:** State highways (SHs) are the roads that connect the state's major cities and district headquarters, as well as those that connect them to the federal highway system or the highways of other states. Provincial highways is another name for these routes. These roads act as main thoroughfares for traffic heading to and from the state's district roads. Two lanes should ideally be present on state highways. They ought to have contemporary surfacing as well. The state governments are in charge of building and maintaining these roads. However, funding for the construction of these roads is provided by the national government.
- **Major district road.** The important roads within a district serving areas of production and markets and connecting these places with each other or with the main highways are known as Major District Roads (MDRs)
- **Other District Roads.** The roads serving rural areas of production and providing them with outlets to market centers, Tehsil headquarters, block development headquarters, railway stations, etc. are known as Other District Roads (ODRs)
- **Village Roads.** The roads connecting villages or groups of villages with each other or with the nearest road of the higher category are known as Village roads.

9.3. A STRATEGY TO IMPROVE THE FEASIBILITY

Highways promote regional connectivity, which boosts a country's economic growth. Additionally, it promotes equitable development across a nation's regions, facilitates better distribution of goods and services, and eases traffic in developing areas. The advantages of building new highways have a direct impact on the movement and accessibility of people, products, and time, in addition to the other advantages of increased regional competitiveness and GDP.

In order to facilitate their economic expansion, nations like Australia, China, South Africa, France, Spain, the United States, and Canada have all invested in highway projects. One argument for raising the degree of competitiveness in pricing logistics is the development of infrastructure. In addition to increasing transactional activity, the strategy will support macroeconomic growth. With roughly 262 million people living there as of 2016, Indonesia is a developing nation in South East Asia. Since 1973, the state has been attempting to build toll roads, and as of right now, 987,444 km of them are in operation throughout the nation. Even though the government has built a lot of infrastructure over the last five years, the entire operated highway lags behind other countries. Japan is the country that has constructed 8358.3 km of expressways; Korea is in charge of 4193 km of toll roads and plans to operate 6160 km by 2020; and Malaysia, its neighbor, already has 1820 km of toll highways. Indonesia must expand the project since the motorway is essential to the country's competitiveness.

In an effort to link raw material resources with the processing sector, the nation is constructing the Trans-Sumatera Toll Road (TSTR), a new highway in the western portion of the nation. Notwithstanding the advantages of developing highway projects, investment costs emerge as the primary challenges, aside from land acquisition. For the purpose of developing 2700 km of highways, the project needs 300 trillion rupiahs, or about US$ 24 billion. Because the Indonesian government lacks the resources to complete the necessary highway construction, it should explore other options, such as assigning a portion of the financing responsibility to private investors. Researchers and academics talked about how an optimal feasibility was the first step toward the project's attractiveness.

TSTR's financial rate of return is relatively low, at roughly 7.79%. The figure is relatively low when compared to the country's minimum attractiveness rate of return (MARR), which is more than 10%, from the perspective of infrastructure and transportation investors. As a result, businesses are reluctant to participate in projects that significantly jeopardize their cash flow. In light of those issues, it becomes imperative to increase the feasibility by accounting for the project's added value.

Value engineering (VE) (Younker D. 2003) will be used in the study in an effort to increase project feasibility. VE has been used in a number of nations to enhance project or product development by lowering costs or providing an extra benefit by highlighting substantial cost savings. Over the past 30 years, the United States has integrated the VE approach into National Highway System (NHS) projects that receive funding and have a total cost of at least $50,000,000. It has been demonstrated that VE enhances transportation project performance. In 2015, VE successfully reduced the project's overall cost by 5 points9%. The study adds to the corpus of information regarding value engineering's application to the construction of highways. The approach provides the best value possible, which may entice investors to fund the project. It is anticipated that the findings of this study will be applied to comparable contexts involving not only transportation but also other value-added creations.

9.3.1. The Trans-Sumatera Highway Project

One of the nation's megaprojects, the Trans-Sumatera Highway Road Project, is situated on Sumatera Island, an island in the western section of the nation. The project's estimated length is 2700 km, and its estimated cost is 330 trillion rupiahs. Out of the ten provinces in the region, it will connect eight provinces on the mainland. The project comprises a Master Plan on ASEAN Connectivity (MPAC) and is anticipated to be a part of the Asian Highway network. Over the course of the project, it is anticipated that the project will produce benefits such as accelerating the logistical flow of goods and services distribution and creating approximately 97,000 new jobs. Additionally, it raises the price of real estate in ways that are competitive and indirectly aid in regional development. The route has been developed by the Ministry of Public Works by using the island's western path's flat topography. The idea is split into two corridors: the main route from Lampung province to Aceh province is the South-North corridor. Conversely, the West-East Corridor, which connects the two regions, is divided into three sections: Medan-Sibolga, Pekanbaru-Padang, and Palembang-Bengkulu. Each section's financial rate of return (FIRR) varies in the results, which adds up to 7 points79% overall. Most investors tend to withhold their involvement because the feasibility is below the transportation companies' minimum rate of return (MARR). Two bidding processes were attempted by the government, but none of them succeeded in reaching a financial close. The government is working to expedite economic development, but infrastructure development is essential for people's access to services and mobility. Therefore, it is necessary to create added value that supports microeconomic and financial growth (Kim E, Bae Y. 2015).

9.3.2. Developing Conceptual design of Trans-Sumatera Highway Road

First, the island of Sumatra's population and economic landscape, as measured by GDP, were taken into account during the conceptual design of the highway. Potential regencies and cities for the highway route were generated using the location quotient (LQ). By considering the industrial portfolio, LQ provides financial flexibility while analyzing and estimating the local economic basis. The approach has been used in many industries for related goals, including food, marine, and carbon emissions.

It is expressed by a mathematical model as follows:

$$LQ = \frac{ei/e}{Ei/E} \quad (1)$$

where LQ is location quotient in a regency/district, ei is GRDP for i sector in a Regency/district, e is GRDP for the whole compared sector in a Regency/district, Ei is GRDP for i sector in the province, and E is GRDP for i whole compared sector in the province.

The LQ evaluates the prospective industries for every city and regency across eight provinces on the island. The outcome will be taken into account when deciding the project's course. The highway ought to link districts and regencies that significantly affect the entire island. Thus, in order to improve the project's viability, a thorough route planning process that considers potential commodities becomes essential. Figure 9.1 depicts the idea of the highway road. The eight provinces that make up Sumatera Island are connected overall. The route is roughly 2427.15 km long overall, with 2102.10 km of lowlands and 325.05 km of hills.

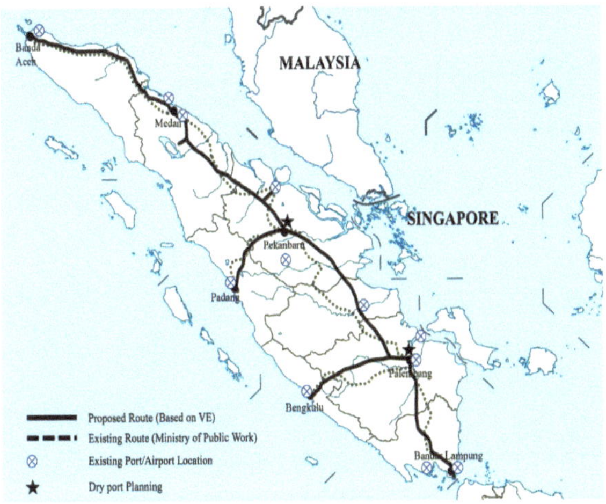

Figure 9.1. *Conceptual design of Trans-Sumatera highway road.*

Source: https://cdnintech.com/media/chapter/57639/1512345123/media/F2.png.

9.3.3. Value Engineering on Trans-Sumatera Highway Road

The value engineering in the study follows the VE job plan and collaborates with varied backgrounds in the process. (Dell'Isola A. 1997)

9.3.3.1. Information phase

Data collection is essentially being used at this point to further define the project. Journal articles, government and research reports, regulations, policies, and other relevant activities are the sources of the data and information. The output was presented but not limited as follows:

Project objectives

The government has made an effort to build the project, which would connect five airports and six major ports in addition to cities, from Bakauheni to Aceh. The island's economic corridor, which is primarily dependent on natural resources, processing industries, and agricultural products, is anticipated to benefit from the project. Included in the project is the ASEAN and Asian Highway Network, which was signed during the 61st UN ESCAP in Shanghai on April 26–28, 2008.

Project benefits

The project intends to enhance the distribution of goods and services, particularly within Java Island and Sumatera. The project's construction will speed up economic growth and lower logistics costs because the area's existing arterial road is overcrowded and generally in poor condition. It is also thought that the project will enable employment from the start of construction to the end of operation.

Scope

The project is in the conceptual design phase which assesses the route and potential added value to increase the project feasibility.

Concept design existing

The project consists of two corridors with a combined planned length of approximately 2700 km and an estimated cost of 330 trillion rupiahs. The central corridor connects Aceh province in the north of Sumatra to Lampung province in its southern region. The connecting corridor, on the other hand, is divided into three sections: Medan-Sibolga in North Sumatera; the city of Pekanbaru in Riau Province to the city of Padang in West Sumatera; and the city of Palembang in South Sumatera to Bengkulu.

9.3.3.2. Function analysis phase

The phase involved using a FAST diagram to set up the problem's scope, order function, primary function, and other functions in order to identify their functions. Using a logical model, it is possible to recognize, categorize, create, and choose functions that add greater value and advantages to the project's advancement. As seen in Figure 9.2., the FAST diagram is visible.

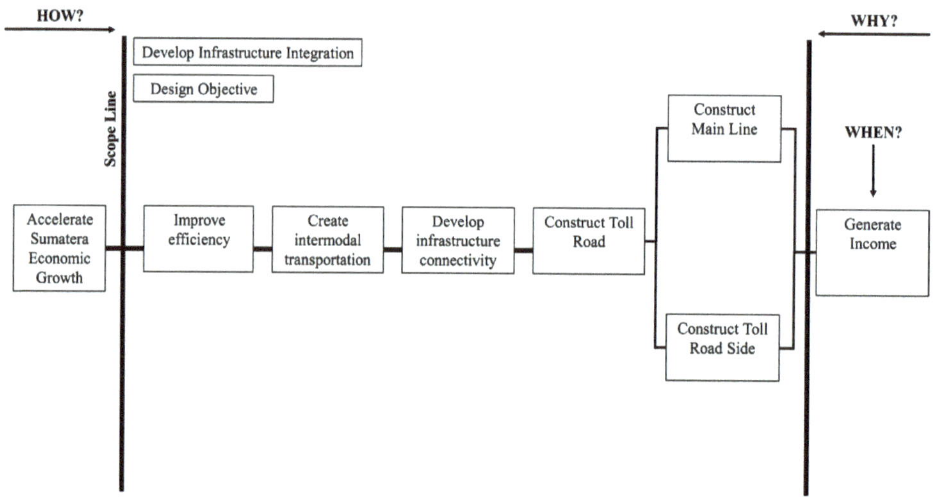

Figure 9.2. *Initial FAST diagram.*

Source: https://cdnintech.com/media/chapter/57639/1512345123/media/F3.png.

9.3.3.3. Creativity phase

This stage assesses the functional analysis from the previous stage by considering the project's potential for greater detail. It will incorporate extended functions and extended processes, updating the previous FAST diagram model. In Figure 9.3., the updated concept design is displayed.

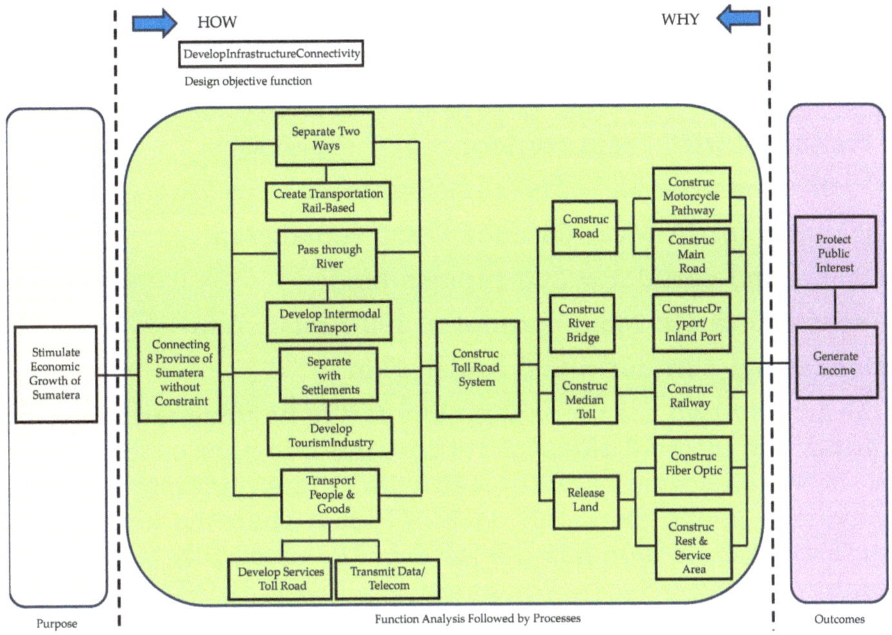

Figure 9.3. *Revised FAST diagram.*

Source: https://cdnintech.com/media/chapter/57639/1512345123/media/F4.png.

9.3.3.4. Evaluation Phase

The project life cycle cost is taken into account when evaluating the study. The identified additional function needs to make the project more feasible in order to draw in private investors. Even in cases where the function failed to produce financial gains, it can still be justified by other benefits like job opportunities and economic growth, among many others. The following scenario discusses project life cycle costs.

9.3.4. Financial Scenario

The life-cycle cost approach is used to evaluate the financial scenario, taking into account the mid-scenario for each function that has been identified. It takes into account revenue, operating and maintenance expenses, and initial costs. For the estimation boundary, a number of hypotheses are put forth, taken from Statistics Indonesia and other trustworthy sources. The project life cycle is approximately 40 years, the discount rate is approximately 6.71%, and the inflation rate varies depending on the industry, ranging from 1.63% to 5.95%.

9.3.4.1. Single function

The new highway concept has 22 sections and 6 corridors covering a total distance of approximately 2527 km. The initial cost is followed by the unit cost from the Ministry of Public Works, which is approximately 106,6 million rupiahs/km for lowland roads and 164 million rupiahs/km for hilly terrain roads. Three tariff scenarios are taken into account and revenue is split between the main and connecting corridors. The situation looks like this:

- **Low scenario tariff:** Main corridor = 150 rupiahs/km
  - Connecting corridor = 300 rupiahs/km
- **Mid scenario tariff:** Main corridor = 300 rupiahs/km
  - Connecting corridor = 300 rupiahs/km
- **High scenario tariff:** Main corridor = 300 rupiahs/km
  - Connecting corridor = 600 rupiahs/km.

Seven of the twenty-two scenarios with low tariffs have negative net present values (NPVs), indicating low internal rates of return (IRR). In this case, the average IRR is roughly 7.25%. Nineteen scenarios with positive NPV are generated by the middle tariff scenario. The average is approximately 11.34% higher than the low tariff scenario. The section Tebing Tinggi-Lima Puluh has the lowest IRR, at roughly 4.88%. The low number results from high construction costs compared to a low anticipated volume trip over the course of the operational stage's 40 years. The 20 scenarios in the high tariff scenario have positive net present value. In this case, the average IRR is roughly 12.61%. The Tebing Tinggi-Lima Puluh and Sigli-Banda Aceh sections of the scenario have negative net present value (NPV) even though they generate higher IRR. More functions were suggested after financial engineering was unable to bring the sections to the desired viability. Table 9.2 displays the scenario's specifics.

Table 9.2. *Life Cycle Cost Analysis of Highway Function*

Highway section	Length	Tariff scenario 1		Tariff scenario 2		Tariff scenario 3	
		NPV	IRR (%)	NPV	IRR (%)	NPV	IRR (%)
Pekanbaru-Medan (main corridor)	**547.2**	**Tariff IDR 150/km**		**Tariff IDR 300/km**		**Tariff IDR 300/km**	
Bakauheni-Bandar Lampung	85.8	Negative	4.31	Positive	8.55	Positive	8.55
Bandar Lampung-Metro-Kota Bumi	102.3	Positive	11.92	Positive	17.32	Positive	17.32
Kota Bumi-Kayu Agung	185.3	Positive	7.21	Positive	12.32	Positive	12.32
Kayu Agung-Palembang	63.8	Positive	7.67	Positive	13.04	Positive	13.04
Palembang-Pangkalan Balai-Sekayu	110	Positive	11.30	Positive	18.63	Positive	18.63

Palembang-Pekanbaru (main corridor)	**494.8**	**Tariff IDR 150/ km**		**Tariff IDR 300/ km**		**Tariff IDR 300/ km**	
Sekayu-Meranti Panjang-Jambi	151.5	Negative	2.87	Positive	7.37	Positive	7.37
Jambi-Rengat	184.9	Positive	7.76	Positive	13.17	Positive	13.17
Rengat-Pangkalan Kerinci-Pekanbaru	158.4	Positive	7.03	Positive	11.99	Positive	11.99
Pekanbaru-Medan (main corridor)	**581.4**	**Tariff IDR 150/ km**		**Tariff IDR 300/ km**		**Tariff IDR 300/ km**	
Pekanbaru-Pematang Duri-Dumai	133.2	Positive	10.22	Positive	16.55	Positive	16.55
Pematang Duri-Rantau Prapat	191	Positive	6.86	Positive	11.99	Positive	11.99
Rantau Prapat-Pematang Siantar-Raya	135	Negative	5.39	Positive	10.35	Positive	10.35
Pematang Siantar-Tebing Tinggi-Medan	93.08	Negative	10.90	Positive	17.60	Positive	17.60
Tebing Tinggi-Lima Puluh	29.12	Negative	1.57	Negative	4.88	Negative	4.88
Banda-Aceh Medan (main corridor)	**480.2**	**Tariff IDR 150/ km**		**Tariff IDR 300/ km**		**Tariff IDR 300/ km**	
Medan-Langsa	118	Positive	9.68	Positive	16.16	Positive	16.16
Langsa-Lhokseumawe	136.8	Negative	6.72	Positive	12.27	Positive	12.27
Lhokseumawe-Sigli	145.5	Negative	3.61	Positive	8.78	Positive	8.78
Sigli-Banda Aceh	79.9	Negative	2.61	Negative	6.58	Negative	6.58
Pekanbaru-Padang (connecting corridor)	**212.9**	**Tariff IDR 300/ km**		**Tariff IDR 300/ km**		**Tariff IDR 600/ km**	
Pekanbaru-Bangkinang	50.4	Positive	7.06	Positive	7.06	Positive	11.96
Bangkinang-Paya Kambuh	72.5	Positive	7.78	Positive	7.78	Positive	12.97

Paya Kambuh-Batusangkar-Padang	90	Positive	12.41	Positive	12.41	Positive	19.65
Palembang-Bengkulu (connecting corridor)	**211.1**	**Tariff IDR 300/ km**		**Tariff IDR 300/ km**		**Tariff IDR 600/ km**	
Sekayu-Lubuklinggau	116.1	Positive	9.55	Positive	9.55	Positive	15.37
Lubuklinggau-Bengkulu	95	Negative	5.13	Negative	5.13	Positive	9.81
Average IRR		**Average IRR**	7.25	**Average IRR**	**11.34**	Average IRR	12.61

Source: https://www.intechopen.com/chapters/57639.

9.3.4.2. Multifunction

The idea combines the use of motorbikes and trains with the integration of highways and other possible uses, including fiber optics, dry ports, rest areas, tourism, and services. Every function's initial cost, operating and maintenance expenses, and revenue over a 40-year operational period are examined. The two most profitable functions were determined to be the development of tourist attractions and rest areas. Fiber optic, dry port, motorbike integration, and railway integration came next. Figure 9.4 shows the estimated revenue for each extra function.

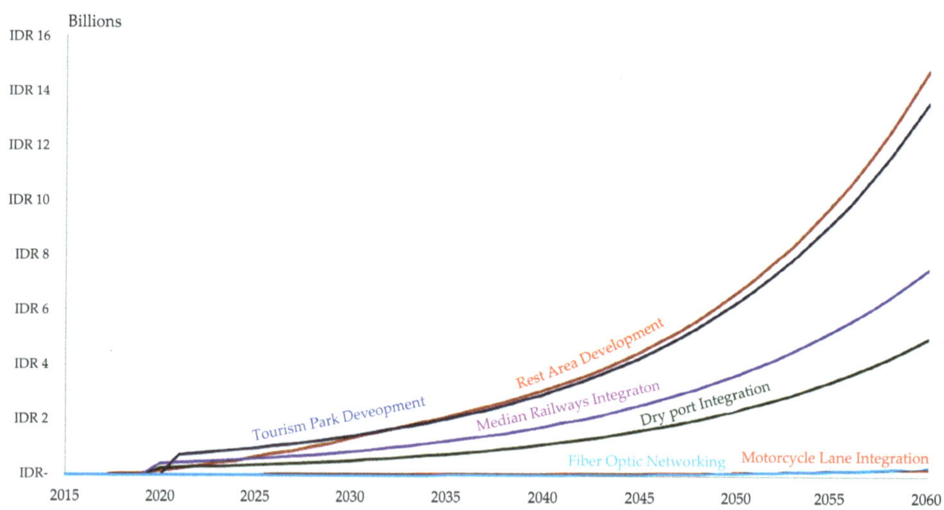

Figure 9.4. *Projected revenue for each additional function on the conceptual design.*

Source: https://cdnintech.com/media/chapter/57639/1512345123/media/F5.png.

The highway with a multifunction concept brings in more money for the project than the single function does. In 2060, the multifunction concept's low tariff generates 200 billion rupiahs, while the single function generates 160 billion rupiahs. Furthermore, the revenue from the multifunction concept's mid-tariff exceeds that of the single-function concept's higher tariff scenario. In contrast to 330 billion and 290 billion rupiahs, respectively, it produces 350 billion rupiahs in 2060. Figure 9.5 shows the expected revenue for both single and multifunction functions.

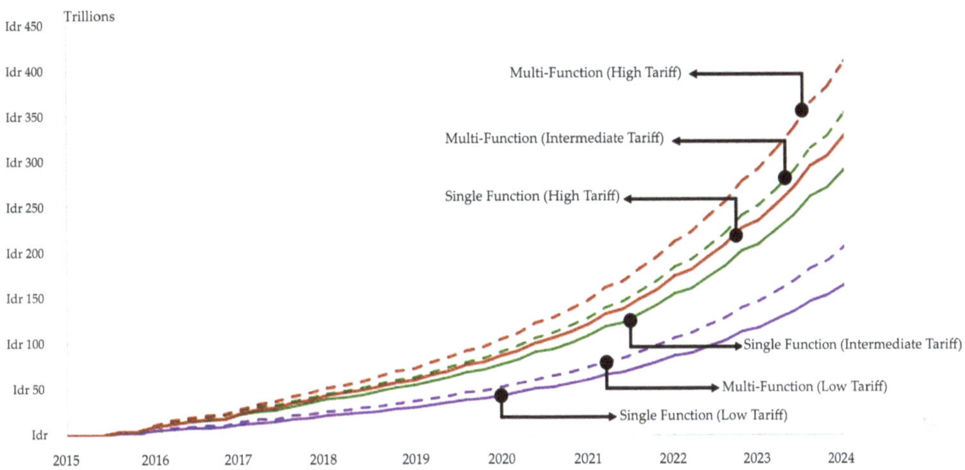

Figure 9.5. *Projected revenue comparison for multifunction and single function.*

Source: https://cdnintech.com/media/chapter/57639/1512345123/media/F6.png.

Life cycle cost in multifunction is simulated using eight different settings. Every alternative was explained with IRR, NPV, and payback period (PbP) and was different from the active function. While some others suggested deactivating the extra function, Alternative A suggested incorporating all of the extra functions. Table 9.3. displays the results of the simulation.

Table 9.3. *Life Cycle Cost Simulation for Multifunction Conceptual Design*

Alternative	Highway road	Rest area	Tourism	Dry port	Railways	Motorcycle	Fiber optic
A	√	√	√	√	√	√	√
B	√	√	√	√	√	x	x
C	√	√	√	x	x	√	√
D	√	√	√	x	x	x	x
E	√	√	x	√	√	√	√
F	√	x	√	√	√	√	√
G	√	x	x	√	√	x	x
H	√	x	x	x	x	√	√

Source: https://www.intechopen.com/chapters/57639.

With the lowest PbP of roughly 8.30 and 34.39%, option E generates a higher internal rate of return (IRR) in light of the low tariff. Alternative A offers greater net present value (NPV) than alternative E, despite having a lower IRR and a higher PbP. A similar problem can be seen for mid- and high-tariff tariffs, where alternative A suggests higher NPV and alternative E suggests higher IRR and PbP. Option E proposes that tourism be left out of the project. Nonetheless, one of the major sources of project income is tourism. It is no longer possible to ignore the sector given its immense appeal. As a result, the best course of action for the project's development is suggested by option A, which incorporates all additional functions. Table 9.4. presents a comprehensive analysis of each option's financial viability comparison.

Table 9.4. *Scenario on Eight Alternatives for Multifunction Conceptual Design*

Alt	Low tariff			Intermediate tariff			**High tariff**		
	NPV (million IDR)	**IRR (%)**	**PBP**	**NPV (million IDR)**	**IRR (%)**	**PBP**	**NPV (million IDR)**	**IRR (%)**	**PBP**
A	83,743,843	8.29	34.47	398,199,357	12.71	20.72	501,149,460	13.77	19.62
B	83,609,123	8.29	34.46	397,025,087	12.70	20.45	498,935,639	13.76	18.65
C	76,617,476	8.22	34.76	379,764,988	12.66	20.52	471,407,089	13.63	18.83
D	76,482,757	8.22	34.76	378,590,718	12.65	20.52	469,193,268	13.61	18.88
E	78,811,227	8.30	34.39	380,774,195	12.84	20.22	471,231,753	13.82	18.59
F	72,121,410	8.13	35.27	375,411,362	12.58	20.74	467,195,903	13.57	20.00
G	67,054,074	8.13	35.35	356,811,930	12.71	20.54	435,064,376	13.60	19.02
H	60,062,427	8.04	35.36	339,551,832	12.66	20.59	407,535,826	13.45	19.19

Source: https://www.intechopen.com/chapters/57639.

9.3.5. Public-Private Partnership

The scheme will evaluate division of responsibility between the government and private on the life cycle of the project from initial cost, operational and maintenance cost, as well as revenue. The financing scheme consists of three scenarios:

- **Scenario A:** The government shares 60% and the private 40% for the initial cost.
- **Scenario B:** The government shares 50% and private 50% for the initial cost.
- **Scenario C:** The government shares 40% and private 60% for the initial cost.

Cost sharing on operating and maintenance expenses follows each scenario, with the private sector contributing 60, 50, and 40% of the total. Additionally, revenue will be split 80/50 with the private sector. Three factors—the revenue weighting score, the operational and maintenance cost weighting score, and the comparison of IRR and MARR—are taken into account when determining the optimal scenario. The private revenue weighting score will be the lowest. It is not required to use higher revenue in the calculation when IRR reaches the expected MARR with the lowest revenue. The O and M cost is likewise subjected to a comparable scenario. Lesser cost is no longer required when the highest O&M cost helps to achieve IRR over MARR.

9.3.5.1. Scenario A

Twelve of the original scenarios from A.1 to A.12. First, a parameter about the minimum IRR is tested, and six cases pass. Second, three possible scenarios are generated from A based on the evaluation of the revenue weighting score for private to gain the expected MARR (12%). 2, A. 6 as well as A. 10. Finally, the two most likely scenarios, both of which include A, are produced by weighting the operational and maintenance cost scores. In both scenarios, private revenue from the project is regulated along with private responsibility for initial, operating, and maintenance costs.

A. Sixth scenario yields a positive net present value (NPV) and an internal rate of return (IRR) of 12.48%. The private sector will contribute roughly 60% of the initial cost and 50% of the operating and maintenance costs. In this case, the private will receive seventy percent of project earnings. Conversely, though, A. Scenario 10 results in a positive net present value (NPV) and an IRR of roughly 12.31%. In the scenario, the private sector contributes a different percentage—roughly 60%—to the costs of operation and maintenance.

9.3.5.2. Scenario B

The scenario B consists of 12 initial scenarios from B.1 to B.12. Similar treatment was also conducted as in scenario A. The scenario generates two most potential scenarios, B.7 and B.11. Scenario B.7 produces a positive NPV of 12.63%. The scenario suggests that the private party contributes 50% of the initial cost and operational and maintenance costs, and reserves the right to gain 60% of the revenue. Moreover, scenario B.11 produces a close IRR to B.7, about 12.43%, and a positive NPV. The private party obtains 60% of the revenue by contributing 50% of the initial cost and 60% of the operational and maintenance costs.

9.3.5.3. Scenario C

The scenario C consists of 12 initial scenarios from C.1 to C.12. Similar treatment was also conducted for scenarios A and B. The scenario generates two most potential scenarios, C.8 and C.12. The C.8 scenario produces a positive NPV of 12.86%. The scenario regulates private responsibility and rights for initial cost, operational and maintenance cost, as well as revenue by 40%, 50%, and 50%, respectively. C.12 generates an IRR of 12.60% and a positive NPV. The private sector should invest 40% of the initial cost and 60% of the operational and maintenance cost. They will gain 50% of revenue from the investment.

9.3.5.4. The Proposed Scenario

As the above scenarios have been generated, the most optimal financing scheme by the division of responsibility between government and private shall be conducted. All the scenarios show a similar IRR of about 12%. The comparison for each scenario can be seen in Table 9.5. It shows the investor's responsibility for the initial cost of revenue.

Table 9.5. *Comparison for Optimal Scenario*

Scenario	Initial cost (%)	OM cost (%)	Revenue (%)	IRR (%)
A.6	60	50	70	12.48
A.10	60	60	70	12.31
B.7	50	50	60	12.63
B.11	50	60	60	12.42
C.8	40	50	50	12.86
C.12	40	60	50	12.60

Source: https://www.intechopen.com/chapters/57639.

Out of the six possible situations, the C. Regarding the allocation of responsibilities in the public-private partnership, scenario 8 is advised. From a public viewpoint, the government bears about 60% of the project's sunk costs but also receives about 50% of the project's revenue, which can be reinvested as capital to fund additional infrastructure projects. In contrast to other scenarios, the private sector contributes higher operational and maintenance costs and splits revenue equally. Since the private sector's initial outlay is only 40%, the IRR generates competitive value. The intended plan seeks to advance public interest without jeopardizing project revenue for private parties.

9.3.6. Institutional Scheme

The financing plan generates an ideal situation that raises the IRR from 10.59 to 12.86%. Given that the project employs multiple disciplines, it is suggested that a consortium that supports the multidisciplinary sector be formed. The joint venture company, investor, consultant, insurance institution, national bank, multilateral bank, contractor, operator, and user are the nine primary stakeholders in the project. Six distinct sectors make up the joint venture company: industry, port, railway, tourism, transportation, and telecommunication. They ought to work with the government to establish a company that will oversee the project. A national bank provides support to the investor, who splits the expense with the joint venture companies. Conversely, the multilateral bank will provide soft loans to the government. Figure 9.6 depicts the institutional scheme in detail.

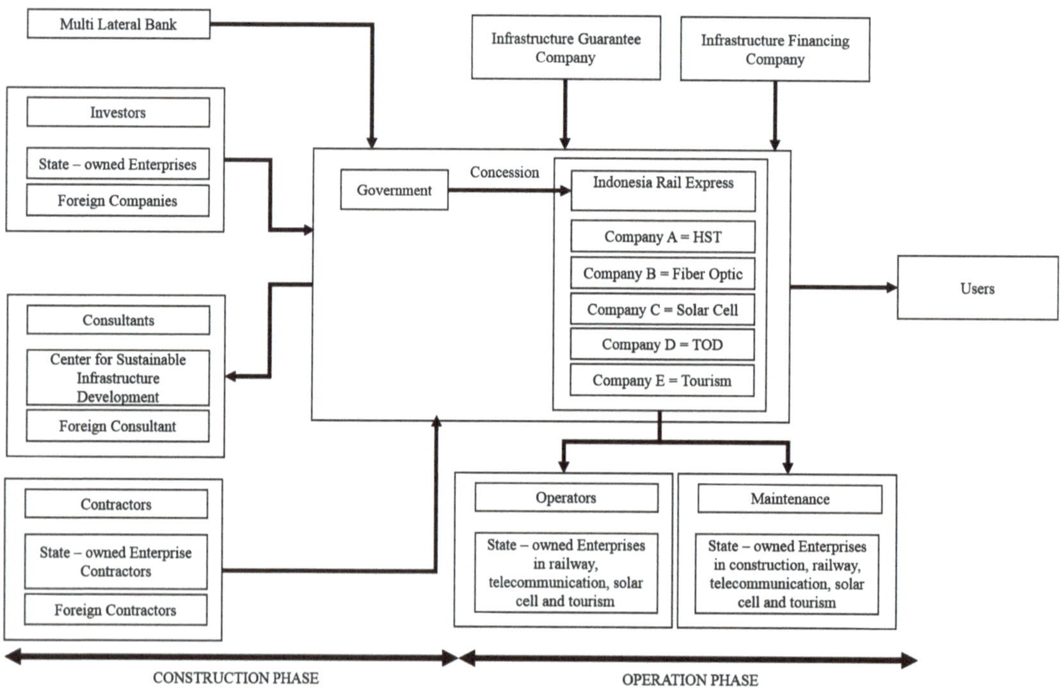

Figure 9.6. *Institutional scheme on the project.*

Source: https://cdnintech.com/media/chapter/57639/1512345123/media/F7.png.

CASE STUDY

UNITED STATES NUMBERED HIGHWAY SYSTEM

In general, U.S. Routes do not have a minimum design standard, unlike the later Interstate Highways, and are not usually built to freeway standards. Some stretches of U.S. Routes do meet those standards. Many are designated using the main streets of the cities and towns through which they run. New additions to the system, however, must "substantially meet the current AASHTO design standards." As of 1989, the United States Numbered Highways system had a total length of 157,724 miles (253,832 km).

Except for toll bridges and tunnels, very few U.S. Routes are toll roads. AASHTO policy says that a toll road may only be included as a special route, and that "a toll-free routing between the same termini shall continue to be retained and marked as a part of the U.S. Numbered System." U.S. Route 3 (US 3) meets this obligation; in New Hampshire, it does not follow tolled portions of the Everett Turnpike. However, US Routes in the system do use parts of five toll roads:

- US 51 uses part of the Jane Addams Memorial Tollw· ay in Illinois; the old road is Illinois Route 251.
- US 278 uses the tolled Cross Island Parkway in South Carolina; the old road is US 278 Business. The tolls were removed in July 2021.
- US 301 is a toll road through Delaware; the former routing is a free road and uses several Delaware state routes.
- US 412 uses the Cimarron Turnpike in Oklahoma; the old road is US 64.
- US 412 also uses the Cherokee Turnpike in Oklahoma; the old road is US 412 Alternate.

Numbering

U.S. Routes in the contiguous United States follow a grid pattern, in which odd-numbered routes run generally north to south and even-numbered routes run generally east to west, though three-digit spur routes can be either-or.[d] Usually, one- and two-digit routes are major routes, and three-digit routes are numbered as shorter spur routes from a main route. Odd numbers generally increase from east to west; U.S. Route 1 (US 1) follows the Atlantic Coast and US 101 follows the Pacific Coast. (US 101 is one of the many exceptions to the standard numbering grid; its first "digit" is "10," and it is a main route on its own and not a spur of US 1) Even numbers tend to increase from north to south; US 2 closely follows the Canadian border, and US 98 hugs the Gulf Coast. The longest routes connecting major cities are generally numbered to end in a 1 or a 0; however, extensions and truncations have made this distinction largely meaningless. These guidelines are very rough, and exceptions to all of the basic numbering rules exist. The numbering system also extended beyond the borders of the United States in an unofficial manner. Many Canadian highways were renumbered in the 1940s and

1950s to adopt the same number as the U.S. Route they connected to – mostly in the western provinces.[citation needed] Examples include British Columbia's highways 93, 95, 97, and 99; Manitoba's highways 59, 75, and 83; or 71. The reverse happened with U.S. Route 57, originally a Texas state highway numbered to match Mexican Federal Highway 57.

In the 1950s, the numbering grid for the new Interstate Highway System was established as intentionally opposite from the US grid insofar as the direction route numbers increased. Interstate Highway numbers increase from west-to-east and south-to-north, to keep identically numbered routes geographically apart in order to keep them from being confused with one another, and it omits 50 and 60 which would potentially conflict with US 50 and US 60. In the US Highway system, three-digit numbers are assigned to spurs of one or two-digit routes. US 201, for example, splits from US 1 at Brunswick, Maine, and runs north to Canada. Not all spurs travel in the same direction as their "parents"; some are connected to their parents only by other spurs, or not at all, instead only traveling near their parents, Also, a spur may travel in different cardinal directions than its parent, such as US 522, which is a north–south route, unlike its parent US 22, which is east–west. As originally assigned, the first digit of the spurs increased from north to south and east to west along the parent; for example, US 60 had spurs, running from east to west, designated as US 160 in Missouri, US 260 in Oklahoma, US 360 in Texas, and US 460 and US 560 in New Mexico. As with the two-digit routes, three-digit routes have been added, removed, extended, and shortened; the "parent-child" relationship is not always present. AASHTO guidelines specifically prohibit Interstate Highways and U.S. Routes from sharing a number within the same state. As with other guidelines, exceptions exist across the U.S. Some two-digit numbers have never been applied to any U.S. Route, including 37, 39, 47, 86, and 88.

Signage

Route numbers are displayed on a distinctively-shaped white shield with large black numerals in the center. Often, the shield is displayed against a black square or rectangular background. Each state manufactures its own signage, and as such subtle variations exist all across the United States. Individual states may use cut-out or rectangular designs, some have black outlines, and California prints the letters "US" above the numerals. One- and two-digit shields generally feature the same large, bold numerals on a square-dimension shield, while 3-digit routes may either use the same shield with a narrower font, or a wider rectangular-dimension shield. Special routes may be indicated with a banner above the route number, or with a letter suffixed to the route number. Signs are generally displayed in several different locations. First, they are shown along the side of the route at regular intervals or after major intersections (called reassurance markers), which shows the route and the nominal direction of travel. Second, they are displayed at intersections with other major roads, so that intersecting traffic can follow their chosen course. Third, they can be displayed on large green guide signs that indicate upcoming interchanges on freeways and expressways.

Divided and special routes

Main articles: List of divided U.S. Routes and List of special routes of the United States Numbered Highway System. Since 1926, some divided routes were designated to serve related areas and designate roughly-equivalent splits of routes. For instance, US 11 splits into US 11E (east) and US 11W (west) in Bristol, Virginia, and the routes rejoin in Knoxville, Tennessee. Occasionally only one of the two routes is suffixed; US 6N in Pennsylvania does not rejoin US 6 at its west end. AASHTO has been trying to eliminate these since 1934; its current policy is to deny approval of new split routes and to eliminate existing ones "as rapidly as the State Highway Department and the Standing Committee on Highways can reach an agreement with reference thereto."

Special routes—those with a banner such as alternate or bypass—are also managed by AASHTO. These are sometimes designated with lettered suffixes, like A for alternate or B for business.

Naming

The official route log, last published by AASHTO in 1989, has been named United States Numbered Highways since its initial publication in 1926. Within the route log, "U.S. Route" is used in the table of contents, while "United States Highway" appears as the heading for each route. All reports of the Special Committee on Route Numbering since 1989 use "U.S. Route," and federal laws relating to highways use "United States Route" or "U.S. Route" more often than the "Highway" variants. The use of U.S. Route or U.S. Highway on a local level depends on the state, with some states such as Delaware using "route" and others such as Colorado using "highway."

CLASS ACTIVITY

ACTIVITY: ROAD DESIGN SIMULATION

Divide the class into groups and assign each group a section of road to design. Provide them with parameters such as traffic volume, terrain, and environmental factors. Have them use software like AutoCAD or Civil 3D to design their section of the highway. Then, each group presents their design to the class, explaining their choices and justifying their decisions.

Objective: To understand and simulate the road design factors.

Materials: Graph paper, rulers, compasses, calculators, and design guidelines.

SUMMARY

- A highway engineer plays a critical role in the development process of transportation infrastructure, encompassing a broad range of responsibilities from planning and design to construction and maintenance.
- The role of a highway engineer is multifaceted, requiring a deep understanding of engineering principles, project management, and regulatory compliance. They must balance technical demands with public safety and environmental stewardship, ensuring that each project contributes positively to the broader transportation network.
- Any major road or highway construction involves a variety of transport professionals, including highway engineers, highway designers, water resources engineers, transport planners, environmental planners, and many more experienced personnel.
- A career in highway engineering involves planning roadway construction and improvement projects. Before a road structure can be designed, the Engineer must determine the exact route the road will take, how many lanes are needed, how this road interfaces with other roads, and how the flow of traffic will be maintained.
- Becoming a highway engineer requires a combination of education, practical experience, and professional certification. The journey typically begins with a strong foundation in mathematics and science during high school, where subjects such as calculus, physics, and chemistry are essential.
- Continuing education is important for highway engineers to stay current with advancements in technology, materials, and methodologies. Many states require PEs to complete a certain number of continuing education hours to maintain their licensure.
- Encompass a broad range of tasks that ensure the planning, design, construction, and maintenance of safe, efficient, and sustainable roadways. At the outset, highway engineers engage in extensive planning and feasibility studies.
- Embarking on a career in highway engineering offers a dynamic and rewarding professional path, combining technical expertise with impactful contributions to society's infrastructure.
- Gaining practical experience is crucial in this field. Internships or cooperative education programs during college offer hands-on training and exposure to real-world projects.
- The workplace environment of a highway engineer is diverse, encompassing both office settings and field locations, each with its own set of tasks and challenges. In the office, highway engineers engage in planning and design activities.
- The career path progression for a highway engineer offers a structured journey from entry-level positions to advanced leadership roles, providing opportunities for professional development, specialization, and increased responsibility.
- The process of transportation planning entails developing a transportation

plan for an urban region. It is an ongoing process that seeks to address the transport needs of the inhabitants of the area, and with the aid of a process of consultation with all relevant groups, strives to identify and implement an appropriate plan to meet these needs.

- The planning process commences with the collection of historical traffic data covering the geographical area of interest. Growth levels in past years act as a strong indicator regarding the volumes one can expect over the chosen future time, be it 15, 20, or 30 years.

REVIEW QUESTIONS

1. How to become a highway engineer?
2. Prepare to bring the highway engineering career.
3. What are the highway planning processes?
4. Which types of decision-making process used in highway and transport planning?
5. Identify the developing conceptual design of trans-Sumatera highway road?

REFERENCES

1. Bel, G. (2011). Infrastructure and nation building: The regulation and financing of network transportation infrastructures in Spain (1720–2010). Business History, 53(5), 688–705.
2. Berawi, M. A., & Susantono, B. (2012). Developing conceptual design of mega infrastructure project: Creating innovation and added value. Value World, 35(1), 12–20.
3. Berawi, M. A., Berawi, A. R. B., Prajitno, I., Nahry, N., Miraj, P., Abdurachman, Y., Tobing, E., & Ivan, A. (2015). Developing conceptual design of high-speed railways using value engineering method: Creating optimum project benefits. International Journal of Technology, 6(4), 670–679.
4. Berawi, M. A., Zagloel, T. M., Berawi, A. R. B., & Abdurachman, Y. (2017). Producing alternative concept for the Trans-Sumatera toll road project development using location quotient method. Procedia Engineering, 171, 265–273.
5. Berawi, M. A., Zagloel, T. M., Berawi, A. R. B., & Abdurachman, Y. (2015). Feasibility analysis of Trans-Sumatera toll road using value engineering method. International Journal of Technology, 6(3), 388–399. http://doi.org/10.14716/ijtech.v6i3.1475.
6. Berawi, M. A., Zagloel, T. Y., Miraj, P., Berawi, A. B., Titaheluw, W., & Karim, S. B. B. A. (2016). Developing road infrastructure route planning: Increasing feasibility of toll road project. Makara Journal of Technology, 20(3), 147–152.
7. Cho, S. M. (2016). Road safety facilities in Korea. Paper presented at expert group meeting on Asian Highway, Bangkok, Thailand. Retrieved from: http://www.unescap.org/sites/default/files/3.4%20Republic%20of%20Korea_Mr.%20Sungmin%20Cho.pdf (accessed on 16 July 2024).

8. Dell'Isola, A. (1997). Value Engineering: Practical Application for Design, Construction, Maintenance and Operations (Vol. 35). RSMeans.
9. Federal Highway Administration, Department of Transportation. (2017). Value Engineering Final Rule. Retrieved from: https://www.fhwa.dot.gov/ve/ (accessed on 16 July 2024).
10. Khmel, V., & Zhao, S. (2016). Arrangement of financing for highway infrastructure projects under the conditions of public–private partnership. IATSS Research, 39(2), 138–145.
11. Kim, E., & Bae, Y. (2015). Economic contribution of the private sector on financing and operation of a highway: Financial computable general equilibrium. Journal of Transport Economics and Policy (JTEP), 49(3), 397–414.
12. Kumar, L., Jindal, A., & Velaga, N. R. (2017). Financial risk assessment and modelling of PPP based Indian highway infrastructure projects. Transport Policy. Advance online publication. http://doi.org/10.1016/j.tranpol.2017.03.010.
13. Leslie, T. F., Frankenfeld, C. L., & Makara, M. A. (2012). The spatial food environment of the DC metropolitan area: Clustering, co-location, and categorical differentiation. Applied Geography, 35(1), 300–307.
14. Li, Z., & Hensher, D. A. (2010). Toll roads in Australia: An overview of characteristics and accuracy of demand forecasts. Transport Reviews, 30(5), 541–569.
15. Liu, J., Love, P. E., Smith, J., Regan, M., & Davis, P. R. (2014). Life cycle critical success factors for public-private partnership infrastructure projects. Journal of Management in Engineering, 31(5), 04014073.
16. Malaysia Highway Authority. (2012). Malaysian Roads. Retrieved from: https://www.piarc.org/ressources/documents/1216,road-network-in-malaysia-v2.pdf (accessed on 16 July 2024).
17. Meijers, E., Hoekstra, J., Leijten, M., Louw, E., & Spaans, M. (2012). Connecting the periphery: Distributive effects of new infrastructure. Journal of Transport Geography, 22, 187–198.
18. Miles, L. D. (2015). Techniques of Value Analysis and Engineering. Miles Value Foundation, USA.
19. Ministry of Land, Infrastructure, Transport and Tourism. (2015). Expressway in Japan. Retrieved from: www.mlit.go.jp/road/road_e/pdf/ROAD2014web.pdf (accessed on 16 July 2024).
20. Morrissey, K. (2014). Producing regional production multipliers for Irish marine sector policy: A location quotient approach. Ocean & Coastal Management, 91, 58–64.
21. Muller, R. H. (1996). Examining toll road feasibility studies. Public Works Financing, NJ USA.
22. Otsuka, A., Goto, M., & Sueyoshi, T. (2010). Industrial agglomeration effects in Japan: Productive efficiency, market access, and public fiscal transfer. Papers in Regional Science, 89(4), 819–840.

23. Shen, Q., & Liu, G. (2007). The selection of benchmarking partners for value management: An analytic approach. International Journal of Construction Management, 7(2), 11–22.

24. Snodgrass, T. J., & Kasi, M. (1986). Function Analysis: The Stepping Stones to Good Value. University of Wisconsin-Madison, Wisconsin, USA.

25. Trappey, A. J., Trappey, C. V., Liu, P. H., Hsiao, C. T., Ou, J. J., & Chen, K. W. (2013). Location quotient EIO-LCA method for carbon emission analysis. In Concurrent Engineering Approaches for Sustainable Product Development in a Multi-Disciplinary Environment (pp. 367–377). London: Springer.

26. Woodhead, R. M., & Berawi, M. A. (2008). An alternative theory of idea generation. International Journal of Management Practice, 3(1), 1–19.

27. World Bank. (2016). Population of Country–Indonesia. Retrieved from: http://data.worldbank.org/indicator/SP.POP.TOTL?locations=ID (accessed on 16 July 2024).

28. Younker, D. (2003). Value Engineering: Analysis and Methodology. CRC Press, Boca Raton, Florida, USA.

CHAPTER 10

Transportation Planning Process

LEARNING OBJECTIVES

After studying this chapter, you will be able to:

- Give an overview of the transportation planning process.
- Discuss the relation between land use and transportation planning in the scope of smart growth strategies.

KEY TERMS FROM THIS CHAPTER

Transportation	Geographic information systems
Infrastructure	Climate change
Economic growth	Density
Metropolitan planning organizations	Collaborative planning

INTRODUCTION

In order to get ready for the need to move people and goods to destinations in the future, transportation planning involves defining future policies, goals, investments, and spatial planning designs. As it is currently implemented, it is a cooperative process that takes into account the opinions of numerous parties, including the public, private, and government sectors. In order to influence positive outcomes, transportation planners analyze the many options and their impacts on the transportation system using a multi-modal and/or comprehensive approach. The process of evaluating, assessing, designing, and siting transportation facilities—typically streets, highways, bike lanes, and public transportation lines—is known as transportation planning or transport planning in other parts of the world.

The first step in creating an area's mobility is transportation planning. It combines disciplines to figure out how to plan the most efficient transportation system possible given the existing and projected transit patterns. Through the balance of investments, technology, land use, and policy, transportation planning offers integrated solutions. These are intended to promote strategic accessibility and long-term growth. Before creating plans, effective transport planning examines travel trends, community needs, and demographics. This entails assessing the infrastructure as it stands, projecting transportation requirements, and making suggestions for upgrades. Any financial restrictions should be considered during the planning process.

Policymakers can define their objectives for a region through transportation planning, which helps analyze travel patterns and demographics. They can foresee how traits will evolve to formulate future regulations. Then, investments and designs for spatial planning can be adjusted to meet present and future demands as well as the demands of urban mobility.

10.1. AN OVERVIEW OF TRANSPORTATION PLANNING PROCESS

The goal of the transportation planning process is to create strategies that will increase the sustainability, safety, and efficiency of transportation networks through a methodical and thorough approach. In order to meet the needs of people and goods for mobility, minimize negative environmental effects, and promote economic growth, this process is essential. Goal-setting, data collection, analysis, alternative evaluation, decision-making, implementation, and monitoring are typically included as important phases.

The first step in goal setting is to identify specific, quantifiable targets that align with larger local, state, or federal priorities. These objectives frequently center on enhancing safety, reducing traffic, improving accessibility, maintaining the environment, and boosting the local economy. In order to make sure that the goals are inclusive and representative of a range of interests and needs, it is crucial to involve stakeholders during this phase, including the public, businesses, and government agencies.

Data collection is done after goal setting in order to obtain pertinent information regarding the state of transportation at the time. This covers information on travel habits, land use patterns, traffic volumes, and environmental effects in addition to demographic data. The accuracy and scope of data collection efforts are improved by the integration of cutting-edge technologies like sensor networks, big data analytics, and geographic information systems (GIS), which also provide a solid basis for further analysis.

The next stage is data analysis, in which gathered information is analyzed to find patterns, obstacles, and opportunities in the transportation network. Planners can better grasp problems like traffic hotspots, security risks, and environmental deterioration with the use of this analysis. Analytical tools are used to forecast future scenarios and assess the possible effects of different interventions, including cost-benefit analysis, traffic simulation, and travel demand modeling.

The next step is the evaluation of alternatives, which entails creating and analyzing various approaches to deal with the recognized transportation problems. Improvements to the infrastructure, advances in technology, modifications to laws, and demand management techniques are some examples of alternatives. Every choice is assessed in light of factors including cost, practicality, efficacy, equity, and environmental impact. During this phase, public participation is frequently encouraged to get feedback and create consensus around preferred solutions, ensuring that the chosen tactics are acceptable in both social and political contexts.

The next stage is called decision-making, during which planners, legislators, and other interested parties examine the data and decide on the best course of action. To get the best results at this point, balancing conflicting priorities and making trade-offs are necessary. Gaining public support and making sure that decisions are fair and justified depends heavily on decision-making processes that are inclusive and transparent. The process of carrying out the selected plans is called implementation. This entails obtaining the required funding, getting the approvals, and liaising with different organizations and agencies. To handle new issues and adjust to shifting circumstances, careful project management, prompt execution, and continuous stakeholder communication are necessary for effective implementation.

Lastly, to determine the success of the strategies put into practice, monitoring and assessment are carried out. This entails monitoring key performance indicators to ascertain whether the predetermined goals and objectives are being fulfilled, such as traffic flow, safety data, and environmental metrics. Adaptive management is made possible by continuous monitoring, which makes it possible to modify and improve strategies in response to changing conditions and observed results.

The goal of the dynamic, iterative transportation planning process is to develop sustainable, safe, and effective transportation networks. Transportation planners strive to satisfy the various and changing needs of society through methodically defining objectives, gathering and analyzing data, assessing alternatives, making defensible decisions, putting plans into action, and keeping an eye on results. This all-encompassing strategy guarantees that transportation networks can adjust to new obstacles and meet the demands of the present.

10.1.1. What is the Transportation Planning Process?

In order to meet future transportation-related needs, future policies, objectives, investments, action plans, and spatial planning projects are defined through a process known as transport planning. It is a collaborative process that takes into account the opinions of many different stakeholders, including state-owned businesses, the private sector, and various local and federal government agencies. This is how it is currently done. In order to contribute to successful outcomes, transport planners, planning organizations, or federal agencies use a multidisciplinary approach to analyze a range of alternatives and their impact on the transportation system.

The inspection, assessment, and design of transportation facilities—typically roads, highways, bike lanes, and public transportation lines—are all part of the transport planning process, also known as international transport planning. Globally, the COVID-19 pandemic has had a significant impact on public transit and service provision. New public transportation networks and transportation regulations are being created as an increasing number of nations start to resume their regular travel schedules. Compared to the previous COVID-19 period, these measures imply a significant reduction in service capacity and an increase in the use of transportation planning resources in the transportation planning process (Figure 10.1).

Figure 10.1. *Transportation planning process.*

Source: https://cms-media.bartleby.com/wp-content/uploads/sites/2/2021/11/26115126/image-90.png.

10.1.1.1. The Four Steps of Transportation Planning

10.1.1.1.1. Step 1: Trip generation

Trip generation examines the overall structure of the current transit system within a specified area. It classifies the amount of time spent traveling in each zone according to its origin, destination, and purpose. Specific segments can be subjected to additional analysis based on land use, population demographics, socioeconomic variables, and other relevant features.

10.1.1.1.2. Step 2: Trip distribution

The goal of trip distribution is to determine the number of trips that take place between a destination and an origin zone. This can help with traffic planning optimization by providing additional insight into regional mobility. Analyzing macroscopic relationships between locations is a part of calculating trip distribution. Relative activity levels at the origin and destination, transit-related travel expenses, traffic, and enhancements to sustainability measures are all included in the analysis.

10.1.1.1.3. Step 3: Mode choice

When evaluating mobility, mode choice considers both trip generation and distribution.

Based on region, it establishes which mode of transportation is most popular for each transit route. The percentage of trips made by each mode of transportation within each zone is calculated to achieve this. The availability, cost, and societal preferences for transit all influence mode choice.

10.1.1.1.4. Step 4: Route choice

Analyzing the favored travel routes between origins and destinations within a given area is the process of route choice. It places passengers on the roads that lead to their destinations. The transportation network is then loaded with the totality of estimated trips based on origin, destination, and mode in order to determine how route choice affects congestion in relation to the time of day. This makes it possible to plan for transportation upgrades based on current transit systems. Reducing travel time and increasing accessibility are the objectives.

10.1.2. What Does Transport Planning Involve?

Planning for transportation encompasses all the necessary elements required to improve the usability of transportation networks. Analyzing and evaluating the situation of urban transportation and traffic patterns at the moment is the first step. Subsequently, plans and policies are proposed using this data. These seek to offer comprehensive solutions that take longevity, accessibility, safety, and sustainability into account.

Transport planning is conducted at various levels in the targeted geographic area. It is crucial to address it at the local level initially, but some regions may require the implementation of international policies to see significant improvements. To create the most effective transportation planning system, policymakers and engineers should start by examining local communities, and then move on to national site design, before considering how to connect international regions. In order to support end-to-end services and achieve net carbon zero, strategic transportation planning must also take safety and decarbonizing transport options into consideration. Both the state of the global environment and the future of the economy depend on this. Planning for sustainable regional transportation can lessen the effects of climate change and cut carbon emissions (Figure 10.2).

Figure 10.2. *Planning for sustainable regional transportation.*

Source: https://www.ptvgroup.com/sites/default/files/styles/1x1_640/public/2023-07/highway-1929866_1920.jpg.webp?h=e5aec6c8&itok=rrEMZpjr.

10.1.3. What Does a Transportation Planner Do?

It is the duty of transportation planners to examine existing trends and create a methodical plan for sustainable transportation in the future. Their goal is to tackle the shortcomings and demands of the existing infrastructures. This entails enhancing the infrastructure for all users, including cyclists and pedestrians, while also taking anticipated and unforeseen safety and environmental concerns into account. Increasing an area's overall efficiency through thorough regional transport planning is the main duty of a transportation planner.

This integral perspective requires an understanding of current infrastructures, as well as the projected growth and evolution of a city, town, or region. Some of the primary responsibilities include:

- Recommending transportation improvements and solutions to ongoing problems, such as congestion, vehicle emissions, inaccessibility of public transport, etc.
- Working to decarbonize transport networks for future sustainability.
- Enhancing user safety within the transportation system.
- Assessing the impact of recent or anticipated building developments. This includes transport systems, housing, or corporate structures.
- Designing efficient research methods and survey techniques.
- Modeling traffic flows using mathematical techniques.
- Running simulated computer applications to determine the impact of recommended transportation improvements.
- Analyzing and interpreting data.
- Preparing reports and publications to share data with policymakers and city officials.
- Giving presentations about proposed transportation improvements and regional plans.
- Acting as the subject matter expert during public inquiries.
- Staying up to date and knowledgeable about ongoing government transport and planning policies.

Depending on the regional requirements, transportation planners can be employed at either a public or private level. At either level, their roles remain vast and varied.

10.1.4. Methods and Sustainability

In the past, logical planning models have been used in transportation planning to establish goals and objectives, pinpoint issues, create and investigate alternatives, and enhance existing systems. Strategic character, transport-focused development, satisfaction, rising planning, organizational process, cooperative planning, and political discourse are some other planning models.

Because of the increasing significance of nature, organizers are expected to take a multifaceted approach. One strategy to encourage drivers to abandon their cars and take public transportation is the application of morality. The transport planner's responsibilities have evolved from providing technical analysis to advocating for sustainability via integrated transportation policies. For instance, the growing number of motorcycles in Hanoi is not only affecting the quality of the air but also decelerating economic growth. The goal is to gradually lessen traffic congestion by implementing changes to urban planning. Experts intend to improve performance metrics and short-term decrease traffic through economic stimulus and other alluring measures.

While value-based approaches to examining transportation patterns are still regarded as the foundation of transportation planning, the importance of qualitative analysis, method combinations, and the application of critical analysis frameworks is becoming more widely acknowledged. When creating, assessing, and choosing project and policy options.

10.1.4.1. Models and sustainability

The rational planning model, which includes defining goals and objectives, identifying problems, generating alternatives, evaluating alternatives, and developing plans, has historically been the basis for transportation planning, also known as transport planning. Rational actors, transit-oriented development, satisficing, incremental planning, organizational process, collaborative planning, and political bargaining are some other planning models. The need for planners to take a multidisciplinary approach is growing, particularly considering environmentalism's increasing significance. As an illustration, behavioral psychology may be used to convince drivers to give up driving and switch to public transportation. The focus of a transport planner's work is evolving from technical analysis to integrated transport policy promotion of sustainability. For instance, the growing motorcycle population in Hanoi is causing both environmental harm and a deceleration of economic growth. Long-term traffic reduction is intended to be achieved through modifications to urban planning. Short-term traffic lightening is the goal that experts hope to achieve through financial incentives and appealing alternatives.

Although quantitative methods are widely regarded as vital in transport planning, the importance of qualitative and mixed-methods analysis, as well as critical analytical frameworks, is increasingly being acknowledged. These approaches play a crucial role in integrating various planning criteria to effectively generate, evaluate, and select policy and project options in transport planning practice.

10.1.5. The Metropolitan Transportation Planning Process

Decisions regarding transportation policies and initiatives are made through the metropolitan transport planning process. By doing this, the editors enhance the data regarding the effects of employing alternate strategies, such as bus lanes, new highways, or parking restrictions, in the context of transportation services. Federal officers and decision-makers use this information to help them choose transportation policies and programs. The goals of the study area and the objectives of the transportation planning process must be followed. To make sure that the goals and objectives reflect the current values of the society, measures to improve communication and local government officials, as well as organs of state and local citizens, must be taken into consideration at the outset of the process.

10.1.5.1. Technical process

The rational planning model is employed by the majority of regional transport planners. According to the model, planning is a methodical and technical process that looks at quantitative data analysis to decide how much money is best to invest in both new and old transportation infrastructure. The sharp rise in the number of cars on the road, the growth of suburban areas, and the notable rise in local and federal government spending on urban transportation could all be contributing factors to the popularity of these models. During the late 1940s, early 1950s, and early 1960s, planning culture was dominated by all of these events.

According to Johnston (2004) and the Federal Transit Administration (FTA) (2007), there are three main categories into which the US process typically falls. The metropolitan planning organizations (MPOs) or agencies should involve the community, address planning questions with financial delay, and take environmental and air quality concerns into account in each of the three phases. The MPO examines the issues and challenges the area is facing as well as the goals and objectives it can set for itself to help address those concerns during the first stage, known as pre-analysis. In addition, the MPO gathers information on various regional indicators during this phase, creates a set of alternate approaches to be taken into consideration during the planning process, and gathers quantifiable results to be used in assessing whether goals and objectives have been met. Johnston observes that while many of these activities seem like truly "unnecessary" planning resources, many MPOs perform poorly in this area.

10.1.5.2. Technical analysis

Technical analysis is the second stage. Although there is a great deal of technical guidance involved in the process, model development can be broadly divided into the following categories. Before launching, MPO gathers a lot of information. These data can be divided into two groups: those pertaining to the local land use and those concerning the transportation system.

10.1.6. Urban Transportation Modeling System (UTMS)

The Urban Transportation Modeling System (UTMS), even though the procedure is frequently described as a four-step one. The four steps of UTMS are travel production, travel distribution, mode selection, and route allocation, to give it a nickname. The area is split up into a great number of tiny analysis units known as traffic analysis areas for motion production. A specific number of trips are generated, which depends on the quantity and nature of the cities in each area. The journey is separated into categories in the second step, distribution of travel, according to its origin and goal. Typically, these categories include homework and other non-domestic work. Using the data gathered, the journey is compared to the origins and areas in each of the three categories.

10.1.6.1. Mode selection

The trip is assigned to a mode during the mode selection process based on local availability, local resources in that area, and the cost and time of each mode. It is assumed to remain in one location and is excluded from the analysis because most trips taken on foot or by bicycle are typically shorter. Lastly, the network provides the trip in the route allocation. The vehicle's speed reduces as more network segments are made available for the trip; consequently, other trips are distributed among other routes so that the total travel time is the same. This is significant because completely integrating the entire system—rather than just a single person—is the main objective. The traffic flow and link speeds in the network are the final results. These models should ideally be able to incorporate all of the various behaviors connected to the transportation planning process, including intricate and inherently more natural transport policy questions. In reality, this is frequently not feasible due to the complexity of transportation-related issues.

10.2. RELATION BETWEEN LAND USE AND TRANSPORTATION PLANNING IN THE SCOPE OF SMART GROWTH STRATEGIES

Within the framework of smart growth strategies, which seek to develop sustainable, livable, and economically vibrant communities, land use and transportation planning are closely related. By encouraging mixed-use development, higher densities, and the preservation of open spaces, smart growth encourages efficient land use and lessens the need for extensive transportation networks. When land use and transportation planning are combined, reliance on cars is lessened and traffic congestion and greenhouse gas emissions are decreased. Walking, bicycling, and public transportation are encouraged. Coordinated planning improves the overall quality of life by ensuring that residential, commercial, and recreational areas are connected by easily accessible and effective transit options.

Cities are expanding and car ownership is rising as a result of a rapidly growing population. As a result of their financial implications, problems with land use and transportation become important concerns. According to Wey and Hsu, urban sprawl and traffic congestion are now unavoidable development trends in the process of economic expansion. Alongside the current trend of development, problems related to land use and transportation are becoming increasingly interwoven.

Numerous studies have shown that density has a major impact on the energy debate. Other important aspects of sustainability include development incentives, balanced parking policies, mixed-use designs, job-housing ratios, and accessibility to high-capacity transit. Land use planning problems and transport-oriented problems are closely related and have an iterative relationship that is highly interdependent.

It is important to conduct traffic impact analyzes to determine if existing link capacities can accommodate increasing traffic volumes. Traditional land use planning typically focuses on social and economic factors when making decisions, but it is crucial to also consider traffic impact analyzes as a key component of planning parameters. The emergence of new planning paradigms, such as new urbanism and smart growth (SG), has brought fresh perspectives on managing growth and incorporating urban design into planning processes. New urbanism and SG have similarities but also significant differences in their origins and influences, with new urbanism being more architecturally and physically driven, while SG was initiated by environmentalists, citizen groups, transportation planners, and policymakers. The integration of comprehensive planning with affordable housing advocacy and public health initiatives has expanded the scope beyond traditional land use, transportation, and environmental concerns.

Urban sprawl is directly impacted by the more intelligent and sophisticated planning trend that has been brought about by numerous urban problems. These issues can be summed up as increased traffic congestion, loss of open space, and pollution of the air and water. The emerging planning movement has been called "SG." SG strategies, which support mixed-use development, transit-oriented development, and environments conducive to bicycling and walking, have been developed to prevent unplanned sprawl and adverse effects.

Due to variations in location requirements, the goals of the SG strategies differ. However, the three main pillars of those goals are open space preservation, continued development of existing urban areas, and protection of environmentally sensitive areas. A truly intelligent site plan ought to outperform all other suggested plans for land development in measurable terms. There is not, however, a numerical definition of the SG.

SG strategies have been used to integrate land use and transportation planning, with the goal of directing new development into already developed urban areas and enhancing the viability of car alternatives. This method can be used to address planning and design issues in order to control growth and maximize the efficiency of land use (e.g. human population control). The planning principles that are promoted by the SG network are widely acknowledged. It also supports complete streets, neighborhood schools, compact, walkable, and bicycle-friendly land use, as well as mixed-use development with a variety of housing options.

Since those paradigms aim to preserve the positive effects of existing land use and transportation patterns and practices while reducing their negative effects, SG strategies may be assessed during the land use planning process. The goal of this research is to create a land use planning model that incorporates both traffic impact analyzes and SG strategies. The study intersection has been chosen to be a major signalized intersection in Denizli, Turkey that handles high traffic volumes and is situated between three major arterials. The Denizli Transportation Master Plan (DTMP) provided the data, which included land use plans and travel demand. Two scenarios were looked into for the projection year. While SG strategies are taken into consideration in the second scenario, conventional land use planning decisions are applied in the first. Performance indicator visual analyzes and quantitative evaluations are provided by VISSIM traffic simulations.

10.2.1. Methodology

This study's primary goal is to assess the accuracy of land use planning decisions from the standpoint of sustainable transportation by applying SG strategies to the process. A four-step land use planning process is suggested in this situation. Figure 10.3 shows the suggested procedure's flow chart.

The suggested process begins with initializing the problem parameters, as shown in Figure 10.3. Step 1 provides the land use pattern of the study area, travel demand between all origin-destination (O-D) pairs, and details about the transportation network, including link capacities, free flow travel times, and signal timings for intersections with signalization.

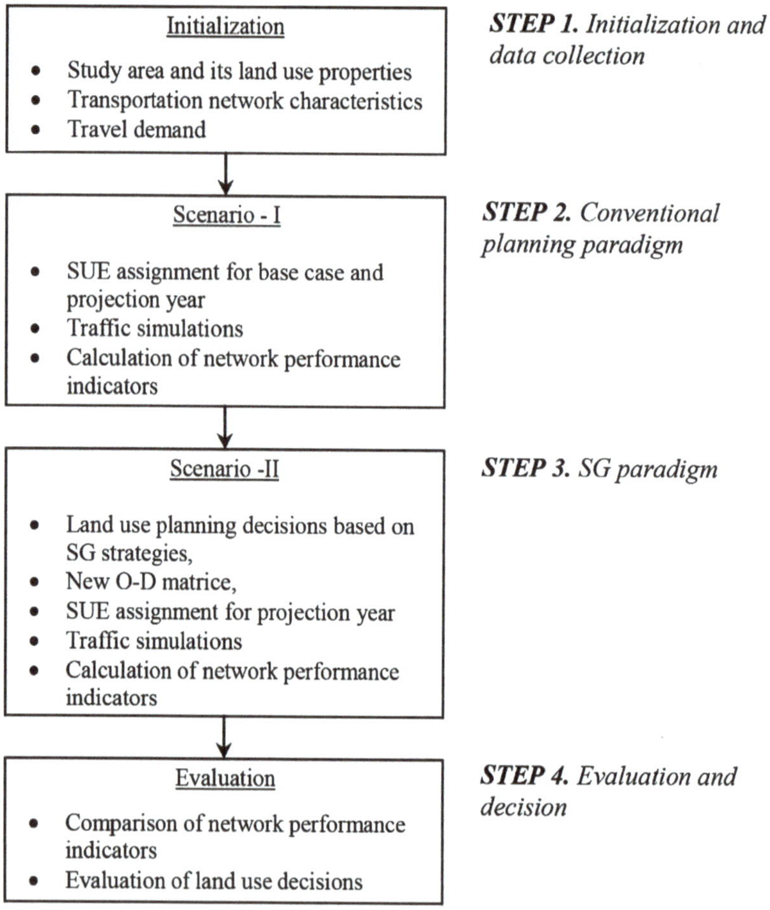

Figure 10.3. *Four-step procedure.*

Source: https://cdnintech.com/media/chapter/50282/1512345123/media/Figure%201.png.

Considering a road network with sets of nodes N, directed links A, O-D pairs W, and routes P, the SUE link traffic volumes may be calculated by solving Eq. (1).

$$\underset{\mathbf{v}(\psi)}{Minimise} \quad Z(\mathbf{v}(\psi),\psi) = -\mathbf{q}^{T}\mathbf{y}(\mathbf{v}(\psi),\psi) + \mathbf{v}^{T}\mathbf{t}(\mathbf{v}(\psi),\psi) - \sum_{a\in A}\int_{0}^{v_a(\psi)} t_a(\psi,x)dx \tag{1}$$

subject to

$$\mathbf{q} = \mathbf{\Lambda h}, \quad \mathbf{v}(\psi) = \mathbf{\delta h}, \quad \mathbf{h} \geq \mathbf{0} \tag{2}$$

where **q** is the vector of O-D demands $[q_w; \forall\, w \in W]$, $\mathbf{v}(\psi)$ represents the vector of link traffic volumes, $\psi$is the vector of signal timings, **h** is the vector of route traffic volumes $[h_p; \forall\, p \in P]$, h_p is the traffic volume on route p, $\mathbf{y}(\mathbf{v}(\psi), \psi)$ represents the vector that consists of travel times on all routes $[y_p; \forall\, p \in P]$, $\mathbf{t}(\mathbf{v}(\psi), \psi)$ is the vector of link travel times, $[y_p; \forall\, p \in P]$ is the travel time along link a, v_a is the flow on link a, while Λ is the O-D/route incidence matrix $[\Lambda_p; \forall\, p \in P]$ and δ represents the link/route incidence matrix where δ_{ap}=1 if link a is on route p and δ_{ap}=0 otherwise $[\delta_{ap}; \forall\, a \in A; \forall\, p \in P]$.

Eq. (1) can be solved by the path flow estimator (PFE) which is a traffic assignment tool using logit route choice model. The solution procedure of PFE is given in Figure 10.4.

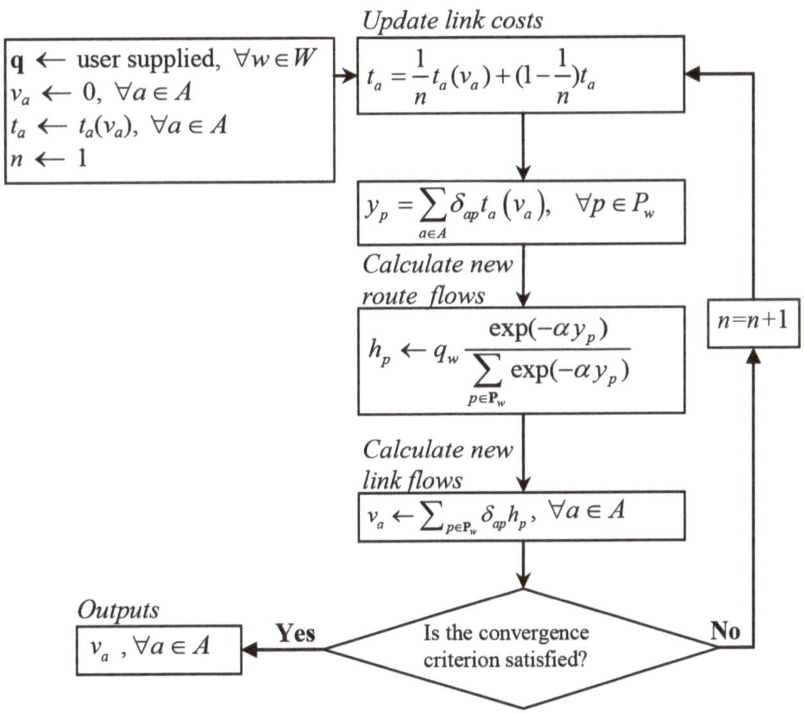

Figure 10.4. *Flowchart of the PFE.*

Source: https://cdnintech.com/media/chapter/50282/1512345123/media/Figure%202.png.

Step 2 involves an analysis of Scenario-I, which stands for the traditional land use planning paradigm. To determine the link traffic volumes in this case, a traffic assignment is performed. Stochastic user equilibrium (SUE) traffic assignment is suggested in the developed land use planning procedure since it accounts for drivers' perception errors when they choose their routes.

$$\frac{\sqrt{\sum_a \left(v_a^{n+1} - v_a^n\right)}}{\sum_a v_a^n} \quad (3)$$

Applications may accept 0.01 as the value of the convergence criterion for the PFE solution. Network performance indicators are computed for base-case and projection year under Scenario-I after obtaining the link traffic volumes. Throughout this study, average speed (km/h), average number of stops per vehicle, average stopped delay per vehicle (seconds), total delay time (hours), number of stops, total stopped delay (hours), and total travel time (hours) are all measured using VISSIM traffic simulation software in addition to visual analyzes of the traffic.

Under Scenario-II, new land use decisions are made in Step 3 based on SG strategies. The city's land use plan is then taken into account when modifying the densities of residential areas. Economic, social, spatial, and cultural factors can all be taken into account during the evaluation process. After that, a SUE assignment is done to determine link traffic volumes for the projection year, and the O-D demand matrix is updated in direct proportion to the new land use decisions. The performance indicators are determined for Scenario-II and the traffic is simulated on the road network, just like it was in Step 2. In Step 4, the new land use decisions are assessed and Scenarios I and II are compared in terms of the network performance indicators.

10.2.2. Study Area

In Turkey's Aegean Region sits the industrial metropolis of Denizli, home to more than 600,000 people in its central district. In addition, it is a tourist destination with 80 traffic analysis zones, which served as the administrative neighborhood districts prior to the implementation of new laws.

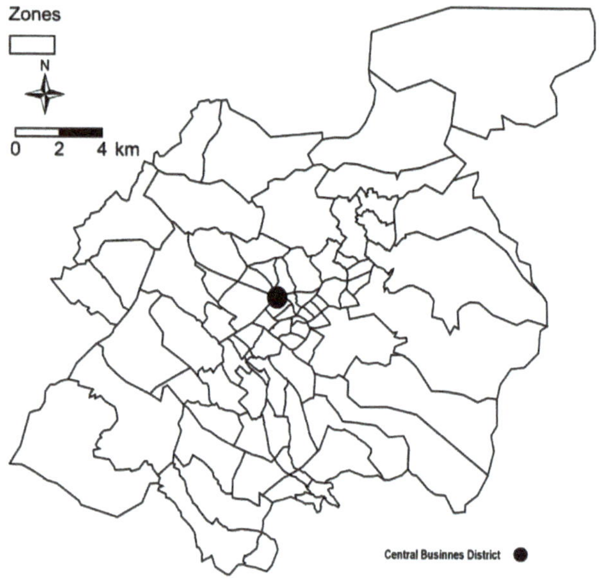

Figure 10.5. *Zonal layout.*

Source: https://cdnintech.com/media/chapter/50282/1512345123/media/Figure%203.png.

Mixed traffic makes up the transportation demand, and it is met by service vehicles, taxis, buses, minibusses, and private automobiles. The high density of private car use in Denizli has led to an increase in traffic problems in recent years. The percentage of people who own a car is roughly 22%, which is roughly twice as high as the average in Turkey. The trips between (7:00 and 09:00 a.m.) account for roughly 30% of all trips, as determined by household surveys. Figure 10.5 shows the city's traffic analysis zones.

The city's zonal structure is depicted in Figure 10.5. At the city's periphery, zone sizes are substantially larger and land use densities are naturally lower. The main cause of traffic issues is delays at intersections. Thus, the field of study has been determined to be a major signalized intersection that serves high traffic volumes and is situated between three major arterials. Figure 10.6 shows the aerial photos of the chosen intersection.

Figure 10.6. *Illustration of the study intersection (a) and queue occurrence (b).*

Source: https://cdnintech.com/media/chapter/50282/1512345123/media/Figure%204.png.

Figure 10.6.a illustrates the signalized roundabout at the study intersection, with four entry lanes on each approach. In Figure 10.6.b, the queue occurrence is depicted on an approach with three isolated lanes merging with the downstream link immediately after the roundabout. It is evident that the intersection's performance will deteriorate and result in significant traffic congestion due to the projected increase in travel demand.

10.2.3. Analyzes

10.2.3.1. Scenario-I: Conventional Transportation Planning Paradigm

The city of Denizli's suggested land use planning process is provided. Keep in mind that DTMP provides the application's necessary data. The projection year is set at 2030 in light of the DTMP's 20-year projection period. The performance indicators of the road network are derived from land use patterns, travel demand between all O-D pairs, and features of the transportation network, including link capacities, free flow travel times, and signal timings for signalized intersections. The link traffic volumes for the base case and the projection year under Scenario-I have been determined by applying an SUE assignment. Keep in mind that the analyzes are done for the morning peak hours of (7:00 and 9:00 a.m.). Figures 10.5 and 10.6 present the traffic volumes that resulted for the road network in 2010 and 2030, respectively (see Figures 10.7 and 10.8).

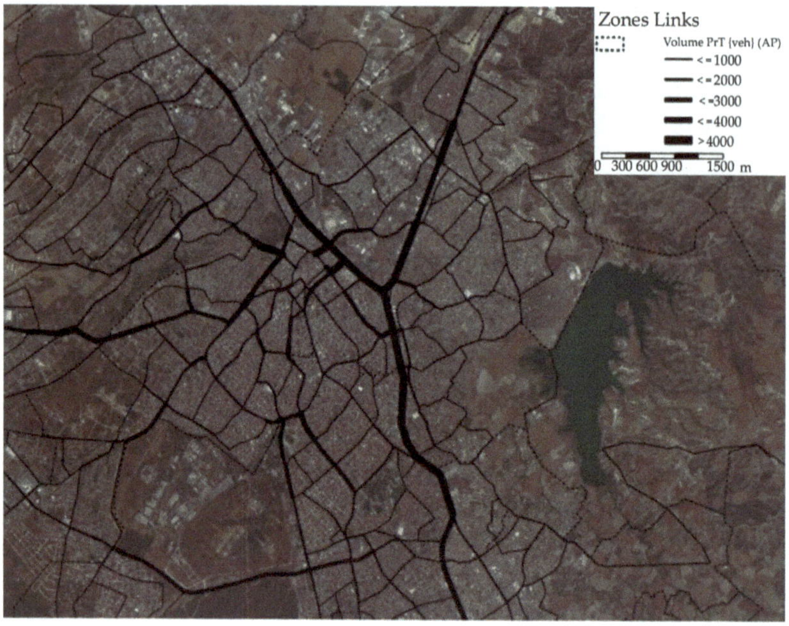

Figure 10.7. *Traffic volumes on the road network for 2010.*

Source: https://cdnintech.com/media/chapter/50282/1512345123/media/Figure%205.png.

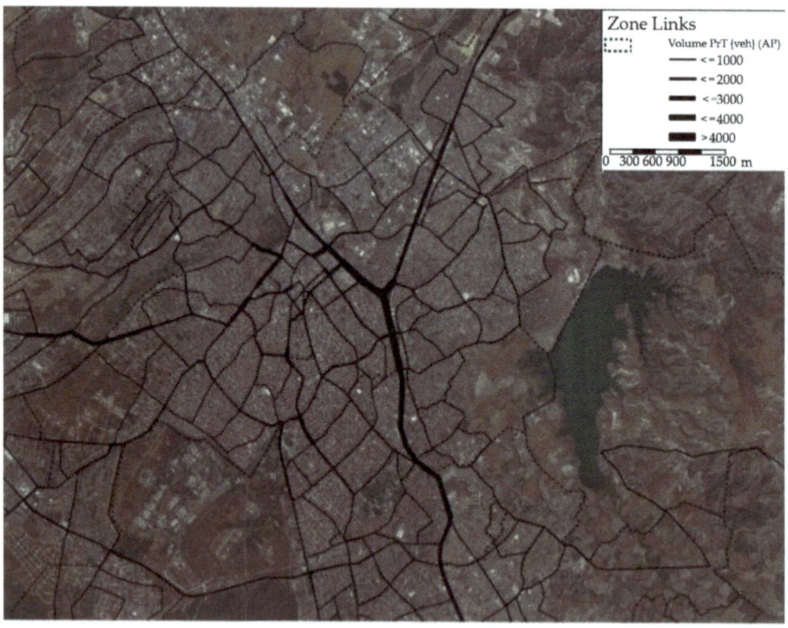

Figure 10.8. *Traffic volumes on the road network for 2030.*

Source: https://cdnintech.com/media/chapter/50282/1512345123/media/Figure%206.png.

The links that meet at the study intersection have the highest traffic volumes, as shown in Figures 10.7 and 10.8. It is also possible to argue that, given the growing volumes of traffic on the road network, the rising demand will result in worsening traffic conditions by 2030. Figure 10.9 and Table 10.1., respectively, provide turn movements and the resulting link traffic volumes in order to examine the performance of the chosen intersection.

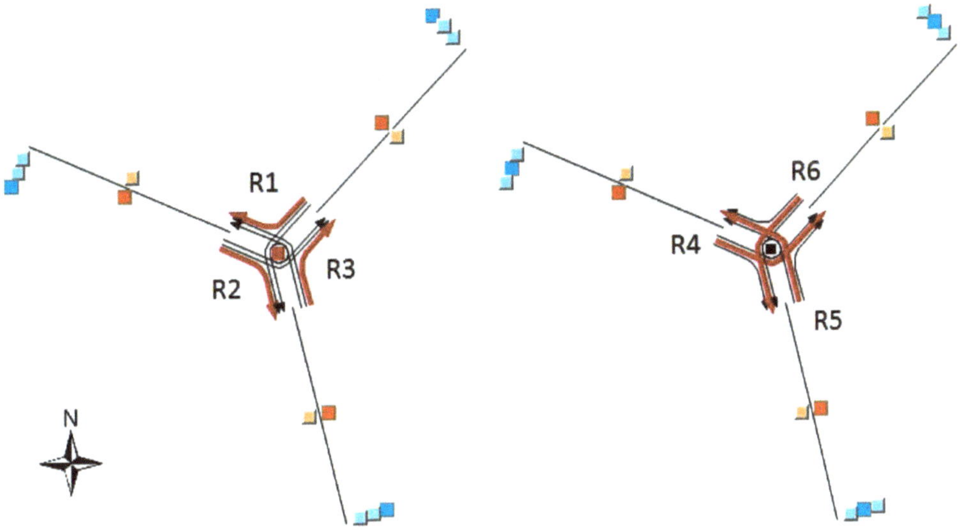

Figure 10.9. *Turning movements in the intersection.*

Source: https://cdnintech.com/media/chapter/50282/1512345123/media/Figure%207.png.

Table 10.1. *SUE Link Flows Under Scenario-I*

Movements	SUE flows for 2010 (veh/h)	SUE flows for 2030 (veh/h)	**Increase (%)**
R1	1549	2410	56
R2	732	1319	80
R3	433	574	33
R4	504	818	62
R5	2282	3819	67
R6	522	788	51

Table 10.1 illustrates how traffic volumes along the intersection's approaches are predicted to rise by 2030 in a variety of ratios. The second movement will see the largest increase, at roughly 80%, while the third movement will see the lowest, at roughly 33%. Currently, base-case and 2030 traffic volumes are taken into account when creating VISSIM traffic simulations for Scenario-I. VISSIM snapshots for Scenario-I are displayed in Figure 10.10.

Figure 10.10a demonstrates that queues form on the upstream links, resembling the pattern seen in Figure 10.10b. The simulation results for the base case show that these queues are manageable thanks to sufficient queue storage on the upstream links.

However, Figure 10.10b predicts that with increasing travel demand, longer queues may form in 2030 under Scenario-I, potentially exceeding the discharge capacity in a single green period. Performance indicator values from the simulations can be found in Table 10.2.

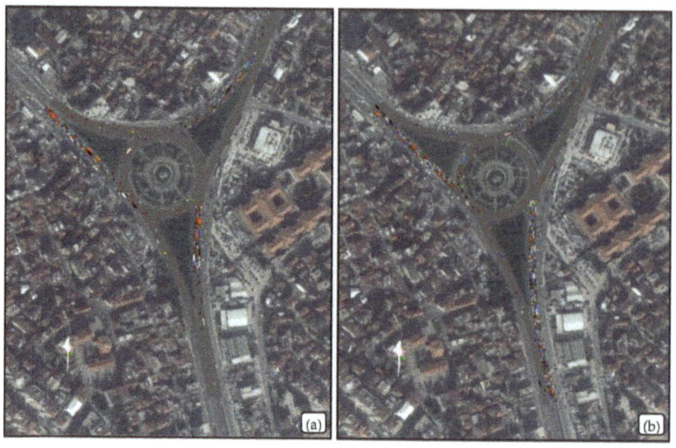

Figure 10.10. *Traffic simulation snapshots for base case (a) and projection year (b).*

Source: https://cdnintech.com/media/chapter/50282/1512345123/media/Figure%208.png.

Table 10.2. *Performance indicators for base-case and Scenario-I.*

	2010	2030	Change (%)
Average delay time per vehicle (s)	85.15	166.21	95
Average number of stops per vehicle	1.55	3.08	99
Average stopped delay per vehicle (s)	60.93	128.77	111
Total delay time (h)	140.17	285.29	104
Number of stops	9210	19033	107
Total stopped delay (h)	100.30	221.02	120
Total travel time (h)	182.58	334.97	83
Average speed (km/h)	17.58	11.24	–36

Table 10.2 shows that the number of stops, delay times, and total travel time increase by over 100% by 2030 considering the traditional land use planning decisions. Meanwhile, the average speed in the intersection decreases by about 36%.

10.2.3.2. Scenario-II: Transportation planning paradigm based on smart growth (SG)

The SG approach may be used to configure the transportation demand, which causes traffic issues when allocated to the road network. In this case, trip attraction and trip generation rates are primarily determined by city block densities. The trip generation and trip attraction increases in the city of Denizli for the zonal case of 2030 are depicted in Figures 10.11. and 10.12.

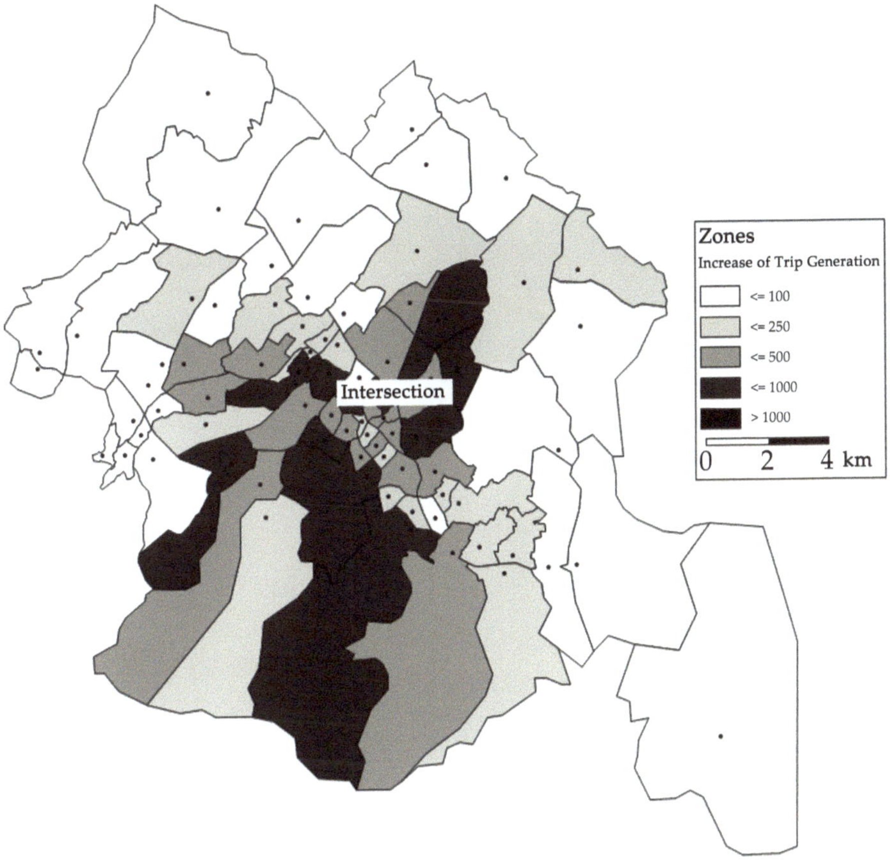

Figure 10.11. *Trip generation increase for 2030.*

Source: https://cdnintech.com/media/chapter/50282/1512345123/media/Figure%209.png.

For the case 2030, Figure 10.12 illustrates that the center of attractive activities is located southwest of the intersection. The zones with higher trip generation values are situated within the same geographic area. In contrast to this type of location selection, the eastern portion of the intersection contains multiple zones with high trip generation values. The western side of the intersection is home to the zones with appealing features. Keep in mind that the study intersection is the only way to go between the urban districts. Given that using the intersection for access may be required in this situation, higher trip generation values at the intersection's eastern district should be questioned. An alternate land use planning paradigm for urban planners is to reduce the trip generation characteristics of the zones located at the eastern part of the intersection. To relieve traffic congestion, residential development areas located on the eastern side of the intersection may be moved to the other side.

Scenario-II in the SG context has a 50% reduction in residential area densities at the eastern portion of the intersection. Trip generation rates therefore decrease in a direct proportion to the O-D matrices. This reduction has been implemented by assessing the city's land use plan.

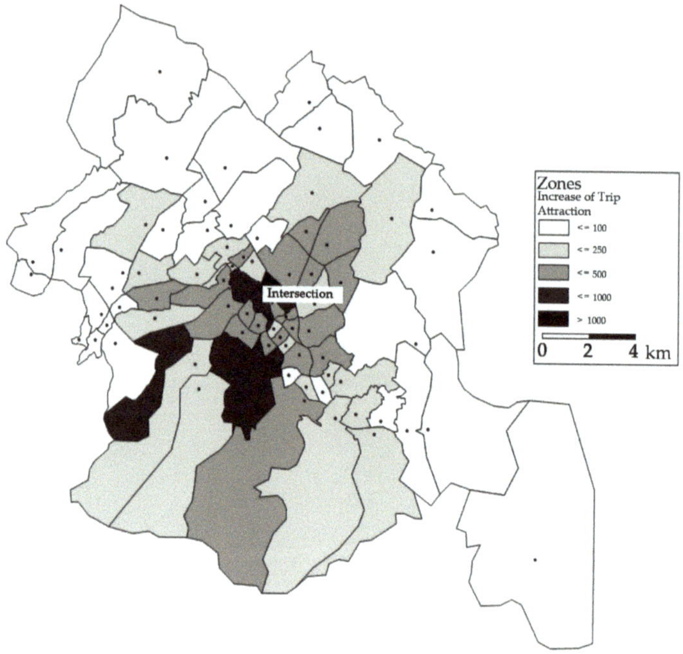

Figure 10.12. *Trip attraction increase for 2030.*

Source: https://cdnintech.com/media/chapter/50282/1512345123/media/Figure%2010.png.

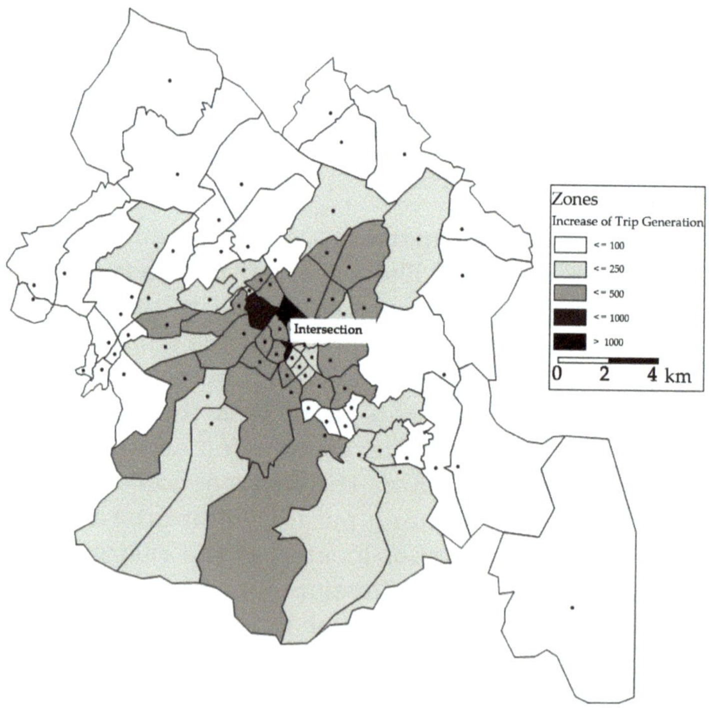

Figure 10.13. *Rearranged trip generation increase for 2030.*

Source: https://cdnintech.com/media/chapter/50282/1512345123/media/Figure%2011.png.

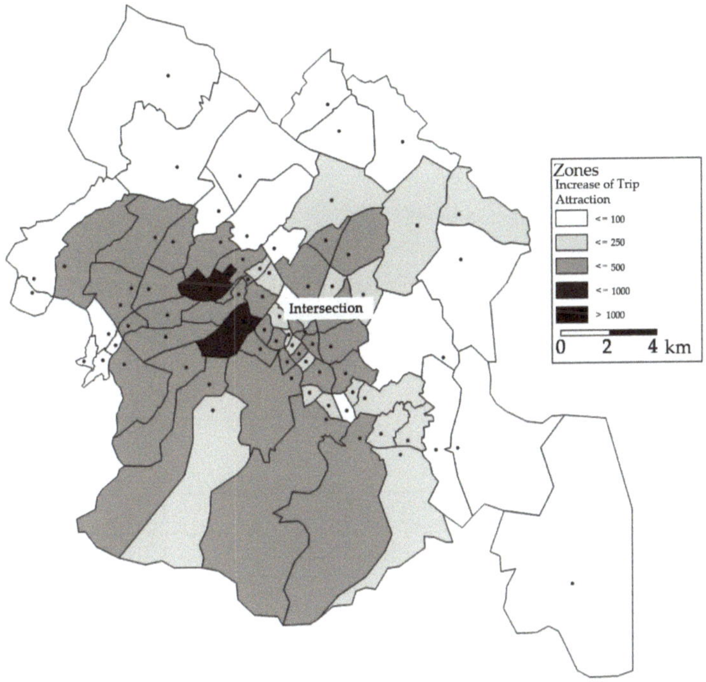

Figure 10.14. *Rearranged trip attraction increase for 2030.*

Source: https://cdnintech.com/media/chapter/50282/1512345123/media/Figure%2012.png.

All decreases and increases have been reflected in the O-D demands, and vacant areas appropriate for residential development have been considered. Figures 10.11–10.14 depict the changes in trip generation and attraction in the zonal case following the implementation of new land use modifications in Scenario-II.

As can be seen in Figures 10.13 and 10.14 the land use densities are sprawled over the area more homogenously in comparison with Scenario-I as shown in Figures 10.11 and 10.12. For Scenario-II, an SUE assignment has been applied with the new O-D travel demand in order to calculate the link traffic volumes. The resulting volumes are given in Table 10.3.

Table 10.3. *SUE Link Flows for the Scenarios*

Movements	Scenario-I (veh/h)	Scenario-II (veh/h)	Decrease (%)
R1	2410	1801	25.27
R2	1319	998	24.34
R3	574	540	5.92
R4	818	753	7.95
R5	3819	2502	34.49
R6	788	711	9.77

As can be seen in Table 3. the traffic volumes along the approaches of the intersection may be decreased from 6% to 35% by applying Scenario-II. In order to evaluate the impacts of the SG strategies in terms of the performance indicators, VISSIM simulations have been made for Scenario-II, and the resulting values of those indicators are provided in Table 10.4.

Table 10.4. *Performance Indicators for Scenarios*

	Scenario-I	**Scenario-II**	**Decrease (%)**
Average delay time per vehicle (s)	166.21	158.96	4.36
Average number of stops per vehicle	3.08	2.82	8.44
Average speed (km/h)	11.24	11.71	–4.18
Average stopped delay per vehicle (s)	128.77	123.32	4.23
Total delay time (h)	285.29	269.75	5.45
Number of stops	19033	17198	9.64
Total stopped delay (h)	221.02	209.26	5.32
Total travel time (h)	334.97	319.12	4.73

Table 10.4 shows that the number of stops in the intersection may be decreased by about 10% while the total delay time decreases by about 5%. Meanwhile, the average travel speed in the study intersection increases by about 4% in comparison with Scenario-I. Therefore, it may be stated that the traffic congestion may be reduced, and the performance of the road network could be improved by applying the SG land use planning strategies.

CASE STUDY

FORT STREET REVIVAL

Background

The City of Victoria, British Columbia, created a plan to "promote a green, clean, walkable and safe city." The plan aims to promote alternate modes of transportation, improve mobility in the center of the city, and reduce motor vehicle pollution. Fort Street is one of three key streets in the city's bikeway network. Improving Fort Street for cycle traffic would help improve the bikeway network as well. Cycling, an excellent alternative to driving, is already the fastest growing mode of transportation in the city. Cycling could represent 1/5th, or 20%, of commuter trips in Victoria by 2026.

Problem

The city made changes in four key areas:

- *Improved bicycle safety*. Creating a dedicated bike lane increased the safety of cyclists along this key bike travel route.
- *Improved pedestrian safety.* Pedestrian safety was improved at a signal crosswalk at Fern Street by creating a center refuge island. By creating boulevards where possible, the city separated pedestrians from vehicle traffic. Bike lanes also buffered pedestrians from cars.
- *Reduced vehicle speeds.* Landscaped medians and fewer vehicle lanes mean slower traffic speeds. Adding center medians and making lanes less wide reduces collisions and vehicle speed.
- *Changes to street parking.* Congestion was reduced by removing limited evening parking on one side of the street. Time-limited parking on both sides of Fort Street was kept to help nearby businesses.

Solution

Through a public planning process, citizens had opportunities to hear proposed plans, provide input and ideas, and ensure that many different views were heard. Inclusive and open processes invite authentic and meaningful conversation and participation by citizens. With citizens' groups assisting in promoting and reviewing new plans, solid input from residents was available to city planners, making new developments more reflective of the growing needs of the city.

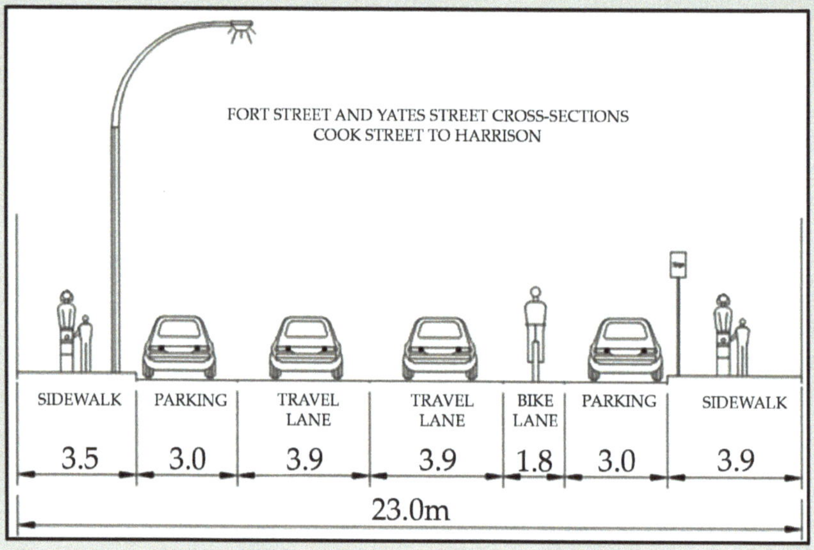

Challenges and Future Possibilities

The City of Victoria is growing and changing. As it does, new initiatives come up and changes are required. Fort Street changes shown here represent the best available resources at the time.

In New York, a recent change along 9th Avenue that includes separated bicycle paths has increased the number of cyclists and reduced collisions by 40%. The City of Vancouver introduced a separate bicycle lane on the Burrard Street bridge in July 2009, which led to a 30% increase in bike traffic.

Separated bicycle lanes may become the norm for all cities someday. Groups such as the Greater Victoria Cycling Coalition and Capital Bike and Walk in Victoria provide cyclists with newsletters on bike culture and celebrate the fun and inspiring aspects of bicycle travel. These and other groups also provide useful information for city planners on bicycle needs, safety, and other concerns.

CLASS ACTIVITY

In this class activity, students will work in groups to design a sustainable transportation plan for a fictional city. Each group will identify key stakeholders, collect, and analyze relevant data, and develop innovative solutions to address transportation issues. They will create a detailed plan including a vision statement, goals, proposed solutions, and an implementation timeline. Groups will present their plans to the class, followed by a discussion on common themes and unique ideas. This activity enhances students' understanding of the transportation planning process, emphasizes the importance of stakeholder engagement, and fosters teamwork and critical thinking skills.

SUMMARY

- Transportation planning is the process of defining future policies, goals, investments, and spatial planning designs to prepare for future needs to move people and goods to destinations.
- The transportation planning process is a systematic and comprehensive approach aimed at developing strategies to improve the efficiency, safety, and sustainability of transportation systems.
- Transportation planners are responsible for analyzing current patterns and developing a systematic approach to the future of sustainable transport.
- The metropolitan transport planning process is a process that leads to decisions about transport policies and programs.
- Urban Transportation Modeling System (UTMS), although it is often referred to as a four-step process. As a nickname suggestion, UTMS has four steps: travel production, travel distribution, mode selection, and route allocation.

REVIEW QUESTIONS

1. What are the main stages of the transportation planning process?
2. Who are the primary stakeholders involved in transportation planning, and why is their involvement important?
3. What types of data are typically collected in the transportation planning process, and how is this data used?
4. How does travel demand forecasting contribute to transportation planning, and what are the common methods used?

5. What is the significance of integrating land use planning with transportation planning?
6. What are some key policies and regulations that influence transportation planning at the federal, state, and local levels?
7. How is cost-benefit analysis used in the transportation planning process, and what are its limitations?
8. What strategies can be employed to promote sustainability in transportation planning?
9. How can technology and innovation improve transportation planning and system management?
10. What are some common challenges faced in the implementation and monitoring of transportation plans, and how can they be addressed?

REFERENCES

1. Behan, K., Maoh, H., & Kanaroglou, P. (2008). Smart growth strategies, transportation and urban sprawl: Simulated futures for Hamilton, Ontario. *The Canadian Geographer / Le Géographe Canadien, 52*, 291–308.
2. Cervero, R., & Duncan, M. (2006). Which reduces vehicle travel more: Jobs-housing balance or retail-housing mixing? *Journal of the American Planning Association, 72*, 475–490.
3. Ceylan, H. (2013). Optimal design of signal-controlled road networks using differential evolution optimization algorithm. *Mathematical Problems in Engineering, Article ID: 696374*, 1–11.
4. Ceylan, H., & Ceylan, H. (2012). A Hybrid Harmony Search and TRANSYT hill climbing algorithm for signalized stochastic equilibrium transportation networks. *Transportation Research Part C: Emerging Technologies, 25*, 152–167.
5. Chapin, T. S. (2012). Introduction: From growth controls, to comprehensive planning, to smart growth: Planning's emerging fourth wave. *Journal of the American Planning Association, 78*(1), 5–15.
6. Chatman, D. G. (2013). Does TOD need the T? On the importance of factors other than rail access. *Journal of the American Planning Association, 79*, 17–31.
7. DBM. (2010). *Denizli Intercity and Immediate Surroundings Transportation Master Plan and Process Management, 1. Phase Final Report* [in Turkish].
8. Echenique, M. H., Hargreaves, A. J., Mitchell, G., & Namdeo, A. (2012). Growing cities sustainably: Does urban form really matter? *Journal of the American Planning Association, 78*, 121–137.
9. Geller, A. L. (2003). Smart growth: A prescription for livable cities. *American Journal of Public Health, 93*(9), 1410–1415.
10. Gulhan, G., Ceylan, H., Baskan, O., & Ceylan, H. (2014). Using potential accessibility measure for urban public transportation planning: A case study of Denizli, Turkey. *Promet-Traffic & Transportation, 26*(2), 129–137.

11. Gulhan, G., Ceylan, H., Ozuysal, M., & Ceylan, H. (2013). Impact of utility-based accessibility measures on urban public transportation planning: A case study of Denizli, Turkey. *Cities, 32*, 102–112.

12. Handy, S. (2005). Smart growth and the transportation-land use connection: What does the research tell us? *International Regional Science Review, 28*, 146.

13. Handy, S., Cao, X., & Mokhtarian, P. (2005). Correlation or causality between the built environment and travel behavior? Evidence from Northern California. *Transportation Research Part D: Transport and Environment, 10*, 427–444.

14. Harris, G. A. (2012). Implementing smart growth approaches in southwest Atlanta neighborhoods. Retrieved from: http://www.smartgrowth.org (accessed on 16 July 2024).

15. Johnston, R. A. (2004). The urban transportation planning process. In S. Hansen, & G. Guliano (Eds.), *The Geography of Urban Transportation* (pp. 115–138). The Guilford Press.

16. Loukaitousideris, A. (2010). A new-found popularity for transit-oriented developments? Lessons from Southern California. *Journal of Urban Design, 15*, 49–68.

17. McLeod, S., Schapper, J. H. M., Curtis, C., & Graham, G. (2019). Conceptualizing freight generation for transport and land use planning: A review and synthesis of the literature. *Transport Policy, 74*, 24–34.

18. Moglen, G. E., Gabriel, S. A., & Faria, J. A. (2003). A framework for quantitative smart growth in land development. *Journal of the American Water Resources Association, 39*(4), 947–959.

19. Southern, A. (2006). Modern-day transport planners need to be both technically proficient and politically astute. *Local Transport Today, 448*.

20. Tumlin, J., & Millard-ball, A. (2003). How to make transit-oriented development work. *Planning, 69*, 14–19.

21. Wey, W. M. (2015). Smart growth and transit-oriented development planning in site selection for a new metro transit station in Taipei, Taiwan. *Habitat International, 47*, 158–168.

22. Wey, W. M., & Hsu, J. (2014). New urbanism and smart growth: Toward achieving a smart National Taipei University District. *Habitat International, 42*, 164–174. DOI: 10.1016/j.habitatint.2013.12.001.

Index

S

T

FSC
www.fsc.org
MIX
Papier | Fördert
gute Waldnutzung
FSC® C083411

Kolibri 360 GmbH
Ferdinand-Jühlke-Straße 7
99095 Erfurt, Deutschland
produktsicherheit@kolibri360.de

FSC
www.fsc.org
MIX
Papier | Fördert
gute Waldnutzung
FSC® C083411

Kolibri 360 GmbH
Ferdinand-Jühlke-Straße 7
99095 Erfurt, Deutschland
produktsicherheit@kolibri360.de